The graph of $y = c + a \sin b(x - d)$ or $y = c + a \cos b(x - d)$ has amplitude $|a|$, period $2\pi/b$, a vertical translation c units up if $c > 0$ or $|c|$ units down if $c < 0$, and a phase shift d units to the right if $d > 0$ or $|d|$ units to the left if $d < 0$. Throughout, we assume $b > 0$. The graph of $y = a \tan bx$ or $y = a \cot bx$ has period π/b.

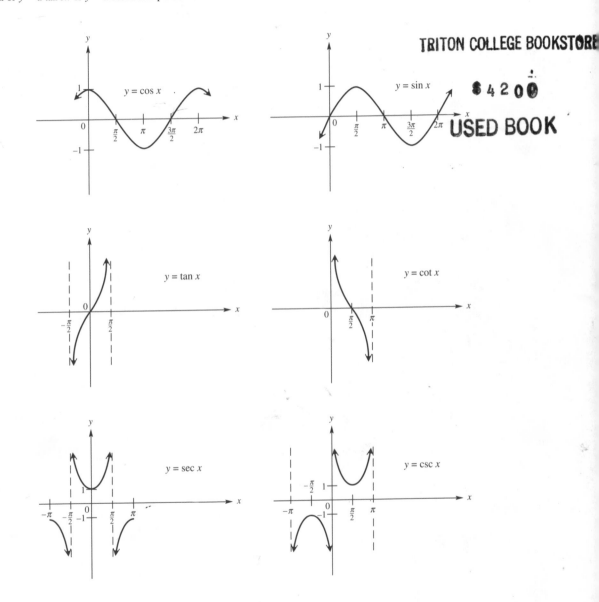

TRIGONOMETRY
FIFTH EDITION

TO THE STUDENT

If you need further help with trigonometry, you may want to obtain a copy of the *Student's Solution Manual* that goes with this textbook. It contains solutions to all the odd-numbered exercises. Your college bookstore either has this book or can order it for you.

Sponsoring Editor: Anne Kelly
Developmental Editor: Linda Youngman
Project Editor: Cathy Wacaser
Art Director: Julie Anderson
Text and Cover Design: Lesiak/Crampton Design Inc: Cynthia Crampton
Cover and Chapter Opener Photos: The Chicago Photographic Company
Photo Researcher: Karen Koblik
Production: Steve Emry, Linda Murray
Compositor: The Clarinda Company
Printer and Binder: R.R. Donnelley & Sons Company
Cover Printer: The Lehigh Press, Inc.

PHOTO ACKNOWLEDGMENTS
Page 22, Charles D. Miller
Page 29, National Park Service, Mount Rushmore National Memorial
Page 105, Comstock
Page 108, Charles D. Miller (top left and bottom left)
Page 119, NASA
Page 127, P. Dugan, M.D. and B. Gorham, R.N.

Trigonometry, Fifth Edition
Copyright © 1993 by HarperCollins College Publishers

Library of Congress Cataloging-in-Publication Data

Lial, Margaret L.
    Trigonometry/Margaret L. Lial, Charles D. Miller, E. John
Hornsby, Jr.—5th ed.
        p.   cm.
    Includes index.
    ISBN 0-673-46647-7
    1. Trigonometry.   I. Miller, Charles David.   II. Hornsby,
E. John.   III. Title.
QA531.L5   1992
516.24—dc20   92-13849
                                               CIP

93   94   95   9   8   7   6   5   4   3

PREFACE

The Fifth Edition of *Trigonometry* is designed for a one-semester or one-quarter course that will prepare students for calculus or for further work in electronics and other technical fields. Applications for both types of study are given throughout the text.

We have written the book assuming a background in algebra. A course in geometry is a desirable prerequisite, but many students reach trigonometry with little or no background in geometry. For this reason, we have included a section on geometry in Chapter 1 and have explained the necessary ideas from geometry in the text as needed.

In this edition of *Trigonometry,* we have attempted to maintain the strengths of past editions while enhancing the pedagogy, readability, usefulness, and attractiveness of the text. Many new features have been added to make the text easier and more enjoyable for both students and teachers to use, including new exercises, increased emphasis on calculator use, and full color. We continue to provide an extensive supplemental package. For students, we offer a solution manual, interactive tutorial software, and videotapes. For instructors, we present an instructor's edition with all answers provided next to the exercises in a special section, overhead transparencies, a computerized test generator, a test manual, and complete solutions to all exercises.

All of the successful features of the previous edition are carried over in the new edition: careful exposition, fully developed examples with comments printed at the side (about 285 examples), and carefully graded section and chapter review exercises (more than 3200 in all). Screened boxes set off important definitions, formulas, rules, and procedures to further aid students in learning and reviewing the course material.

NEW FEATURES

Several new features, designed to assist students in the learning process, have been integrated into this edition. The use of full color and changes in format create a fresh look for the book. The design has been crafted to enhance the book's pedagogical features and increase its accessibility. The next three pages illustrate these features.

NEW FEATURES

PROBLEM-SOLVING STRATEGIES
Special paragraphs labeled "Problem Solving" expand our discussion of strategies to include connections to techniques learned earlier.

CAUTIONARY REMARKS
Common student errors and difficulties are now highlighted graphically and identified with the heading "Caution." Important comments are similarly highlighted with the heading "Note."

EXAMPLE TITLES
Each example now has a title to help students see the purpose of the example. The titles also facilitate working the exercises and studying for examinations.

CHAPTER SUMMARIES
Key ideas are presented in a new "grid" format that makes it easier to review the chapter section by section.

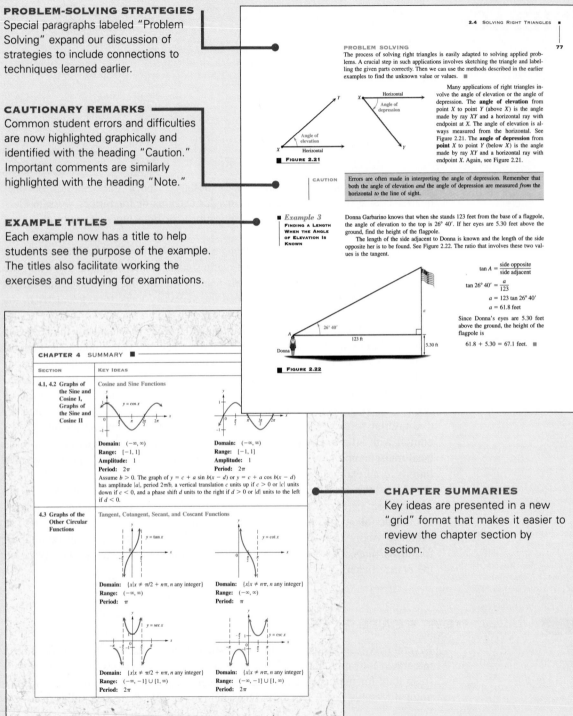

PROBLEM SOLVING
The process of solving right triangles is easily adapted to solving applied problems. A crucial step in such applications involves sketching the triangle and labelling the given parts correctly. Then we can use the methods described in the earlier examples to find the unknown value or values. ■

Many applications of right triangles involve the angle of elevation or the angle of depression. The **angle of elevation** from point X to point Y (above X) is the angle made by ray XY and a horizontal ray with endpoint at X. The angle of elevation is always measured from the horizontal. See Figure 2.21. The **angle of depression** from **point** X to point Y (below X) is the angle made by ray XY and a horizontal ray with endpoint X. Again, see Figure 2.21.

■ FIGURE 2.21

CAUTION Errors are often made in interpreting the angle of depression. Remember that both the angle of elevation *and* the angle of depression are measured *from* the horizontal *to* the line of sight.

■ *Example 3*
FINDING A LENGTH WHEN THE ANGLE OF ELEVATION IS KNOWN

Donna Garbarino knows that when she stands 123 feet from the base of a flagpole, the angle of elevation to the top is 26° 40′. If her eyes are 5.30 feet above the ground, find the height of the flagpole.

The length of the side adjacent to Donna is known and the length of the side opposite her is to be found. See Figure 2.22. The ratio that involves these two values is the tangent.

$$\tan A = \frac{\text{side opposite}}{\text{side adjacent}}$$

$$\tan 26° 40′ = \frac{a}{123}$$

$$a = 123 \tan 26° 40′$$

$$a = 61.8 \text{ feet}$$

Since Donna's eyes are 5.30 feet above the ground, the height of the flagpole is

$$61.8 + 5.30 = 67.1 \text{ feet.} \quad ■$$

■ FIGURE 2.22

CHAPTER 4 SUMMARY ■

SECTION	KEY IDEAS
4.1, 4.2 Graphs of the Sine and Cosine I, Graphs of the Sine and Cosine II	**Cosine and Sine Functions**

$y = \cos x$

Domain: $(-\infty, \infty)$
Range: $[-1, 1]$
Amplitude: 1
Period: 2π

Domain: $(-\infty, \infty)$
Range: $[-1, 1]$
Amplitude: 1
Period: 2π

Assume $b > 0$. The graph of $y = c + a \sin b(x - d)$ or $y = c + a \cos b(x - d)$ has amplitude $|a|$, period $2\pi/b$, a vertical translation c units up if $c > 0$ or $|c|$ units down if $c < 0$, and a phase shift d units to the right if $d > 0$ or $|d|$ units to the left if $d < 0$.

4.3 Graphs of the Other Circular Functions **Tangent, Cotangent, Secant, and Cosecant Functions**

$y = \tan x$

$y = \cot x$

Domain: $\{x | x \neq \pi/2 + n\pi, n \text{ any integer}\}$
Range: $(-\infty, \infty)$
Period: π

Domain: $\{x | x \neq n\pi, n \text{ any integer}\}$
Range: $(-\infty, \infty)$
Period: π

$y = \sec x$

$y = \csc x$

Domain: $\{x | x \neq \pi/2 + n\pi, n \text{ any integer}\}$
Range: $(-\infty, -1] \cup [1, \infty)$
Period: 2π

Domain: $\{x | x \neq n\pi, n \text{ any integer}\}$
Range: $(-\infty, -1] \cup [1, \infty)$
Period: 2π

Graph the following functions over the interval $[-2\pi, 2\pi]$. Give the amplitude. See Example 1.

1. $y = 2 \cos x$ **2.** $y = 3 \sin x$ **3.** $y = \frac{2}{3} \sin x$ **4.** $y = \frac{3}{4} \cos x$

5. $y = -\cos x$ **6.** $y = -\sin x$ **7.** $y = -2 \sin x$ **8.** $y = -3 \cos x$

Graph each of the following functions over a two-period interval. Give the period and the amplitude. See Examples 2–5.

9. $y = \sin \frac{1}{2} x$ **10.** $y = \sin \frac{2}{3} x$ **11.** $y = \cos \frac{1}{3} x$ **12.** $y = \cos \frac{3}{4} x$

13. $y = \sin 3x$ **14.** $y = \sin 2x$ **15.** $y = \cos 2x$ **16.** $y = \cos 3x$

17. $y = -\sin 4x$ **18.** $y = -\cos 6x$ **19.** $y = 2 \sin \frac{1}{4} x$ **20.** $y = 3 \sin 2x$

21. $y = -2 \cos 3x$ **22.** $y = -5 \cos 2x$ **23.** $y = \cos \pi x$ **24.** $y = -\sin \pi x$

Match each function with its graph in Exercises 25–32.

25. $y = \sin x$

26. $y = \cos x$

A   B   C

27. $y = -\sin x$

28. $y = -\cos x$

D   E   F

29. $y = \sin 2x$

30. $y = \cos 2x$

G   H

31. $y = 2 \sin x$

32. $y = 2 \cos x$

appropriate responses.

s amplitude _____, period _____, _____ and has vertical translation _____

horizontally to the _____, it will coincide

period, any vertical translation, and any phase

$= \frac{2}{3} \sin \left(x + \frac{\pi}{2} \right)$ **13.** $y = 4 \cos \left(\frac{x}{2} + \frac{\pi}{2} \right)$

$= 3 \cos 2 \left(x - \frac{\pi}{4} \right)$ **16.** $y = \frac{1}{2} \sin \left(\frac{x}{2} + \pi \right)$

17. $y = 2 - \sin \left(3x - \frac{\pi}{5} \right)$ **18.** $y = -1 + \frac{1}{2} \cos (2x - 3\pi)$

Graph each of the following functions over a two-period interval. See Examples 1 and 2.

19. $y = \cos \left(x - \frac{\pi}{2} \right)$ **20.** $y = \sin \left(x - \frac{\pi}{4} \right)$ **21.** $y = \sin \left(x + \frac{\pi}{4} \right)$

22. $y = \cos \left(x - \frac{\pi}{3} \right)$ **23.** $y = 2 \cos \left(x - \frac{\pi}{3} \right)$ **24.** $y = 3 \sin \left(x - \frac{3\pi}{2} \right)$

Graph each of the following functions over a one-period interval. See Example 3.

25. $y = \frac{3}{2} \sin 2 \left(x + \frac{\pi}{4} \right)$ **26.** $y = -\frac{1}{2} \cos 4 \left(x + \frac{\pi}{2} \right)$ **27.** $y = -4 \sin (2x - \pi)$

28. $y = 3 \cos (4x + \pi)$ **29.** $y = \frac{1}{2} \cos \left(\frac{1}{2} x - \frac{\pi}{4} \right)$ **30.** $y = -\frac{1}{4} \sin \left(\frac{3}{4} x + \frac{\pi}{8} \right)$

Graph each of the following functions over a two-period interval. See Example 4.

31. $y = -3 + 2 \sin x$ **32.** $y = 2 - 3 \cos x$ **33.** $y = 1 - \frac{2}{3} \sin \frac{3}{4} x$ **34.** $y = -1 - 2 \cos 5x$

35. $y = 2 - \cos x$ **36.** $y = 1 + \sin x$ **37.** $y = 1 - 2 \cos \frac{1}{2} x$ **38.** $y = -3 + 3 \sin \frac{1}{2} x$

39. $y = -2 + \frac{1}{2} \sin 3x$ **40.** $y = 1 + \frac{2}{3} \cos \frac{1}{2} x$

Graph each of the following functions over a one-period interval. See Example 5.

41. $y = -3 + 2 \sin \left(x + \frac{\pi}{2} \right)$ **42.** $y = 4 - 3 \cos (x - \pi)$

43. $y = \frac{1}{2} + \sin 2 \left(x + \frac{\pi}{4} \right)$ **44.** $y = -\frac{5}{2} + \cos 3 \left(x - \frac{\pi}{6} \right)$

45. Consider the function $y = -4 - 3 \sin 2(x - \pi/6)$. Without actually graphing the function, write an explanation of how the constants -4, -3, 2, and $\pi/6$ affect the graph, using the graph of $y = \sin x$ as a basis for comparison.

46. Explain how, by an appropriate translation, the graph of $y = \sin x$ can be made to coincide with the graph of $y = \cos x$.

CONCEPTUAL AND WRITING EXERCISES ■

Several exercises requiring an understanding of the concepts introduced in a section are included in almost every exercise set (more than 180 in all). Further, approximately 100 exercises require the student to respond by writing a few sentences. (There is some overlap of the two categories.)

GRAPHING CALCULATOR

Selected chapters include a brief section at the end about the use of graphing calculators. We hope this may enable students to experiment with a graphing calculator and understand some of the concepts in a fresh way.

SCIENTIFIC CALCULATOR EMPHASIS

Based on the assumption that all students have access to scientific calculators, all exposition and examples are written accordingly. References to tables have been eliminated in this edition, although the standard tables are included in the back of the book.

■ THE GRAPHING CALCULATOR ■

When we verify a trigonometric identity, we are essentially showing that two trigonometric functions have the same function values for all numbers in their domains. In Section 5.2 we discussed how trigonometric identities can be verified by using algebraic manipulations and trigonometric substitutions.

A graphing calculator can help us determine whether a given equation is an identity—that is, whether two functions are identical. Consider the equation

$$\tan^2 x(1 + \cot^2 x) = \frac{1}{1 - \sin^2 x},$$

which was verified as an identity in Example 2 of Section 5.2. We can also use a popular model graphing calculator, the TI-81 by Texas Instruments, to verify this identity. First, be sure that the calculator is set in the radian and sequential graphing modes, using the MODE key. Then, use the ZOOM key to set the limits for x and y based on the "Trig" domain and range. Now enter the function on the left side of the equation as Y_1, using the Y= key. Care should be taken here to use parentheses as necessary. One way of entering the function will result in the following display.

$$Y_1 = (\tan x)_\wedge 2 * \left(1 + \left(\frac{1}{\tan x}\right)_\wedge 2\right)$$

Notice how cot x is entered as 1/tan x. The function on the right side of the equation can now be entered as Y_2, resulting in the following display.

$$Y_2 = 1/(1 - (\sin x)_\wedge 2)$$

Now press the GRAPH key. The calculator will begin by graphing Y_1. See the figure.

Notice that a small square appears in the upper right corner of the screen, indicating that graphing is taking place. After Y_1 is graphed, the graph of Y_2 will follow. Since this equation is indeed an identity, the graph of Y_2 will coincide with that of Y_1, as indicated by the prolonged presence of the small square. Had this not been an identity, the graph of Y_2 would have been different from that of Y_1.

You may wish to experiment with the following identities, using your graphing calculator as described above. They are identities that appear in the exercises for Section 5.2. For simplicity, we will use the variable x in all of them.

1. $\dfrac{\cot x}{\csc x} = \cos x$

2. $\cos^2 x (\tan^2 x + 1) = 1$

3. $\dfrac{\sin^2 x}{\cos x} = \sec x - \cos x$

4. $\cot x + \tan x = \sec x \csc x$

5. $\dfrac{\cos x}{\sin x \cot x} = 1$

6. $\dfrac{\cos x + 1}{\tan^2 x} = \dfrac{\cos x}{\sec x - 1}$

7. $\dfrac{\tan^2 x - 1}{\sec^2 x} = \dfrac{\tan x - \cot x}{\tan x + \cot x}$

8. $\sin x (1 + \tan x) + \cos x (1 + \cot x) = \sec x + \csc x$

9. $\dfrac{1}{\cos^2 x} + \dfrac{1}{\sin^2 x} = \dfrac{1}{\cos^2 x - \cos^4 x}$

10. $\dfrac{\sec x}{1 + \tan x} = \dfrac{\csc x}{1 + \cot x}$

11. A force of 176 lb makes an angle of 78° 50′ with a second force. The resultant of the two forces makes an angle of 41° 10′ with the first force. Find the magnitude of the second force and of the resultant.

12. A force of 28.7 lb makes an angle of 42° 10′ with a second force. The resultant of the two forces makes an angle of 32° 40′ with the first force. Find the magnitudes of the second force and of the resultant.

13. A plane flies 650 mph on a bearing of 175.3°. A 25 mph wind, from a direction of 266.6°, blows against the plane. Find the resulting bearing of the plane.

14. A pilot wants to fly on a bearing of 74.9°. By flying due east, he finds that a 42 mph wind, blowing from the south, puts him on course. Find the airspeed and the groundspeed.

15. Starting at point A, a ship sails 18.5 km on a bearing of 189°, then turns and sails 47.8 km on a bearing of 317°. Find the distance of the ship from point A.

16. Two towns 21 mi apart are separated by a dense forest. (See the figure.) To travel from town A to town B, a person must go 17 mi on a bearing of 325°, then turn and continue for 9 mi to reach town B. Find the bearing of B from A.

17. An airline route from San [...] a bearing of 233°. A jet fly [...] ing flies into a wind blow [...] tion of 114°. Find the re [...] speed of the plane.

18. A pilot is flying at 168 m [...] to be on a bearing of 57° [...] the south at 27.1 mph. [...] should fly, and find the p [...]

19. What bearing and airspeed are required for a plane to fly 400 mi due north in 2.5 hr if the wind is blowing from a direction of 328° at 11 mph?

20. A plane is headed due south with an airspeed of 192 mph. A wind from a direction of 78° is blowing at 23 mph. Find the groundspeed and resulting bearing of the plane.

21. An airplane is headed on a bearing of 174° at an airspeed of 240 km per hr. A 30 km per hr wind is blowing from a direction of 245°. Find the groundspeed and resulting bearing of the plane.

22. A ship sailing due east in the North Atlantic has been warned to change course to avoid a group of icebergs. The captain turns and sails on a bearing of 62° for a while, then changes course again to a bearing of 115° until the ship reaches its original course. (See the figure.) How much farther did the ship have to travel to avoid the icebergs?

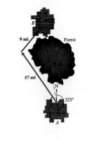

Forest

9 mi

21 mi

87 mi

325°

B

A

N

Icebergs

50 mi

E

23. The aircraft carrier *Tallahassee* is travelling at sea on a steady course with a bearing of 30° at 32 mph. Patrol planes on the carrier have enough fuel for 2.6 hr of flight when travelling at a speed of 520 mph. One of the pilots takes off on a bearing of 338° and then turns and heads in a straight line, so as to be able to catch

CHALLENGING EXERCISES

Most sections include a few challenging exercises that extend the ideas presented in the section. These are identified in the instructor's edition, although not in the student's edition.

NEW CONTENT HIGHLIGHTS

- In Chapter 1 we introduce interval notation, and it is used thereafter throughout the text.
- Section 1.3 includes extra material on geometry, because many students enter this course with little background in geometry.
- Chapter 2 introduces reference angles earlier than in the previous edition and includes more calculator-based explanations. We also have included a special section on distinguishing between exact and approximate values for angles, a frequent trouble spot for trigonometry students.
- In Chapter 3 we have added more calculator discussion in the third section and have placed more emphasis on the unit circle concept.
- Chapter 4 on graphing circular functions has been reorganized so the graphs of the sine and cosine functions are introduced in Section 4.1, with an accompanying discussion of amplitude and period changes. Section 4.2 continues with an explanation of the horizontal and vertical shifts of the sine and cosine. The remaining circular functions are discussed in Section 4.3.
- Chapter 6 includes more discussion on how to find inverse trigonometric function values using a calculator.
- In Chapter 7 we place more emphasis on the basic triangle congruence axioms from geometry to accompany the discussion of solving triangles. We have rewritten the summary on how to solve a triangle for each possible situation (Section 7.3).
- Chapter 8 now begins with a review of the quadratic formula to motivate the study of complex numbers. The cis θ notation (for $\cos \theta + i \sin \theta$) is used more extensively in this edition, in both the examples and the exercises.
- In many sections of the text, the exercises have been reorganized to more closely reflect the development of the material in the section. All exercises involving calculator approximation have answers that are based on actual calculator verification instead of tables, as in the previous edition.

SUPPLEMENTS

Our extensive supplemental package includes an annotated instructor's edition that contains answers to all exercises in a special exercise section at the back of the book, testing materials, solution manuals, software, and videotapes.

FOR THE INSTRUCTOR

ANNOTATED INSTRUCTOR'S EDITION With this volume, instructors have immediate access to the answers to every exercise in the text, excluding writing exercises and proofs. In a special section at the end of the book, each answer is printed in color next to the corresponding text exercise. In addition, challenging exercises, which will require most students to stretch beyond the concepts discussed in the text, are marked with the symbol ▲. The conceptual (◉) and writing (✎) exercises are also marked in this edition so instructors may assign these problems at their discretion. Each section of the instructor's edition also includes a list of "Resources," containing cross-references to relevant sections in each of the supplements for *Trigonometry*.

INSTRUCTOR'S TEST MANUAL   Included here are four versions of a chapter test for each chapter, two versions of a final examination, and a set of 100 to 125 additional exercises per chapter, which can be used as an additional source of questions for tests, quizzes, or student review of difficult topics. Answers to all tests and additional exercises also are provided.

INSTRUCTOR'S SOLUTION MANUAL   This manual includes complete, worked-out solutions to every exercise in the textbook (excluding most writing exercises).

HARPERCOLLINS TEST GENERATOR FOR MATHEMATICS   The HarperCollins Test Generator is one of the top testing programs on the market for IBM and Macintosh computers. It enables instructors to select questions for any section in the text or to use a ready-made test for each chapter. Instructors may generate tests in multiple-choice or open-response formats, scramble the order of questions while printing, and produce up to twenty-five versions of each test. The system features printed graphics and accurate mathematical symbols. The program also allows instructors to choose problems randomly from a section or problem type or to choose questions manually while viewing them on the screen, with the option to regenerate variables if desired. The editing feature allows instructors to customize the chapter data disks by adding their own problems.

QUIZMASTER ON-LINE TESTING SYSTEM   The QuizMaster program, available in both IBM and Macintosh formats, coordinates with the HarperCollins Test Generator and allows instructors to create tests for students to take at the computer. The test results are stored on disk so the instructor can view or print test results for a student, a class section, or an entire course.

TRANSPARENCIES   Approximately sixty color overhead transparencies of figures, examples, definitions, procedures, properties, and problem-solving methods are available to assist instructors in presenting important points during their lectures.

FOR THE STUDENT

STUDENT'S SOLUTION MANUAL   Complete, worked-out solutions are given for odd-numbered exercises and chapter review exercises in a volume available for purchase by students. In addition, a practice chapter test is provided for each chapter.

COLLEGE ALGEBRA WITH TRIGONOMETRY: GRAPHING CALCULATOR INVESTIGATIONS This new supplemental text, written by Dennis Ebersole of Northampton County Area Community College, provides investigations that help students visualize and explore key concepts, generalize and apply concepts, and identify patterns.

VIDEOTAPES A new videotape series has been developed to accompany *Trigonometry,* Fifth Edition. In a separate lesson for each section of the book, the series covers all objectives, topics, and problem-solving techniques within the text.

COMPUTER-ASSISTED TUTORIALS These tutorials offer self-paced, interactive review in IBM, Apple, and Macintosh formats. Solutions are given for all examples and exercises, as needed.

GRAPHEXPLORER With this sophisticated software, available in IBM and Macintosh versions, students can graph rectangular, conic, polar, and parametric equations; zoom; transform functions; and experiment with families of equations quickly and easily.

ACKNOWLEDGMENTS

We wish to thank the many users of the fourth edition for their insightful comments and suggestions for improvements to this book. We also wish to thank our reviewers for their contributions:

Randall Brian, *Vincennes University*
Kathleen B. Burk, *Pensacola Junior College*
Lou Cleveland, *Chipola Junior College*
John W. Coburn, *St. Louis Community College at Florissant Valley*
Sally Copeland, *Johnson County Community College*
Elaine Deutschman, *Oregon Institute of Technology*
Al Giambrone, *Sinclair Community College*
Holly E. Hake, *De Anza College*
William A. Hemme, *St. Petersburg Junior College*
Norma F. James, *New Mexico State University*
Kenneth C. Kochey, *Northampton County Area Community College*
Gerald R. Krusinski, *College of DuPage*
Carolyn Nelson, *North Dakota State University*
Julienne K. Pendleton, *Brookhaven College*
David Price, *Tarrant County Junior College*
Bonnie Smith, *Chipola Junior College*
C. Donald Smith, *Louisiana State University, Shreveport*
Lisa M. Sowinski, *Cuesta College*
John Spellmann, *Southwest Texas State University*

Theresa Stalder, *University of Illinois at Chicago*
Lowell Stultz, *Kalamazoo Valley Community College*
James J. Symons, *De Anza College*
Mary K. Vaughn, *Texas State Technical Institute, Waco*
Benjamin W. Volker, *Bucks County Community College*
Richard C. Weimer, *Frostburg State University*

Paul Eldersveld, College of DuPage, has our gratitude for coordinating the print supplements, an enormous and time-consuming task. We wish to thank Kitty Pellissier, of the University of New Orleans, for doing an outstanding job checking all answers that appear in the Annotated Instructor's Edition and in the answer section of the student text. Paul Van Erden, of American River College, has done his usual fine job in creating the index.

Special thanks go to those at HarperCollins who have been so supportive of this project: Anne Kelly, Linda Youngman, Cathy Wacaser, Julie Anderson, and Ellen Keith. Finally, we wish to recognize our good friend Charles Dawkins, whose support and encouragement over many years has helped to make the series what it is today.

Margaret L. Lial
E. John Hornsby, Jr.

CONTENTS

An Introduction to Scientific Calculators

There is little doubt that the appearance of hand-held calculators twenty years ago and the later development of scientific and graphing calculators have changed the methods of learning and studying mathematics forever. Where the study of computations with tables of logarithms and slide rules made up an important part of mathematics courses prior to 1970, today the widespread availability of calculators makes their study a topic only of historical significance.

In the past two decades, the hand-held calculator has become an integral part of our everyday existence. Today calculators come in a large array of different types, sizes, and prices. For the course for which this textbook is intended, the most appropriate type is the *scientific calculator,* which costs between ten and twenty dollars. While some scientific calculators have advanced features such as programmability and graphing capability, these two features are not essential for the study of the material in this text.

In this introduction, we explain some of the features of scientific calculators. However, remember that calculators vary among manufacturers and models, and that while the methods explained here apply to many of them, they may not apply to your specific calculator. For this reason, it is important to remember that *this is only a guide, and is not intended to take the place of your owner's manual.* Always refer to the manual when you need an explanation of how to perform a particular operation.

The explanations that follow apply to *basic* scientific calculators. Modern graphing calculators follow different sequences of keystrokes.

FEATURES AND FUNCTIONS OF MOST SCIENTIFIC CALCULATORS

Most scientific calculators use *algebraic logic.* (Models sold by Texas Instruments, Sharp, Casio, and Radio Shack, for example, use algebraic logic.) A notable exception is Hewlett Packard, a company whose calculators use *Reverse Polish Notation* (RPN). In this introduction, we discuss calculators that use algebraic logic.

ARITHMETIC OPERATIONS   To perform an operation of arithmetic, simply enter the first number, touch the operation key ([+], [−], [×], or [÷]), enter the second number, and then touch the [=] key. For example, to add 4 and 3, use the following keystrokes.

(The final answer is displayed in color.)

CHANGE SIGN KEY The key marked $\boxed{+/-}$ allows you to change the sign of a display. This is particularly useful when you wish to enter a negative number. For example, to enter -3, use the following keystrokes.

MEMORY KEY Scientific calculators can hold a number in memory for later use. The label of the memory key varies among models; two of these are $\boxed{\text{M}}$ and $\boxed{\text{STO}}$. $\boxed{\text{M+}}$ and $\boxed{\text{M-}}$ allow you to add to or subtract from the value currently in memory. The memory recall key, labeled $\boxed{\text{MR}}$, $\boxed{\text{RM}}$, or $\boxed{\text{RCL}}$, allows you to retrieve the value stored in memory.

Suppose that you wish to store the number 5 in memory. Enter 5, then touch the key for memory. You can then perform other calculations. When you need to retrieve the 5, touch the key for memory recall.

If a calculator has a constant memory feature, the value in memory will be retained even after the power is turned off. Some advanced calculators have more than one memory. It is best to read the owner's manual for your model to see exactly how memory is activated.

CLEARING/CLEAR ENTRY KEYS These keys allow you to clear the display or clear the last entry entered into the display. They are usually marked $\boxed{\text{C}}$ and $\boxed{\text{CE}}$. In some models, touching the $\boxed{\text{C}}$ key once will clear the last entry, while touching it twice will clear the entire operation in progress.

SECOND FUNCTION KEY This key is used in conjunction with another key to activate a function that is printed *above* an operation key (and not on the key itself). It is usually marked $\boxed{\text{2nd}}$. For example, suppose you wish to find the square of a number, and the squaring function (explained in more detail later) is printed above another key. You would need to touch $\boxed{\text{2nd}}$ before the desired squaring function can be activated.

SQUARE ROOT KEY Touching the square root key, $\boxed{\sqrt{x}}$, will give the square root (or an approximation of the square root) of the number in the display. For example, to find the square root of 36, use the following keystrokes.

The square root of 2 is an example of an irrational number. The calculator will give an approximation of its value, since the decimal for $\sqrt{2}$ never terminates and never repeats. The number of digits shown will vary among models. To find an approximation of $\sqrt{2}$, use the following keystrokes.

An approximation

SQUARING KEY This key, $\boxed{x^2}$, allows you to square the entry in the display. For example, to square 35.7, use the following keystrokes.

$$\boxed{3}\ \boxed{5}\ \boxed{.}\ \boxed{7}\ \boxed{x^2}\ \boxed{\quad 1274.49\quad}$$

The squaring key and the square root key are often found on the same key, with one of them being a second function (that is, activated by the second function key, described above).

RECIPROCAL KEY   The key marked ⌊1/x⌋ (or ⌊x⁻¹⌋) is the reciprocal key. (When two numbers have a product of 1, they are called *reciprocals.*) Suppose that you wish to find the reciprocal of 5. Use the following keystrokes.

INVERSE KEY   Some calculators have an inverse key, marked ⌊INV⌋. Inverse operations are operations that "undo" each other. For example, the operations of squaring and taking the square root are inverse operations. The use of the ⌊INV⌋ key varies among different models of calculators, so read your owner's manual carefully.

EXPONENTIAL KEY   This key, marked ⌊xʸ⌋ or ⌊yˣ⌋, allows you to raise a number to a power. For example, if you wish to raise 4 to the fifth power (that is, find 4^5), use the following keystrokes.

ROOT KEY   Some calculators have this key specifically marked ⌊√x⌋ or ⌊ʸ√y⌋; with others, the operation of taking roots is accomplished by using the inverse key in conjunction with the exponential key. Suppose, for example, your calculator is of the latter type and you wish to find the fifth root of 1024. Use the following keystrokes.

Notice how this "undoes" the operation explained in the exponential key discussion earlier.

PI KEY   The number π is an important number in mathematics. It occurs, for example, in the area and circumference formulas for a circle. By touching the ⌊π⌋ key, you can get in the display the first few digits of π. (Because π is irrational, the display shows only an approximation.) One popular model gives the following display when the ⌊π⌋ key is activated: [*3.1415927*].

log AND ln KEYS   Many students taking this course have never studied logarithms. Logarithms are covered in Chapter 9 in this book. In order to find the common logarithm (base ten logarithm) of a number, enter the number and touch the ⌊log⌋ key. To find the natural logarithm, enter the number and touch the ⌊ln⌋ key. For example, to find these logarithms of 10, use the following keystrokes.

Common logarithm: ⌊1⌋ ⌊0⌋ ⌊log⌋ [*1*]
Natural logarithm: ⌊1⌋ ⌊0⌋ ⌊ln⌋ [*2.3025851*]   An approximation

The foundations of trigonometry go back at least three thousand years. The ancient Egyptians, Babylonians, and Greeks developed trigonometry to find the lengths of the sides of triangles and the measures of their angles. In Egypt trigonometry was used to reestablish land boundaries after the annual flood of the Nile River. In Babylonia it was used in astronomy. The very word *trigonometry* comes from the Greek words for triangle *(trigon)* and measurement *(metry).* Today trigonometry is used in electronics, surveying, and other engineering areas, and is necessary for further courses in mathematics, such as calculus.

Over the years, the progress of technology has changed the way in which trigonometry is studied. Where our predecessors were required to use tables to find values of trigonometric functions, interpolation to obtain more accuracy, and logarithms to perform computations, today we have scientific calculators that perform these tasks with the simple touching of keys. With this progress in mind, examples and text discussion in this book are based on the assumption that all students have access to scientific calculators.

1.1 BASIC CONCEPTS

Many ideas in trigonometry are best explained with a graph. Each point in the plane corresponds to an **ordered pair,** two numbers written inside parentheses, such as $(-2, 4)$. Graphs are set up with two axes, one for each number in an ordered pair. The horizontal axis is called the **x-axis,** and the vertical axis is the **y-axis.** The two axes intersect at a point called the **origin.** To locate the point that corresponds to the ordered pair $(-2, 4)$, start at the origin, and move 2 units left and 4 units up. The point $(-2, 4)$ and other sample points are shown in Figure 1.1.

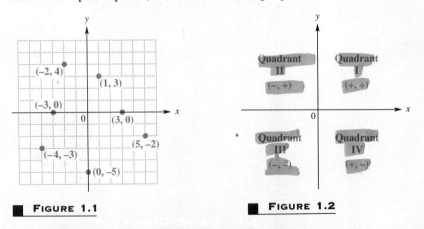

■ **FIGURE 1.1**

■ **FIGURE 1.2**

The axes divide the plane into four regions called **quadrants.** The quadrants are numbered in a counterclockwise direction, as shown in Figure 1.2. The points on the axes themselves belong to none of the quadrants. Figure 1.2 also shows that in quadrant I both the x-coordinate and the y-coordinate are positive; in quadrant II the value of x is negative while y is positive, and so on.

The distance between any two points on a plane can be found by using a formula derived from the **Pythagorean theorem.**

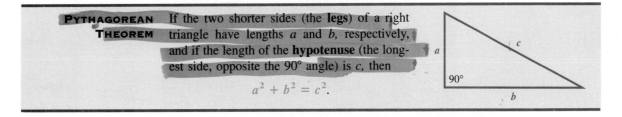

| PYTHAGOREAN THEOREM | If the two shorter sides (the **legs**) of a right triangle have lengths a and b, respectively, and if the length of the **hypotenuse** (the longest side, opposite the 90° angle) is c, then $$a^2 + b^2 = c^2.$$ |

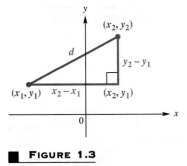

FIGURE 1.3

A proof of the Pythagorean theorem is outlined in Exercise 110.

To find the distance between the two points (x_1, y_1) and (x_2, y_2), start by drawing the line segment connecting the points, as shown in Figure 1.3. Complete a right triangle by drawing a line through (x_1, y_1) parallel to the x-axis and a line through (x_2, y_2) parallel to the y-axis. The ordered pair at the right angle of this triangle is (x_2, y_1).

The horizontal side of the right triangle in Figure 1.3 has length $x_2 - x_1$, while the vertical side has length $y_2 - y_1$. If d represents the distance between the two original points, then by the Pythagorean theorem,

$$d^2 = (x_2 - x_1)^2 + (y_2 - y_1)^2.$$

Upon solving for d, we obtain the *distance formula.*

| DISTANCE FORMULA | The distance between the points (x_1, y_1) and (x_2, y_2) is given by the **distance formula,** $$d = \sqrt{(x_2 - x_1)^2 + (y_2 - y_1)^2}.$$ |

■ *Example 1*
USING THE
DISTANCE FORMULA

Use the distance formula to find the distance, d, between each of the following pairs of points.

(a) $(2, 6)$ and $(5, 10)$

Either point can be used as (x_1, y_1). If we choose $(2, 6)$ as (x_1, y_1) and $(5, 10)$ as (x_2, y_2), then $x_1 = 2$, $y_1 = 6$, $x_2 = 5$, and $y_2 = 10$.

$$
\begin{aligned}
d &= \sqrt{(x_2 - x_1)^2 + (y_2 - y_1)^2} \\
&= \sqrt{(5 - 2)^2 + (10 - 6)^2} \qquad x_1 = 2,\ y_1 = 6,\ x_2 = 5,\ y_2 = 10 \\
&= \sqrt{3^2 + 4^2} \\
&= \sqrt{9 + 16} \\
&= \sqrt{25} \\
&= 5
\end{aligned}
$$

(b) $(-7, 2)$ and $(3, -8)$

$$d = \sqrt{[3 - (-7)]^2 + (-8 - 2)^2}$$
$$= \sqrt{10^2 + (-10)^2}$$
$$= \sqrt{100 + 100}$$
$$= \sqrt{200}$$
$$= 10\sqrt{2}$$

Here $\sqrt{200}$ was simplified as $\sqrt{200} = \sqrt{100} \cdot \sqrt{2} = 10\sqrt{2}$. ■

INTERVAL NOTATION It is often necessary to specify sets of numbers defined by inequalities. One way of doing this is by using **set-builder notation,** such as $\{x|x < 5\}$ (read "the set of all x such that x is less than 5"). Another type of notation, called **interval notation,** can be used for writing intervals. For example, using this notation, the interval $\{x|x < 5\}$ is written as $(-\infty, 5)$. The symbol $-\infty$ does not indicate a number; it is used to show that the interval includes all real numbers less than 5. The parenthesis indicates that 5 is not included. The interval $(-\infty, 5)$ is an example of an **open interval** since the endpoints are not included. Intervals that include the endpoints, as in the example $\{x|0 \le x \le 5\}$, are **closed intervals.** Closed intervals are indicated with square brackets. The interval $\{x|0 \le x \le 5\}$ is written as $[0, 5]$. An interval like $(2, 5]$, that is open on one end and closed on the other, is a **half-open interval.** Examples of other sets written in interval notation are shown in the following chart. In these intervals, assume that $a < b$. Note that a parenthesis is always used with the symbols $-\infty$ or ∞.

Type of Interval	Set	Interval Notation	Graph
Open interval	$\{x|x > a\}$	(a, ∞)	
	$\{x|a < x < b\}$	(a, b)	
	$\{x|x < b\}$	$(-\infty, b)$	
Half-open interval	$\{x|x \ge a\}$	$[a, \infty)$	
	$\{x|a < x \le b\}$	$(a, b]$	
	$\{x|a \le x < b\}$	$[a, b)$	
	$\{x|x \le b\}$	$(-\infty, b]$	
Closed interval	$\{x|a \le x \le b\}$	$[a, b]$	

It is also customary to use $(-\infty, \infty)$ to represent the set of all real numbers.

RELATIONS AND FUNCTIONS   A **relation** is defined as a set of ordered pairs. Many relations have a rule or formula showing the connection between the two components of the ordered pairs. For example, the formula

$$y = -5x + 6$$

shows that a value of y can be found from a given value x by multiplying the value of x by -5 and then adding 6. According to this formula, if $x = 2$, then $y = -5 \cdot 2 + 6 = -4$, so that $(2, -4)$ belongs to the relation. In the relation $y = -5x + 6$, the value of y depends on the value of x, so that y is the **dependent variable** and x is the **independent variable.**

NOTE   A relation is a set of points, often defined by an equation such as $y = -5x + 6$. While precise language would require that we say "the relation defined by the equation $y = -5x + 6$," we will often use the less cumbersome language "the relation $y = -5x + 6$."

Most of the relations in trigonometry are also *functions.*

FUNCTION   A relation is a **function** if each value of the independent variable leads to exactly one value of the dependent variable.

It is customary for x to be considered the independent variable and y the dependent variable, and we shall follow that convention.

For example, $y = -5x + 6$ defines a function. For any one value of x that might be chosen, $y = -5x + 6$ gives exactly one value of y. In contrast, $y^2 = x$ defines a relation that is not a function. If we choose the value $x = 16$, then $y^2 = x$ becomes $y^2 = 16$, from which $y = 4$ or $y = -4$. The one x-value, 16, leads to two y-values, 4 and -4, so that $y^2 = x$ does not define a function.

Functions are often named with letters such as f, g, or h. For example, the function $y = -5x + 6$ can be written as

$$f(x) = -5x + 6,$$

where $f(x)$ is read "f of x." For the function $f(x) = -5x + 6$, if $x = 3$ then $f(x) = f(3) = -5 \cdot 3 + 6 = -15 + 6 = -9$, or

$$f(3) = -9.$$

Also,

$$f(-7) = -5(-7) + 6 = 41.$$

Recall that $|a|$ represents the absolute value of a. By definition, $|a| = a$ if $a \geq 0$ and $|a| = -a$ if $a < 0$. Thus $|4| = 4$ and $|-4| = 4$.

■ *Example 2*

USING FUNCTION
NOTATION

Let $f(x) = -x^2 + |x - 5|$. Find each of the following.

(a) $f(0)$

Use $f(x)$ and replace x with 0.

$$f(0) = -0^2 + |0 - 5| = -0 + |-5| = 5$$

(b) $f(-4) = -(-4)^2 + |-4 - 5| = -16 + |-9| = -16 + 9 = -7$

(c) $f(a) = -a^2 + |a - 5|$ Each x was replaced with a.

(d) Why does f define a function?

For each value of x, there is exactly one value of $f(x)$; therefore f defines a function. ■

The set of all possible values that can be used as a replacement for the independent variable in a relation is called the **domain** of the relation. The set of all possible values for the dependent variable is the **range** of the relation. By observing the graph of a relation, it is often easy to determine the domain and the range. For example, in Figure 1.4 the domain is the set of real numbers between -6 and 6, inclusive, and the range is the set of real numbers between -2 and 2, inclusive. Using interval notation, these sets are written $[-6, 6]$ and $[-2, 2]$, respectively.

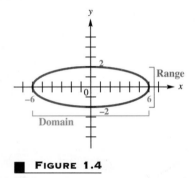

■ **FIGURE 1.4**

For a relation to be a function, each value of x in the domain of the function must lead to exactly one value of y. Figure 1.5 shows the graph of a relation. A point x_1 has been chosen on the x-axis. A vertical line drawn through x_1 intersects the graph in more than one point. Since the x-value x_1 leads to more than one value of y, this graph is not the graph of a function. This example suggests the *vertical line test* for a function.

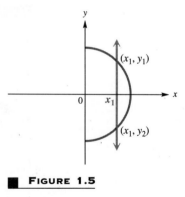

■ **FIGURE 1.5**

VERTICAL LINE TEST   If any vertical line intersects the graph of a relation in more than one point, then the graph is not the graph of a function.

■ *Example 3*

IDENTIFYING
DOMAINS, RANGES,
AND FUNCTIONS

Find the domain and range for the following relations. Identify any functions.

(a) $y = x^2$

Here x, the independent variable, can take on any value, so the domain is the set of all real numbers, $(-\infty, \infty)$. Since the dependent variable y equals the square of x, and since a square is never negative, the range is the set of all nonnegative numbers, $[0, \infty)$.

Each value of x leads to exactly one value of y, so $y = x^2$ defines a function. The graph of $y = x^2$ in Figure 1.6 shows that it satisfies the conditions of the vertical line test.

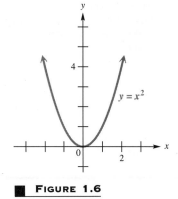

■ FIGURE 1.6

(b) $3x + 2y = 6$

In this relation x and y can take on any value at all. Both the domain and range are $(-\infty, \infty)$. For any value of x that might be chosen, the equation $3x + 2y = 6$ would lead to exactly one value of y. Therefore, $3x + 2y = 6$ defines a function.

(c) $x = y^2 + 2$ (Figure 1.7)

For any value of y, the square of y is nonnegative; that is, $y^2 \geq 0$. Since $x = y^2 + 2$, this means that $x \geq 0 + 2 = 2$, making the domain of the relation $[2, \infty)$. Any real number may be squared, so the range is the set of all real numbers, $(-\infty, \infty)$. To decide whether the relation is a function, choose a sample value of x greater than 2 from the domain. Choosing 6 for x gives

$$6 = y^2 + 2$$
$$4 = y^2$$
$$y = 2 \quad \text{or} \quad y = -2.$$

Since one x-value, 6, leads to two y-values, 2 and -2, the relation $x = y^2 + 2$ does not define a function. As can be seen in Figure 1.7, the graph does not satisfy the requirements of the vertical line test.

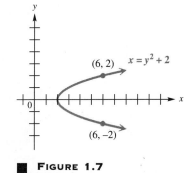

■ FIGURE 1.7

(d) $y = \sqrt{1 - x}$

The domain is found from the requirement that the quantity under the radical, $1 - x$, must be greater than or equal to 0 for y to be a real number.

$$1 - x \geq 0$$
$$x \leq 1$$

The domain is $(-\infty, 1]$. To determine the range, note that the given radical is nonnegative, so the range is $[0, \infty)$.

Since each value of x in the domain leads to a single value of y, this relation is a function. ■

■ *Example 4*

**FINDING THE
DOMAIN OF A
FUNCTION FROM ITS
RULE**

Find the domain for each of the following functions.

(a) $y = \dfrac{1}{x - 2}$

Since division by 0 is undefined, x cannot equal 2. (This value of x would make the denominator become $2 - 2$, or 0.) Any other value of x is acceptable, so the domain is all values of x other than 2, $(-\infty, 2) \cup (2, \infty)$.

(b) $y = \dfrac{8 + x}{(2x - 3)(4x - 1)}$

This denominator is 0 if either

$$2x - 3 = 0 \qquad \text{or} \qquad 4x - 1 = 0.$$

Solve each of these equations.

$$
\begin{aligned}
2x - 3 &= 0 & 4x - 1 &= 0 \\
2x &= 3 & 4x &= 1 \\
x &= \frac{3}{2} & x &= \frac{1}{4}
\end{aligned}
$$

The domain here includes all real numbers x such that $x \neq 3/2$ and $x \neq 1/4$. Using interval notation, this is written $(-\infty, 1/4) \cup (1/4, 3/2) \cup (3/2, \infty)$.

(c) $y = \dfrac{1}{\sqrt{x^2 + 16}}$

For the expression to be defined, the denominator cannot be zero, and the radicand, $x^2 + 16$, must be positive. To find the domain we must solve $x^2 + 16 > 0$.

$$
\begin{aligned}
x^2 + 16 &> 0 \\
x^2 &> -16 \qquad \text{Subtract 16.}
\end{aligned}
$$

Since this last inequality is true for all real numbers, the domain is $(-\infty, \infty)$. ■

NOTE   In later work we will encounter trigonometric functions that are defined in such a way that restrictions are necessary, since their denominators cannot equal zero. Example 4 illustrates this idea with algebraic functions.

1.1 EXERCISES ■

1- 109 odd 1/5/95

Give the quadrant in which each of the following points lies.

1. $(-4, 2)$	**2.** $(-3, -5)$	**3.** $(-9, -11)$	**4.** $(8, -5)$
5. $(9, 0)$	**6.** $(-2, 0)$	**7.** $(0, -2)$	**8.** $(0, 6)$
9. $(-5, \pi)$	**10.** $(\pi, -3)$	**11.** $(3\pi, -1)$	**12.** $(-2\sqrt{2}, 2\sqrt{2})$

13. If $xy = 1$ is graphed, in which quadrants will the points of the graph lie?

14. If (a, b) represents a point that lies in quadrant II, in which quadrant will each of the following points lie?
    (a) $(-a, b)$ (b) $(-a, -b)$ (c) $(a, -b)$

Use the distance formula to find the distance between each of the following pairs of points. See Example 1.

15. $(-2, 7)$ and $(1, 4)$ 16. $(8, -2)$ and $(4, -5)$ 17. $(2, 1)$ and $(-3, -4)$

18. $(-5, 2)$ and $(3, -7)$ 19. $(-1, 0)$ and $(-4, -5)$ 20. $(-2, -3)$ and $(-6, 4)$

21. $(3, -7)$ and $(-2, -5)$ 22. $(-5, 8)$ and $(-3, -7)$ 23. $(-3, 6)$ and $(-3, 2)$

24. $(5, -2)$ and $(5, -4)$ 25. $(\sqrt{2}, -\sqrt{5})$ and $(3\sqrt{2}, 4\sqrt{5})$ 26. $(5\sqrt{7}, -\sqrt{3})$ and $(-\sqrt{7}, 8\sqrt{3})$

A triple of positive integers (a, b, c) is called a Pythagorean triple if it satisfies the equation of the Pythagorean theorem, $a^2 + b^2 = c^2$. Determine whether each of the following is a Pythagorean triple.

27. $(3, 4, 5)$ 28. $(6, 8, 10)$ 29. $(5, 12, 13)$ 30. $(10, 24, 26)$ 31. $(4, 6, 10)$

32. $(5, 10, 15)$ 33. $(7, 24, 25)$ 34. $(9, 40, 41)$ 35. $(8, 15, 17)$

36. Show by an example that the following is a true statement: If, for the positive integers a, b, and c, $a^2 + b^2 = c^2$, then it is not necessarily true that $a + b = c$.

37. Show that $(3, 4, 5)$ is a Pythagorean triple. Then, show that $(3k, 4k, 5k)$ is a Pythagorean triple for $k = 2$, $k = 3$, and $k = 4$. What general conclusion seems likely from this observation? (Although this is not a proof, the conclusion is indeed true, and can be proved using more advanced techniques.)

If we choose positive integers r and s, with $r > s$, the equations
$$a = r^2 - s^2, \qquad b = 2rs, \qquad c = r^2 + s^2$$
will generate a Pythagorean triple. Use the values of r and s given to find a Pythagorean triple, and verify it as such.

38. $r = 2$, $s = 1$ 39. $r = 3$, $s = 2$ 40. $r = 4$, $s = 3$

41. $r = 3$, $s = 1$ 42. $r = 4$, $s = 2$ 43. $r = 4$, $s = 1$

The converse of the Pythagorean theorem says that if a triangle has sides of lengths a, b, and c, where c is the longest side, and $a^2 + b^2 = c^2$, then the triangle is a right triangle. Use this result and the distance formula to decide if the following points are the vertices of right triangles.

44. $(-2, 5)$, $(1, 5)$, $(1, 9)$ 45. $(-9, -2)$, $(-1, -2)$, $(-9, 11)$

46. $(-4, 0)$, $(1, 3)$, $(-6, -2)$ 47. $(-8, 2)$, $(5, -7)$, $(3, -9)$

48. $(\sqrt{3}, 2\sqrt{3} + 3)$, $(\sqrt{3} + 4, -\sqrt{3} + 3)$, $(2\sqrt{3}, 2\sqrt{3} + 4)$

49. $(4 - \sqrt{3}, -2\sqrt{3})$, $(2 - \sqrt{3}, -\sqrt{3})$, $(3 - \sqrt{3}, -2\sqrt{3})$

Find all values of x or y such that the distance between the given points is as indicated.

50. $(x, 7)$ and $(2, 3)$ is 5   51. $(5, y)$ and $(8, -1)$ is 5

52. $(3, y)$ and $(-2, 9)$ is 12   53. $(x, 11)$ and $(5, -4)$ is 17

54. Use the distance formula to write an equation for all points that are 5 units from $(0, 0)$. Sketch a graph showing these points.

55. Write an equation for all points 3 units from $(-5, 6)$. Sketch a graph showing these points.

*The following four exercises use the Pythagorean theorem.**

56. A 1000-ft section of railroad track expands 6 in because the day is very hot. This causes end C (see the figure) to break off and move to position B, forming right triangle ABC. Find BC. (The surprising answer to this simple problem explains why railroad tracks, bridges, and similar structures must be designed to allow for expansion.)

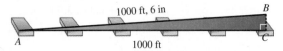

1000 ft, 6 in B

A

1000 ft C

57. Clothing manufacturers sometimes cut their material "on the bias" (that is, at 45° to the direction the threads run) to give it more elasticity. A tie maker

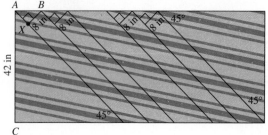

wants to cut twenty 8-in strips of silk on the bias from material that costs $10 per (linear) yd of material 42 in wide (see the figure). Find the total cost of the material. *Note:* This unappealing combination of units—inches, yards, and dollars—is typical of many practical problems, not just in the clothing industry. (*Hint:* First find length AB, using isosceles triangle ABX.)

58. The height (h) of the Great Pyramid of Egypt is 144 m. The apothem (a in the figure) measures 184.7 m. Assuming the base is a square, find the length l of a side of the base.

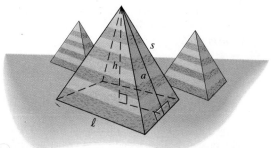

59. Use the result of Exercise 58 to find the length of the edge of the pyramid labeled s in the figure.

Write each of the following sets using interval notation.

60. $\{x|x > 6\}$

61. $\{y|y > 9\}$

62. $\{t|t < -3\}$

63. $\{k|k < -8\}$

64. $\{r|r \geq 1\}$

65. $\{p|p \geq 10\}$

66. $\{x|-3 < x < 7\}$

67. $\{y|8 \leq y \leq 13\}$

68. $\{y|-1 \leq y \leq 1\}$

69.

70.

71.

72. Explain why the set $\{y\,||y| \leq 1\}$ is the same as $[-1, 1]$.

73. Explain why the set $\{y\,||y| \geq 1\}$ is the same as $(-\infty, -1] \cup [1, \infty)$.

Let $f(x) = -2x^2 + 4x + 6$. Find each of the following. See Example 2.

74. $f(0)$

75. $f(-2)$

76. $f(-1)$

77. $f(3)$

78. $f(-3)$

79. $f(-5)$

80. $f(a)$

81. $f(-m)$

82. $f(1 + a)$

83. $f(2 - p)$

Find the domain and range of each of the following. Identify any that are functions. See Example 3.

84. $y = 4x - 3$

85. $2x + 5y = 10$

86. $y = x^2 + 4$

87. $y = 2x^2 - 5$

88. $y = -2(x - 3)^2 + 4$

89. $y = 3(x + 1)^2 - 5$

*From Trigonometry with Calculators by Lawrence S. Levy. Reprinted by permission of the author.

90. $x = y^2$

93. $y = \sqrt{x - 2}$

96.

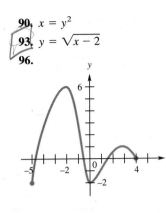

91. $-x = y^2$

94. $y = \sqrt{x^2 + 1}$

97.

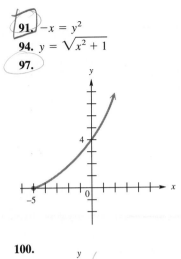

92. $y = \sqrt{4 + x}$

95. $y = \sqrt{1 - x^2}$

98.

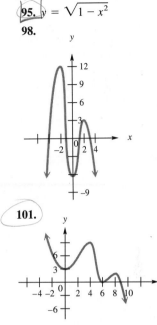

99.

100.

101.

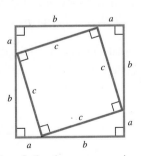

Find the domain for each of the following. See Example 4.

102. $y = \dfrac{1}{x}$

103. $y = \dfrac{-2}{x + 1}$

104. $y = \dfrac{3 + x}{(3x - 7)(2x + 1)}$

105. $y = \dfrac{4 + x^2}{(5x + 1)(3x + 8)}$

106. $y = \dfrac{7 + 5x}{x^2 + 4}$

107. $y = \dfrac{10 + 3x}{-9 - x^2}$

108. $y = \dfrac{-1}{\sqrt{x^2 + 25}}$

109. $y = \dfrac{3 + x}{\sqrt{2x - 5}}$

110. The figure shown is a square made up of four right triangles and a smaller square. By use of the method of *equal areas,* the Pythagorean theorem may be proved. Fill in the blanks with the missing information.

(a) The length of a side of the large square is _____, so its area is (_____)^2 or _____.

(b) The area of the large square may also be found by obtaining the sum of the areas of the four right triangles and the smaller square. The area of each right triangle is _____, so the sum of the areas of the four right triangles is _____. The area of the smaller square is _____.

(c) The sum of the areas of the four right triangles and the smaller square is _____.

(d) Since the areas in (a) and (c) represent the area of the same figure, the expressions there must be equal. Setting them equal to each other we obtain _____ = _____.

(e) Subtract $2ab$ from each side of the equation in (d) to obtain the desired result _____ = _____.

111. The figure shown is a trapezoid. By a method similar to the one used in Exercise 110, the Pythagorean theorem may again be proved. This figure and the accompanying proof are attributed to James A. Garfield, the twentieth president of the United States, who gave the

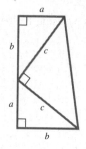

proof in 1876, five years before he was elected president. Prove the Pythagorean theorem using this figure.

112. State the Pythagorean theorem in your own words.

1.2 ——— ANGLES

A line may be drawn through the two distinct points A and B. This line is called **line AB.** The portion of the line between A and B, including points A and B themselves, is **segment AB.** The portion of line AB that starts at A and continues through B, and on past B, is called **ray AB.** Point A is the endpoint of the ray. (See Figure 1.8.)

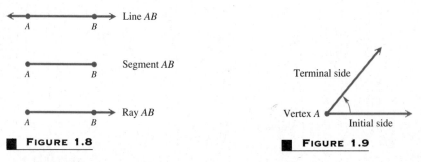

■ **FIGURE 1.8**

■ **FIGURE 1.9**

An **angle** is formed by rotating a ray around its endpoint. The ray in its initial position is called the **initial side** of the angle, while the ray in its location after the rotation is the **terminal side** of the angle. The endpoint of the ray is the **vertex** of the angle. Figure 1.9 shows the initial and terminal sides of an angle with vertex A.

If the rotation of the terminal side is counterclockwise, the angle is **positive.** If the rotation is clockwise, the angle is **negative.** Figure 1.10 shows two angles, one positive and one negative.

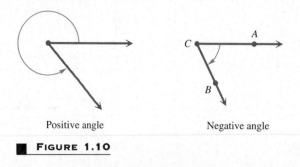

Positive angle   Negative angle

■ **FIGURE 1.10**

An angle can be named by using the name of its vertex. For example, the angle on the right in Figure 1.10 can be called angle *C*. Alternatively, an angle can be named using three letters, with the vertex letter in the middle. For example, the angle on the right also could be named angle *ACB* or angle *BCA*.

There are two systems in common use for measuring the size of angles. The most common unit of measure is the **degree.** (The other common unit of measure is called the *radian*, which is discussed in Chapter 3.) Degree measure was developed by the Babylonians, four thousand years ago. To use degree measure, we assign 360 degrees to a complete rotation of a ray. In Figure 1.11, notice that the terminal side of the angle corresponds to its initial side when it makes a complete rotation.

One degree, written 1°, represents 1/360 of a rotation. For example, 90° represents 90/360 = 1/4 of a complete rotation, and 180° represents 180/360 = 1/2 of a complete rotation. Angles of measure 1°, 90°, and 180° are shown in Figure 1.12.

A complete rotation of a ray gives an angle whose measure is 360°.

■ **FIGURE 1.11**

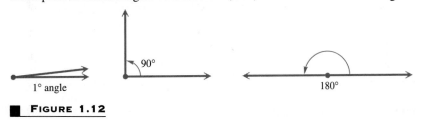

■ **FIGURE 1.12**

Angles are named as shown in the following chart.

TYPES OF ANGLES	Name	Angle Measure	Example
	Acute angle	Between 0° and 90°	60° 82°
	Right angle	Exactly 90°	90°
	Obtuse angle	Between 90° and 180°	97° 138°
	Straight angle	Exactly 180°	180°

If the sum of the measures of two angles is 90°, the angles are called **complementary.** Two angles with measures whose sum is 180° are **supplementary.**

■ *Example 1*
FINDING MEASURES OF COMPLEMENTARY AND SUPPLEMENTARY ANGLES

Find the measure of each angle in Figures 1.13 and 1.14.

(a) In Figure 1.13, since the two angles form a right angle (as indicated by the symbol),

$$6m + 3m = 90°$$
$$9m = 90°$$
$$m = 10°.$$

The two angles have measures of $6 \cdot 10° = 60°$ and $3 \cdot 10° = 30°$.

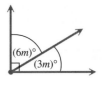

■ **FIGURE 1.13**

(b) The angles in Figure 1.14 are supplementary, so

$$4k + 6k = 180°$$
$$10k = 180°$$
$$k = 18°.$$

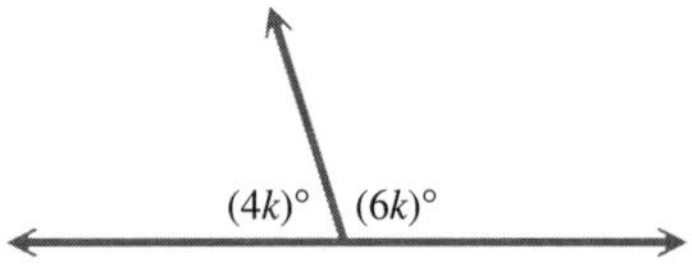

These angle measures are $4(18°) = 72°$ and $6(18°) = 108°.$ ■

■ **FIGURE 1.14**

Angles can be measured with an instrument called a **protractor.** Figure 1.15 shows a protractor measuring an angle of 35°.

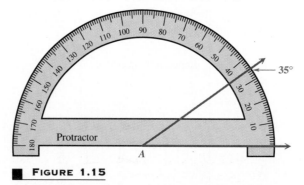

Protractor

■ **FIGURE 1.15**

NOTE Much of the study of trigonometry involves finding angle measures. We will see later in this book that trigonometry does not rely on using protractors to find these measures. For example, if we know the lengths of three sides of a triangle, we can use the Law of Cosines (Section 7.3) to find the angle measures mathematically, and with more precision than we could get from using a protractor.

Do not confuse an angle with its measure. Angle *A* of Figure 1.15 is a rotation; the measure of the rotation is 35°. This measure is often expressed by saying that *m*(angle *A*) is 35°, where *m*(angle *A*) is read "the measure of angle *A*." It saves a lot of work, however, to abbreviate *m*(angle *A*) = 35° as simply angle *A* = 35°.

Traditionally, portions of a degree have been measured with minutes and seconds. One **minute,** written 1′, is 1/60 of a degree.

$$1' = \frac{1}{60}° \quad \text{or} \quad 60' = 1°$$

One **second,** 1″, is 1/60 of a minute.

$$1'' = \frac{1}{60}' = \frac{1}{3600}° \quad \text{or} \quad 60'' = 1'$$

The measure 12° 42′ 38″ represents 12 degrees, 42 minutes, 38 seconds.

The next example shows how to perform calculations with degrees, minutes, and seconds.

■ *Example 2*

CALCULATING WITH DEGREES, MINUTES, AND SECONDS

Perform each calculation.

(a) 51° 29′ + 32° 46′

Add the degrees and the minutes separately.

$$51° 29' + 32° 46' = (51° + 32°) + (29' + 46') = 83° 75'$$

Since 75′ = 60′ + 15′ = 1° 15′, the sum is written

$$83° 75' = 83° + (1° 15') = 84° 15'.$$

(b) 90° − 73° 12′

Write 90° as 89° 60′. Then

$$90° - 73° 12' = 89° 60' - 73° 12' = 16° 48'. ■$$

NOTE **Scientific calculators play a large role in the study of trigonometry as we know it today. In this text we will often give sequences of keystrokes for typical scientific calculators. However, certain scientific calculators may differ in keystroke sequence, and the graphing calculators now available use different keystroke routines than those described here. Therefore, *you should consult your owner's manual if any questions arise concerning calculator use during the study of this text.***

Because calculators are an integral part of our world today, it is now common to measure angles in **decimal degrees.** For example, 12.4238° represents

$$12.4238° = 12\frac{4238}{10,000}°.$$

The next example shows how to change between decimal degrees and degrees, minutes, and seconds. Some calculators will make these conversions automatically, often with a key labeled DMS.

16 ■ *Example 3*

CONVERTING
BETWEEN DECIMAL
DEGREES AND
DEGREES, MINUTES,
SECONDS

(a) Convert 74° 8′ 14″ to decimal degrees. Round to the nearest thousandth of a degree.

Since $1' = \dfrac{1}{60}^{\circ}$ and $1'' = \dfrac{1}{3600}^{\circ}$,

$$74° \; 8' \; 14'' = 74° + \frac{8}{60}^{\circ} + \frac{14}{3600}^{\circ}$$
$$= 74° + .1333° + .0039°$$
$$= 74.137° \text{ (rounded).}$$

One popular scientific calculator converts this example as follows.

Enter	**Press**	**Display**
74.0814	DMS-DD	74.137222

Deg Min Sec

(b) Convert 34.817° to degrees, minutes, and seconds.

$$34.817° = 34° + .817°$$
$$= 34° + (.817)(60')$$
$$= 34° + 49.02'$$
$$= 34° + 49' + .02'$$
$$= 34° + 49' + (.02)(60'')$$
$$= 34° + 49' + 1'' \quad \text{(rounded)}$$
$$= 34° \; 49' \; 1''$$

1 degree = 60 minutes

1 minute = 60 seconds

To convert 34.817° to degrees, minutes, and seconds, use the same calculator as follows. (The column headed "Result" shows how to interpret the display.)

Enter	**Press**	**Display**	**Result**
34.817	INV DMS-DD	34.4901	34° 49′ 1″ ■

NOTE The INV key reverses the conversion. This key will be used extensively later in the text.

An angle is in **standard position** if its vertex is at the origin and its initial side is along the positive *x*-axis. The two angles in Figure 1.16 are in standard position. An angle in standard position is said to lie in the quadrant in which its terminal side lies. For example, an acute angle is in quadrant I and an obtuse angle is in quadrant II. Angles in standard position having their terminal sides along the *x*-axis or *y*-axis, such as angles with measures 90°, 180°, 270°, and so on, are called **quadrantal angles.**

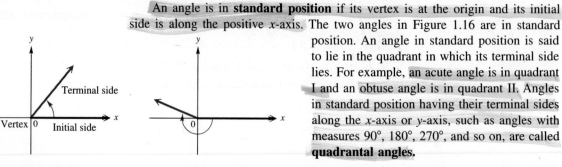

■ **FIGURE 1.16**

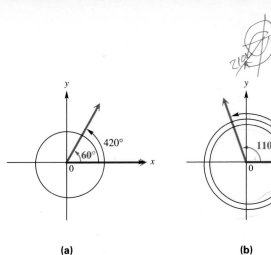

$210°$
$-510°$

420°

60°

830°

110°

360

$110°$

$470°$

(a)

(b)

FIGURE 1.17

A complete rotation of a ray results in an angle of measure 360°. But there is no reason why the rotation need stop at 360°. By continuing the rotation, angles of measure larger than 360° can be produced. The angles in Figure 1.17(a) have measures 60° and 420°. These two angles have the same initial side and the same terminal side, but different amounts of rotation. Angles that have the same initial side and the same terminal side are called **coterminal angles.** As shown in Figure 1.17(b), angles with measures 110° and 830° are coterminal.

■ *Example 4*

FINDING MEASURES OF COTERMINAL ANGLES

Find the angles of smallest possible positive measure coterminal with the following angles.

(a) 908°

Add or subtract 360° as many times as needed to get an angle with measure greater than 0° but less than 360°. Since $908° - 2 \cdot 360° = 908° - 720° = 188°$, an angle of 188° is coterminal with an angle of 908°. See Figure 1.18.

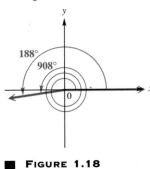

■ **FIGURE 1.18**

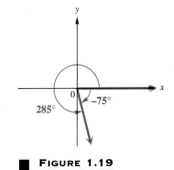

■ **FIGURE 1.19**

(b) −75°

Use a rotation of $360° + (-75°) = 285°$. See Figure 1.19. ■

Sometimes it is necessary to find an expression that will generate all angles coterminal with a given angle. For example, suppose that we wish to do this for a 60° angle. Since any angle coterminal with 60° can be obtained by adding an appropriate integer multiple of 360° to 60°, we can let n represent any integer, and the expression

$$60° + n \cdot 360°$$

will represent all such coterminal angles. The table below shows a few of these.

Value of n	Angle Coterminal with 60°
2	$60° + 2 \cdot 360° = 780°$
1	$60° + 1 \cdot 360° = 420°$
0	$60° + 0 \cdot 360° = 60°$ (the angle itself)
−1	$60° + (-1) \cdot 360° = -300°$

18

■ *Example 5*

**ANALYZING THE
REVOLUTIONS OF A
PHONOGRAPH
RECORD**

A phonograph record makes 45 revolutions per minute. Through how many degrees will a point on the edge of the record move in 2 seconds?

The record revolves 45 times per minute or 45/60 = 3/4 times per second (since there are 60 seconds in a minute). In 2 seconds, the record will revolve 2 · (3/4) = 3/2 times. Each revolution is 360°, so a point on the edge will revolve (3/2) · 360° = 540° in 2 seconds. ■

1.2 EXERCISES ■

Find the measure of each angle in Exercises 1–6. See Example 1.

1.

$(7x)°$ $(11x)°$

2.

$(2y)°$

$(4y)°$

3.

$(5k + 5)°$

$(3k + 5)°$

4. Supplementary angles with measures $10m + 7$ and $7m + 3$ degrees

5. Supplementary angles with measures $6x - 4$ and $8x - 12$ degrees

6. Complementary angles with measures $9z + 6$ and $3z$ degrees

7. If an angle measures x degrees, how can we represent its complement?

8. If an angle measures x degrees, how can we represent its supplement?

9. What angle is its own complement?

10. What angle is its own supplement?

Perform each calculation. See Example 2.

11. $75° \ 15' + 83° \ 32'$ **12.** $62° \ 18' + 21° \ 41'$ **13.** $89° + 23° \ 42'$ **14.** $71° \ 58' + 47° \ 29'$

15. $90° - 51° \ 28'$ **16.** $90° - 73° \ 48'$ **17.** $180° - 152° \ 43'$ **18.** $180° - 124° \ 51'$

19. $90° - 36° \ 18' \ 47''$ **20.** $90° - 72° \ 58' \ 11''$ **21.** $180° - 120° \ 42' \ 37''$ **22.** $180° - 86° \ 39' \ 54''$

Convert each angle measure to decimal degrees. Use a calculator, and round to the nearest thousandth of a degree. See Example 3.

23. $20° \ 54'$ **24.** $38° \ 42'$ **25.** $91° \ 35' \ 54''$

26. $34° \ 51' \ 35''$ **27.** $274° \ 18' \ 59''$ **28.** $165° \ 51' \ 09''$

Convert each angle measure to degrees, minutes, and seconds. Use a calculator as necessary. See Example 3.

29. $31.4296°$ **30.** $59.0854°$ **31.** $89.9004°$

32. $102.3771°$ **33.** $178.5994°$ **34.** $122.6853°$

Find the angles of smallest possible positive measure coterminal with the following angles. Use a calculator in Exercises 45 and 46. See Example 4.

35. $-40°$ **36.** $-98°$ **37.** $-125°$ **38.** $-203°$ **39.** $450°$ **40.** $489°$

41. $539°$ **42.** $699°$ **43.** $850°$ **44.** $1000°$ **45.** $-985.4063°$ **46.** $-1762.3974°$

Give an expression that generates all angles coterminal with the given angle. Let n represent any integer.

47. 30°   **48.** 45°   **49.** 60°   **50.** 90°   **51.** 135°   **52.** 270°   **53.** −90°   **54.** −135°

55. Explain why the answers to Exercises 52 and 53 give the same set of angles.

56. Which two of the following are not coterminal with $r°$?

   (a) $360° + r°$ **(b)** $r° − 360°$ **(c)** $360° − r°$ **(d)** $r° + 180°$

Sketch the following angles in standard position. Draw an arrow representing the correct amount of rotation. Find the measure of two other angles, one positive and one negative, that are coterminal with the given angle. Give the quadrant of each angle.

57. 75°   **58.** 89°   **59.** 122°   **60.** 174°   **61.** 234°   **62.** 250°   **63.** 300°   **64.** 324°

65. 438°   **66.** 593°   **67.** 512°   **68.** 624°   **69.** −52°   **70.** −61°   **71.** −159°   **72.** −214°

Locate the following points in a coordinate system. Draw a ray from the origin through the given point. Indicate with an arrow the angle in standard position having smallest positive measure. Then find the distance r from the origin to the point, using the distance formula of Section 1.1.

73. $(-3, -3)$ **74.** $(-5, 2)$ **75.** $(-3, -5)$ **76.** $(\sqrt{3}, 1)$ **77.** $(-2, 2\sqrt{3})$ **78.** $(4\sqrt{3}, -4)$

Solve each of the following. See Example 5.

79. A tire is rotating 600 times per minute. Through how many degrees does a point on the edge of the tire move in 1/2 second?

80. An airplane propeller rotates 1000 times per minute. Find the number of degrees that a point on the edge of the propeller will rotate in 1 second.

81. A pulley rotates through 75° in one minute. How many rotations does the pulley make in an hour?

82. One student in a surveying class measures an angle as 74.25°, while another student measures the same angle as 74° 20′. Find the difference in these measurements, both to the nearest minute and to the nearest hundredth of a degree.

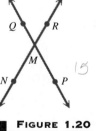

1.3 ─── ANGLE RELATIONSHIPS AND SIMILAR TRIANGLES

In this section we look at some geometric properties that will be used in the study of trigonometry.

   In Figure 1.20, the sides of angle *NMP* have been extended to form another angle, *RMQ*. The pair of angles *NMP* and *RMQ* are called **vertical angles.** Another pair of vertical angles, *NMQ* and *PMR,* are formed at the same time. Vertical angles have the following important property.

■ **FIGURE 1.20**

VERTICAL ANGLES   Vertical angles have equal measures.

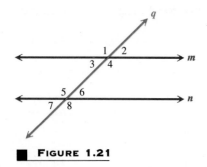

Parallel lines are lines that lie in the same plane and do not intersect. Figure 1.21 shows parallel lines *m* and *n*. When a line *q* intersects two parallel lines, *q* is called a **transversal.** In Figure 1.21, the transversal intersecting the parallel lines forms eight angles, indicated by numbers. Angles 1 through 8 in Figure 1.21 possess some special properties regarding their degree measures. The following chart gives their names with respect to each other, and rules regarding their measures.

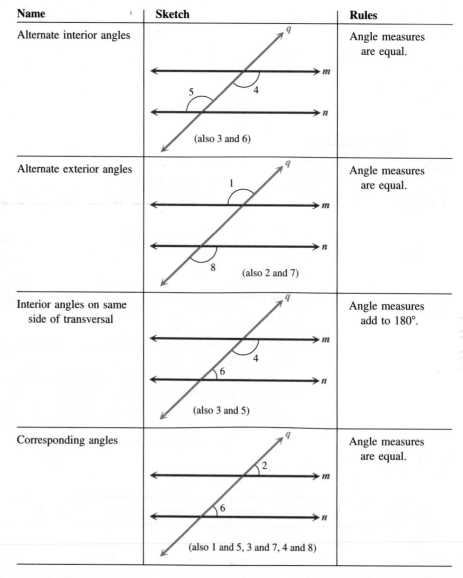

Name	Sketch	Rules
Alternate interior angles	(also 3 and 6)	Angle measures are equal.
Alternate exterior angles	(also 2 and 7)	Angle measures are equal.
Interior angles on same side of transversal	(also 3 and 5)	Angle measures add to 180°.
Corresponding angles	(also 1 and 5, 3 and 7, 4 and 8)	Angle measures are equal.

■ *Example 1*

FINDING ANGLE
MEASURES

Find the measure of each marked angle, given that lines m and n are parallel. (See Figure 1.22.)

The marked angles are alternate exterior angles, which are equal. This gives

$$3x + 2 = 5x - 40$$

$$42 = 2x$$

$$21 = x.$$

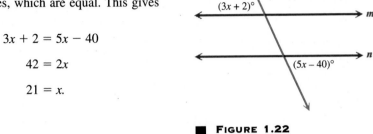

■ FIGURE 1.22

One angle has a measure of $3 \cdot 21 + 2 = 65$ degrees, and the other has a measure of $5x - 40 = 5 \cdot 21 - 40 = 65$ degrees. ■

TRIANGLES An important property of triangles that was first proved by the Greek geometers deals with the sum of the measures of the angles of any triangle.

ANGLE SUM OF A TRIANGLE	The sum of the measures of the angles of any triangle is 180°.

While it is not an actual proof, a rather convincing argument for the truth of this statement can be given using any size triangle cut from a piece of paper. Tear each corner from the triangle, as suggested in Figure 1.23(a). You should be able to rearrange the pieces so that the three angles form a straight angle, as shown in Figure 1.23(b).

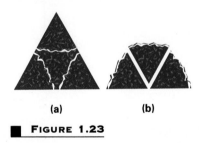

(a)   (b)

■ FIGURE 1.23

Suppose that two of the angles of a triangle are 48° and 61°. Find the measure of the third angle, x, by using the fact that all three angle measures add to 180°.

$$48° + 61° + x = 180°$$
$$109° + x = 180°$$
$$x = 71°$$

The third angle of the triangle measures 71°.

Triangles are classified according to angles and sides as shown in the following chart.

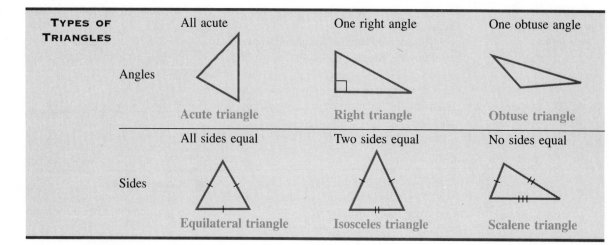

TYPES OF TRIANGLES	All acute	One right angle	One obtuse angle
Angles	Acute triangle	Right triangle	Obtuse triangle
	All sides equal	Two sides equal	No sides equal
Sides	Equilateral triangle	Isosceles triangle	Scalene triangle

Many of the key ideas of trigonometry depend on **similar triangles,** which are triangles of exactly the same shape but not necessarily the same size. Figure 1.24 shows three pairs of similar triangles.

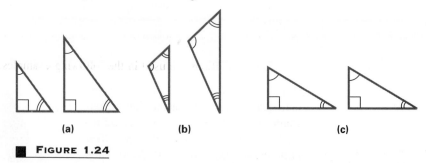

(a)   (b)   (c)

■ **FIGURE 1.24**

The two triangles in the third pair have not only the same shape but also the same size. Triangles that are both the same size and the same shape are called **congruent triangles.** If two triangles are congruent, then it is possible to pick one of

These rock carvings on the island of Hawaii were carved by Polynesians who settled the islands beginning about the year 500. These carvings suggest the ideas of similarity and congruence.

them up and place it on top of the other so that they coincide. If two triangles are congruent, then they must be similar. However, two similar triangles need not be congruent.

The triangle supports for a child's swing are congruent triangles, machine-produced with exactly the same dimensions each time. These supports are just one example of similar triangles. The supports of a long bridge, all the same shape but decreasing in size toward the center of the bridge, are another example of similar triangles.

Suppose that a correspondence between two triangles ABC and DEF is set up as follows.

Angle A corresponds to angle D.   Side AB corresponds to side DE.

Angle B corresponds to angle E.   Side BC corresponds to side EF.

Angle C corresponds to angle F.   Side AC corresponds to side DF.

For triangle ABC to be similar to triangle DEF, the following conditions must hold.

CONDITIONS FOR SIMILAR TRIANGLES	1. Corresponding angles must have the same measure. 2. Corresponding sides must be proportional. (That is, their ratios must be equal.)

These conditions are used in the following examples.

■ *Example 2*

FINDING ANGLE MEASURES IN SIMILAR TRIANGLES

In Figure 1.25, triangles ABC and NMP are similar. Find the measures of angles B and C.

Since the triangles are similar, corresponding angles have the same measure. Since C corresponds to P and P measures 104°, angle C also measures 104°. Since angles B and M correspond, B measures 31°. ■

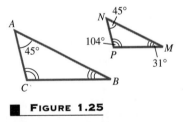

■ **FIGURE 1.25**

NOTE | The small arcs found at the angles in Figure 1.25 are used to denote the corresponding angles in the triangle. This symbolism will be used when appropriate in this book. We will also use △ to denote "triangle."

■ *Example 3*

FINDING SIDE LENGTHS IN SIMILAR TRIANGLES

Given that $\triangle ABC$ and $\triangle DFE$ in Figure 1.26 are similar, find the lengths of the unknown sides of $\triangle DFE$.

As mentioned before, similar triangles have corresponding sides in proportion. Use this fact to find the unknown side lengths in

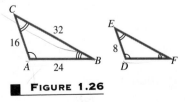

■ **FIGURE 1.26**

△*DFE*. Side *DF* of △*DFE* corresponds to side *AB* of △*ABC*, and sides *DE* and *AC* correspond. This leads to the proportion

$$\frac{8}{16} = \frac{DF}{24}.$$

Recall the following property of proportions.

If $\dfrac{a}{b} = \dfrac{c}{d}$, then $ad = bc$.

This process is called *cross-multiplication*. Use cross-multiplication to solve the equation for *DF*.

$$\frac{8}{16} = \frac{DF}{24}$$

$$8 \cdot 24 = 16 \cdot DF \qquad \text{Cross-multiply.}$$

$$192 = 16 \cdot DF$$

$$12 = DF$$

Side *DF* has a length of 12.

Side *EF* corresponds to *CB*. This leads to another proportion:

$$\frac{8}{16} = \frac{EF}{32}.$$

Cross-multiplication gives

$$8 \cdot 32 = 16 \cdot EF$$

$$16 = EF.$$

Side *EF* has a length of 16. ■

Applied problems can sometimes be solved using properties of similar triangles.

■ *Example 4*

SOLVING A
PROBLEM USING
SIMILAR TRIANGLES

The people at the Arcade Fire Station need to measure the height of the station flagpole. They notice that at the instant when the shadow of the station is 18 feet long, the shadow of the flagpole is 99 feet long. The station is 10 feet high. Find the height of the flagpole.

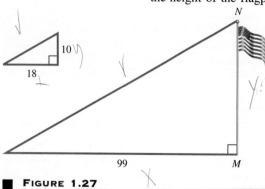

Figure 1.27 shows the information given in the problem. The two triangles shown there are similar, so that corresponding sides are in proportion, with

$$\frac{MN}{10} = \frac{99}{18}$$

or

$$\frac{MN}{10} = \frac{11}{2}$$

$$2 \cdot MN = 110$$

$$MN = 55.$$

The flagpole is 55 feet high. ■

1.3 EXERCISES ■

Use the properties of angle measures given in this section to find the measure of each marked angle. In Exercises 9–12, m and n are parallel. See Example 1.

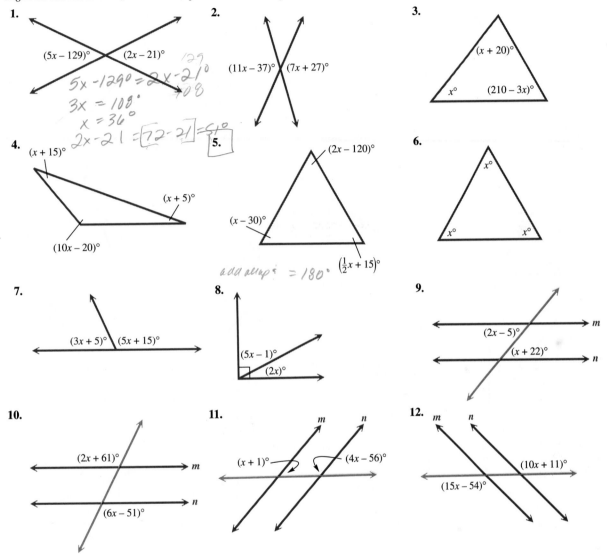

1.

$(5x - 129)°$ $(2x - 21)°$

$5x - 129° = 2x - 21°$

$3x = 108°$

$x = 36°$

$2x - 21 = \boxed{72 - 21} = 51°$

2.

$(11x - 37)°$ $(7x + 27)°$

3.

$(x + 20)°$

$x°$ $(210 - 3x)°$

4.

$(x + 15)°$

$(x + 5)°$

$(10x - 20)°$

5.

$(2x - 120)°$

$(x - 30)°$

add allap $= 180°$

$\left(\frac{1}{2}x + 15\right)°$

6.

$x°$

$x°$ $x°$

7.

$(3x + 5)°$ $(5x + 15)°$

8.

$(5x - 1)°$

$(2x)°$

9.

$(2x - 5)°$ m

$(x + 22)°$ n

10.

$(2x + 61)°$ m

$(6x - 51)°$ n

11.

m   n

$(x + 1)°$ $(4x - 56)°$

12.

m   n

$(10x + 11)°$

$(15x - 54)°$

The measures of two angles of a triangle are given. Find the measure of the third angle.

13. 37°, 52°

14. 29°, 104°

15. 147° 12′, 30° 19′

16. 136° 50′, 41° 38′

17. 74° 12′ 59″, 80° 58′ 05″

18. 29° 51′ 37″, 49° 28′ 50″

19. Can a triangle have two angles of measures 85° and 100°? Explain.

20. Can a triangle have two obtuse angles? Explain.

21. Use the given figure to find the measures of the numbered angles, given that lines *m* and *n* are parallel.

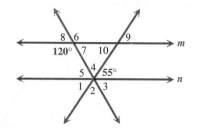

22. Find the measures of the marked angles, given that *x* + *y* = 40. (*Hint:* You must solve a system of equations.)

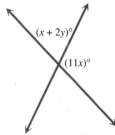

Classify each triangle in Exercises 23–34 as either acute, right, *or* obtuse. *Also classify each as either* equilateral, isosceles, *or* scalene.

23.   **24.**   **25.**   **26.**

27.   **28.**   **29.**   **30.**

31.   **32.**   **33.**   **34.**

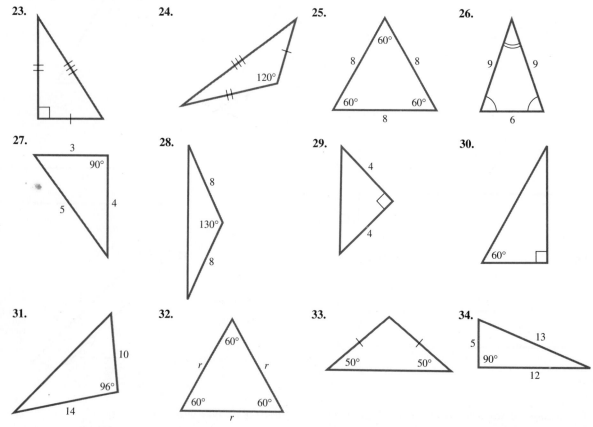

35. Write a definition of *isosceles right triangle*.

36. Explain why the sum of the lengths of any two sides of a triangle must be greater than the length of the third side.

37. Must all equilateral triangles be similar? Explain.

38. In the classic 1939 movie *The Wizard of Oz*, the scarecrow, upon getting a brain, says the following: "The sum of the square roots of any two sides of an isosceles triangle is equal to the square root of the remaining side." Give an example to show that his statement is incorrect.

Name the corresponding angles and the corresponding sides for each of the following pairs of similar triangles.

39.

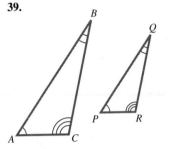

40.

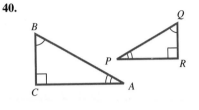

41. (*HK* is parallel to *EF*.)

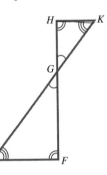

42. (*EA* is parallel to *CD*.)

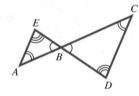

Find all unknown angle measures in each pair of similar triangles. See Example 2.

43.

44.

45.

46.

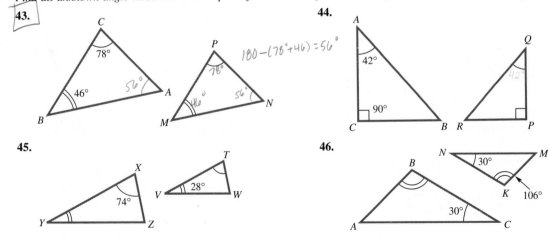
180 − (78° + 46) = 56°

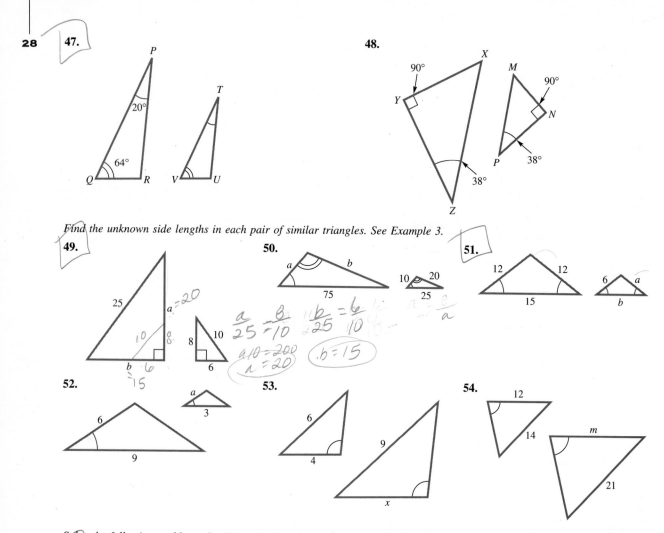

Find the unknown side lengths in each pair of similar triangles. See Example 3.

Solve the following problems. See Example 4.

55. A tree casts a shadow 45 m long. At the same time, the shadow cast by a vertical 2-m stick is 3 m long. Find the height of the tree.

56. A forest fire lookout tower casts a shadow 180 ft long at the same time that the shadow of a 9-ft truck is 15 ft long. Find the height of the tower.

57. On a photograph of a triangular piece of land, the lengths of the three sides are 4 cm, 5 cm, and 7 cm, respectively. The shortest side of the actual piece of land is 400 m long. Find the lengths of the other two sides.

58. The Santa Cruz lighthouse is 14 m tall and casts a shadow 28 m long at 7 P.M. At the same time, the shadow of the lighthouse keeper is 3.5 m long. How tall is she?

59. A house is 15 ft tall. Its shadow is 40 ft long at the same time the shadow of a nearby building is 300 ft long. Find the height of the building.

60. By drawing lines on a map, a triangle can be formed by the cities of Phoenix, Tucson, and Yuma. On the map, the distance between Phoenix and Tucson is 8 cm, the distance between Phoenix and Yuma is 12 cm, and the distance between Tucson and Yuma is 17 cm. The actual straight-line distance from Phoenix to Yuma is 230 km. Find the distances between the other pairs of cities.

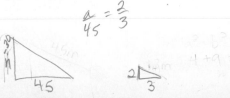

In each diagram, there are two similar triangles. Find the unknown measurement in each. (Hint: In the sketch for Exercise 61, the side of length 100 in the small triangle corresponds to a side of length 100 + 120 = 220 in the larger triangle.)

61.

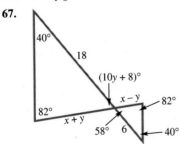

62.

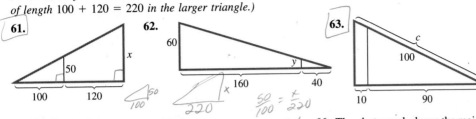

$\frac{50}{100} = \frac{x}{220}$

$110 = x$

63.

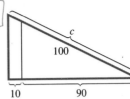

64.

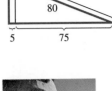

Work each of the following problems.

65. Two quadrilaterals (four-sided figures) are similar. The lengths of the three shortest sides of the first quadrilateral are 18 cm, 24 cm, and 32 cm. The lengths of the two longest sides of the second quadrilateral are 48 cm and 60 cm. Find the unknown lengths of the sides of these two figures.

66. The photograph shows the maintenance of the Mount Rushmore head of Lincoln. Assume that Lincoln was 6 1/3 ft tall and his head 3/4 ft long. Knowing that the carved head of Lincoln is 60 ft tall, find out how tall his entire body would be if it were carved into the mountain.

In each of the following figures, two similar triangles are present. Find the value of each variable in the figures.

67.

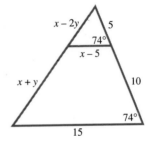

68.

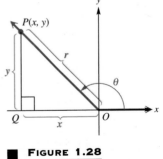

1.4 DEFINITIONS OF THE TRIGONOMETRIC FUNCTIONS

The study of trigonometry covers the six trigonometric functions defined in this section. Most sections in the remainder of this book involve at least one of these functions. To define these six basic functions, start with an angle θ (the Greek letter *theta**) in standard position. Choose any point P having coordinates (x, y) on the terminal side of angle θ. (The point P must not be the vertex of the angle.) See Figure 1.28.

∎ **FIGURE 1.28**

*Greek letters are often used to name angles. A list of Greek letters appears inside the back cover of the book.

A perpendicular from P to the x-axis at point Q determines a triangle having vertices at O, P, and Q. (More will be said about such triangles in Section 1.5.) The distance r from $P(x, y)$ to the origin, $(0, 0)$, can be found from the distance formula.

$$r = \sqrt{(x - 0)^2 + (y - 0)^2}$$
$$r = \sqrt{x^2 + y^2}$$

Notice that $r > 0$, since distance is never negative.

The six trigonometric functions of angle θ are called **sine, cosine, tangent, cotangent, secant,** and **cosecant.**

TRIGONOMETRIC FUNCTIONS

Let (x, y) be a point other than the origin on the terminal side of an angle θ in standard position. The distance from the point to the origin is $r = \sqrt{x^2 + y^2}$. The six trigonometric functions of θ are:

$$\sin \theta = \frac{y}{r} \qquad\qquad \csc \theta = \frac{r}{y} \quad (y \neq 0)$$

$$\cos \theta = \frac{x}{r} \qquad\qquad \sec \theta = \frac{r}{x} \quad (x \neq 0)$$

$$\tan \theta = \frac{y}{x} \quad (x \neq 0) \qquad \cot \theta = \frac{x}{y} \quad (y \neq 0).$$

NOTE Because of the restrictions on the denominators in the definitions of tangent, cotangent, secant, and cosecant, some angles will have undefined function values. This will be discussed in more detail later.

■ Example 1

FINDING THE FUNCTION VALUES OF AN ANGLE

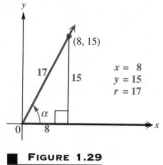

■ FIGURE 1.29

The terminal side of an angle α in standard position goes through the point $(8, 15)$. Find the values of the six trigonometric functions of angle α.

Figure 1.29 shows angle α and the triangle formed by dropping a perpendicular from the point $(8, 15)$ to the x-axis. The point $(8, 15)$ is 8 units to the right of the y-axis and 15 units above the x-axis, so that $x = 8$ and $y = 15$. Since $r = \sqrt{x^2 + y^2}$,

$$r = \sqrt{8^2 + 15^2}$$
$$= \sqrt{64 + 225}$$
$$= \sqrt{289}$$
$$= 17.$$

The values of the six trigonometric functions of angle α can now be found with the definitions given above.

$$\sin \alpha = \frac{y}{r} = \frac{15}{17} \qquad \csc \alpha = \frac{r}{y} = \frac{17}{15}$$

$$\cos \alpha = \frac{x}{r} = \frac{8}{17} \qquad \sec \alpha = \frac{r}{x} = \frac{17}{8}$$

$$\tan \alpha = \frac{y}{x} = \frac{15}{8} \qquad \cot \alpha = \frac{x}{y} = \frac{8}{15} \quad \blacksquare$$

■ **Example 2**

FINDING THE FUNCTION VALUES OF AN ANGLE

The terminal side of angle β in standard position goes through $(-3, -4)$. Find the values of the six trigonometric functions of β.

As shown in Figure 1.30, $x = -3$ and $y = -4$. The value of r is

$$r = \sqrt{(-3)^2 + (-4)^2}$$

$$r = \sqrt{25}$$

$$r = 5.$$

(Remember that $r > 0$.) Then by the definitions of the trigonometric functions,

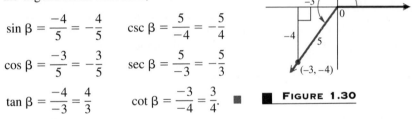

$$\sin \beta = \frac{-4}{5} = -\frac{4}{5} \qquad \csc \beta = \frac{5}{-4} = -\frac{5}{4}$$

$$\cos \beta = \frac{-3}{5} = -\frac{3}{5} \qquad \sec \beta = \frac{5}{-3} = -\frac{5}{3}$$

$$\tan \beta = \frac{-4}{-3} = \frac{4}{3} \qquad \cot \beta = \frac{-3}{-4} = \frac{3}{4}. \quad \blacksquare$$

■ **FIGURE 1.30**

The six trigonometric functions can be found from *any* point on the terminal side of the angle other than the origin. To see why any point may be used, refer to Figure 1.31, which shows an angle θ and two distinct points on its terminal side. Point P has coordinates (x, y) and point P' (read "P-prime") has coordinates (x', y'). Let r be the length of the hypotenuse of triangle OPQ, and let r' be the length of the hypotenuse of triangle $OP'Q'$. Since corresponding sides of similar triangles are in proportion,

$$\frac{y}{r} = \frac{y'}{r'},$$

so that $\sin \theta = y/r$ is the same no matter which point is used to find it. A similar result holds for the other five functions.

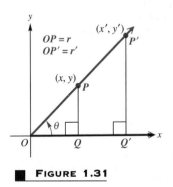

■ **FIGURE 1.31**

Recall from algebra that the graph of the equation

$$Ax + By = 0$$

is a line that passes through the origin. If we restrict x to have only nonpositive or only nonnegative values, we obtain as the graph a ray with endpoint at the origin. For example, the graph of $x + 2y = 0$, $x \geq 0$, is shown in Figure 1.32. A ray such as the one described above can serve as the terminal side of an angle in standard position. By finding a point on the ray, the trigonometric function values of the angle can be found.

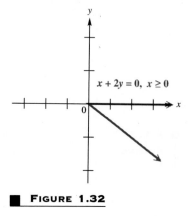

■ **FIGURE 1.32**

■ *Example 3*
FINDING THE
FUNCTION VALUES
OF AN ANGLE

Find the six trigonometric function values of the angle θ in standard position, if the terminal side of θ is defined by $x + 2y = 0$, $x \geq 0$.

The angle is shown in Figure 1.33. We can use *any* point except $(0, 0)$ on the terminal side of θ to find the trigonometric function values, so if we let $x = 2$, we can find the corresponding value of y.

$$x + 2y = 0, x \geq 0$$
$$2 + 2y = 0 \qquad \text{Arbitrarily choose } x = 2.$$
$$2y = -2$$
$$y = -1$$

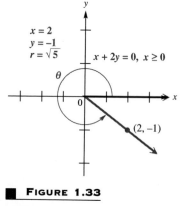

■ **FIGURE 1.33**

The point $(2, -1)$ lies on the terminal side, and the corresponding value of r is $r = \sqrt{2^2 + (-1)^2} = \sqrt{5}$. Now use the definitions of the trigonometric functions.

$$\sin \theta = \frac{y}{r} = \frac{-1}{\sqrt{5}} = -\frac{\sqrt{5}}{5} \qquad \csc \theta = \frac{r}{y} = \frac{\sqrt{5}}{-1} = -\sqrt{5}$$

$$\cos \theta = \frac{x}{r} = \frac{2}{\sqrt{5}} = \frac{2\sqrt{5}}{5} \qquad \sec \theta = \frac{r}{x} = \frac{\sqrt{5}}{2}$$

$$\tan \theta = \frac{y}{x} = \frac{-1}{2} = -\frac{1}{2} \qquad \cot \theta = \frac{x}{y} = \frac{2}{-1} = -2 \quad ■$$

If the terminal side of an angle in standard position lies along the y-axis, any point on this terminal side has x-coordinate 0. Similarly, any angle with terminal side on the x-axis has y-coordinate 0 for any point on the terminal side. Since the values of x and y appear in the denominators of some of the trigonometric functions, and since a fraction is undefined if its denominator is 0, some of the trigonometric function values of quadrantal angles will be undefined.

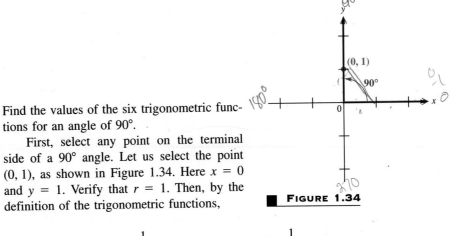

■ **FIGURE 1.34**

■ *Example 4*

FINDING TRIGONOMETRIC FUNCTION VALUES OF A QUADRANTAL ANGLE

Find the values of the six trigonometric functions for an angle of 90°.

First, select any point on the terminal side of a 90° angle. Let us select the point $(0, 1)$, as shown in Figure 1.34. Here $x = 0$ and $y = 1$. Verify that $r = 1$. Then, by the definition of the trigonometric functions,

$$\sin 90° = \frac{1}{1} = 1 \qquad \csc 90° = \frac{1}{1} = 1$$

$$\cos 90° = \frac{0}{1} = 0 \qquad \sec 90° = \frac{1}{0} \text{ (undefined)}$$

$$\tan 90° = \frac{1}{0} \text{ (undefined)} \qquad \cot 90° = \frac{0}{1} = 0. \quad ■$$

The conditions under which the trigonometric function values of quadrantal angles are undefined are summarized here.

If the terminal side of a quadrantal angle lies along the y-axis, the tangent and secant functions are undefined. If it lies along the x-axis, the cotangent and cosecant functions are undefined.

Since the most commonly used quadrantal angles are 0°, 90°, 180°, 270°, and 360°, the values of the functions of these angles are summarized in the following table. This table is for reference only; you should be able to reproduce it quickly.

QUADRANTAL ANGLES	θ	$\sin \theta$	$\cos \theta$	$\tan \theta$	$\cot \theta$	$\sec \theta$	$\csc \theta$
	0°	0	1	0	Undefined	1	Undefined
	90°	1	0	Undefined	0	Undefined	1
	180°	0	−1	0	Undefined	−1	Undefined
	270°	−1	0	Undefined	0	Undefined	−1
	360°	0	1	0	Undefined	1	Undefined

The values given in this table can also be found with a calculator that has trigonometric function keys. First, make sure the calculator is set for *degree mode.* Then, for example, cos 90° can be found by entering 90 and pressing the $\boxed{\text{cos}}$ key.

Enter: 90 $\boxed{\text{cos}}$ **Display:** $\boxed{ 0}$

Trying to find tan 90° would produce the following.

Enter: 90 $\boxed{\text{tan}}$ **Display:** $\boxed{ Error}$

The type of error message displayed depends upon the model of the calculator.

CAUTION | One of the most common errors involving calculators in trigonometry occurs when the calculator is set for *radian measure,* rather than *degree measure.* (Radian measure of angles is studied in Chapter 3.) For this reason, be sure that you know how to set your calculator to *degree mode.*

There are no calculator keys for finding the function values of cotangent, secant, or cosecant. The next section shows how to find these function values with a calculator.

1.4 EXERCISES ■

In Exercises 1–4, sketch an angle θ in standard position such that θ has the smallest possible positive measure, and the given point is on the terminal side of θ.

1. $(-3, 4)$ **2.** $(-4, -3)$ **3.** $(5, -12)$ **4.** $(-12, -5)$

Find the values of the six trigonometric functions for the angles in standard position having the following points on their terminal sides. Rationalize denominators when applicable. Use a calculator in Exercises 29 and 30. See Examples 1, 2, and 4.

5. $(-3, 4)$ **6.** $(-4, -3)$ **7.** $(5, -12)$ **8.** $(-12, -5)$

9. $(6, 8)$ **10.** $(-9, -12)$ **11.** $(-7, 24)$ **12.** $(24, 7)$

13. $(0, 2)$ **14.** $(-4, 0)$ **15.** $(8, 0)$ **16.** $(0, -9)$

17. $(1, \sqrt{3})$ **18.** $(-2\sqrt{3}, -2)$ **19.** $(5\sqrt{3}, -5)$ **20.** $(8, -8\sqrt{3})$

21. $(2\sqrt{2}, -2\sqrt{2})$ **22.** $(-2\sqrt{2}, 2\sqrt{2})$ **23.** $(\sqrt{5}, -2)$ **24.** $(-\sqrt{7}, \sqrt{2})$

25. $(-\sqrt{13}, \sqrt{3})$ **26.** $(-\sqrt{11}, -\sqrt{5})$ **27.** $(\sqrt{15}, -\sqrt{10})$ **28.** $(-\sqrt{12}, \sqrt{13})$

29. $(8.7691, -3.2473)$ **30.** $(-5.1021, 7.6132)$

31. For any nonquadrantal angle θ, sin θ and csc θ will have the same sign. Explain why this is so.

32. If cot θ is undefined, what is the value of tan θ?

33. How is the value of *r* interpreted geometrically in the definitions of the sine, cosine, secant, and cosecant functions?

34. If the terminal side of an angle β is in quadrant III, what is the sign of each of the trigonometric function values of β?

Suppose that the point (x, y) is in the indicated quadrant. Decide whether the given ratio is positive or negative. (Hint: It may be helpful to draw a sketch.)

35. II, $\dfrac{y}{r}$ **36.** II, $\dfrac{x}{r}$ **37.** III, $\dfrac{y}{r}$ **38.** III, $\dfrac{x}{r}$ **39.** III, $\dfrac{y}{x}$ **40.** III, $\dfrac{x}{y}$

41. IV, $\dfrac{x}{r}$ **42.** IV, $\dfrac{y}{r}$ **43.** IV, $\dfrac{y}{x}$ **44.** IV, $\dfrac{x}{y}$ **45.** III, $\dfrac{r}{x}$ **46.** II, $\dfrac{r}{y}$

In Exercises 47–52, an equation with a restriction on x is given. This is an equation of the terminal side of an angle θ in standard position. Sketch the smallest positive such angle θ, and find the values of the six trigonometric functions of θ. See Example 3.

47. $2x + y = 0, \quad x \geq 0$ **48.** $3x + 5y = 0, \quad x \geq 0$ **49.** $-4x + 7y = 0, \quad x \leq 0$

50. $-6x - y = 0, \quad x \leq 0$ **51.** $-5x - 3y = 0, \quad x \leq 0$ **52.** $6x - 5y = 0, \quad x \geq 0$

Use the trigonometric function values of quadrantal angles given in this section to evaluate each of the following. An expression such as $\cot^2 90°$ means $(\cot 90°)^2$ which is equal to $0^2 = 0$.

53. $\cos 90° + 3 \sin 270°$ $0 + -3 = -3$ **54.** $\tan 0° - 6 \sin 90°$

55. $3 \sec 180° - 5 \tan 360°$ **56.** $4 \csc 270° + 3 \cos 180°$

57. $\tan 360° + 4 \sin 180° + 5 \cos^2 180°$ $5(1) = 5$ **58.** $2 \sec 0° + 4 \cot^2 90° + \cos 360°$

59. $\sin^2 180° + \cos^2 180°$ $5(\cos 180°)^2$ **60.** $\sin^2 360° + \cos^2 360°$

61. $\sec^2 180° - 3 \sin^2 360° + 2 \cos 180°$ **62.** $5 \sin^2 90° + 2 \cos^2 270° - 7 \tan^2 360°$

If n is an integer, $n \cdot 180°$ represents an integer multiple of $180°$, and $(2n + 1) \cdot 90°$ represents an odd integer multiple of $90°$. Decide whether each of the following is equal to 0, 1, -1, or is undefined.

63. $\sin [n \cdot 180°]$ **64.** $\cos [(2n + 1) \cdot 90°]$ **65.** $\tan [(2n + 1) \cdot 90°]$ **66.** $\tan [n \cdot 180°]$

| **1.5** | ——————— | # USING THE DEFINITIONS OF THE TRIGONOMETRIC FUNCTIONS |

In this section several useful results are derived from the definitions of the trigonometric functions given in the previous section. First, recall the definition of a reciprocal: the **reciprocal** of the nonzero number x is $1/x$. For example, the reciprocal of 2 is $1/2$, and the reciprocal of $8/11$ is $11/8$. There is no reciprocal for 0.

Scientific calculators have a reciprocal key, usually labeled $\boxed{1/x}$. Using this key gives the reciprocal of any nonzero number entered in the display, as shown in the following examples.

Enter: 5.4 $\boxed{1/x}$ **Display:** .185185185
       .25 $\boxed{1/x}$ 4
       0 $\boxed{1/x}$ *Error*

The definitions of the trigonometric functions in the previous section were written so that functions on the same line are reciprocals of each other. Since $\sin \theta = y/r$ and $\csc \theta = r/y$,

$$\sin \theta = \frac{1}{\csc \theta} \qquad \text{and} \qquad \csc \theta = \frac{1}{\sin \theta}.$$

Also, $\cos \theta$ and $\sec \theta$ are reciprocals, as are $\tan \theta$ and $\cot \theta$. In summary, we have the **reciprocal identities** that hold for any angle θ that does not lead to a zero denominator.

RECIPROCAL IDENTITIES	
$\sin \theta = \dfrac{1}{\csc \theta}$	$\csc \theta = \dfrac{1}{\sin \theta}$
$\cos \theta = \dfrac{1}{\sec \theta}$	$\sec \theta = \dfrac{1}{\cos \theta}$
$\tan \theta = \dfrac{1}{\cot \theta}$	$\cot \theta = \dfrac{1}{\tan \theta}$

NOTE When studying identities, be aware that various forms exist. For example,

$$\sin \theta = \frac{1}{\csc \theta}$$

can also be written

$$\csc \theta = \frac{1}{\sin \theta} \qquad \text{and} \qquad (\sin \theta)(\csc \theta) = 1.$$

You should become familiar with all forms of these identities.

Identities are equations that are true for all meaningful values of the variable. For example, both $(x + y)^2 = x^2 + 2xy + y^2$ and $2(x + 3) = 2x + 6$ are identities. Identities are studied in more detail in Chapter 5.

■ *Example 1*
USING THE
RECIPROCAL
IDENTITIES

Find each function value.

(a) $\cos \theta$, if $\sec \theta = \dfrac{5}{3}$

Since $\cos \theta = 1/\sec \theta$,

$$\cos \theta = \frac{1}{5/3} = \frac{3}{5}.$$

(b) $\sin \theta$, if $\csc \theta = -\dfrac{\sqrt{12}}{2}$

$$\sin \theta = \frac{1}{-\sqrt{12}/2}$$

$$= \frac{-2}{\sqrt{12}}$$

$$= \frac{-2}{2\sqrt{3}} \qquad \sqrt{12} = \sqrt{4 \cdot 3} = 2\sqrt{3}$$

$$= \frac{-1}{\sqrt{3}}$$

$$= \frac{-\sqrt{3}}{3} \qquad \text{Multiply by } \frac{\sqrt{3}}{\sqrt{3}} \text{ to rationalize the denominator.} \quad \blacksquare$$

In the definition of the trigonometric functions, r is the distance from the origin to the point (x, y). Distance is never negative, so $r > 0$. If we choose a point (x, y) in quadrant I, then both x and y will be positive. Since $r > 0$, all six of the fractions used in the definitions of the trigonometric functions will be positive, so that the values of all six functions will be positive in quadrant I.

A point (x, y) in quadrant II has $x < 0$ and $y > 0$. This makes the values of sine and cosecant positive for quadrant II angles, while the other four functions take on negative values. Similar results can be obtained for the other quadrants, as summarized below.

SIGNS OF FUNCTION VALUES	θ in quadrant	$\sin \theta$	$\cos \theta$	$\tan \theta$	$\cot \theta$	$\sec \theta$	$\csc \theta$
	I	+	+	+	+	+	+
	II	+	−	−	−	−	+
	III	−	−	+	+	−	−
	IV	−	+	−	−	+	−

$$
\begin{array}{ccc}
\begin{array}{c} x<0 \\ y>0 \\ r>0 \end{array} & \quad & \begin{array}{c} x>0 \\ y>0 \\ r>0 \end{array} \\
\textbf{II} & & \textbf{I} \\
\text{Sine and cosecant} & & \text{All functions} \\
\text{positive} & & \text{positive} \\
\hline
\begin{array}{c} x<0 \\ y<0 \\ r>0 \end{array} & & \begin{array}{c} x>0 \\ y<0 \\ r>0 \end{array} \\
\textbf{III} & & \textbf{IV} \\
\text{Tangent and cotangent} & & \text{Cosine and secant} \\
\text{positive} & & \text{positive}
\end{array}
$$

38

■ *Example 2*

IDENTIFYING THE
QUADRANT OF AN
ANGLE

Identify the quadrant (or quadrants) for any angle θ that satisfies $\sin \theta > 0$, $\tan \theta < 0$.

Since $\sin \theta > 0$ in quadrants I and II, while $\tan \theta < 0$ in quadrants II and IV, both conditions are met only in quadrant II. ■

Figure 1.35 shows an angle θ as it increases in measure from near $0°$ toward $90°$. In each case, the value of r is the same. As the measure of the angle increases, y increases but never exceeds r, so that $y \leq r$. Dividing both sides by the positive number r gives

$$y \leq r$$

$$\frac{y}{r} \leq 1.$$

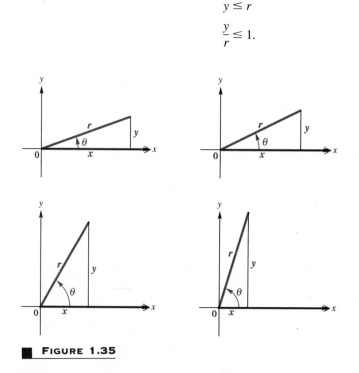

■ FIGURE 1.35

In a similar way, angles in the fourth quadrant suggest that

$$-1 \leq \frac{y}{r},$$

so

$$-1 \leq \frac{y}{r} \leq 1.$$

Since $y/r = \sin \theta$,

$$-1 \leq \sin \theta \leq 1$$

for any angle θ. In the same way,

$$-1 \leq \cos \theta \leq 1.$$

The tangent of an angle is defined as y/x. It is possible that $x < y$, that $x = y$, or that $x > y$. For this reason y/x can take on any value at all, so $\tan \theta$ can be any real number, as can $\cot \theta$.

The functions $\sec \theta$ and $\csc \theta$ are reciprocals of the functions $\cos \theta$ and $\sin \theta$, respectively, making

$$\sec \theta \leq -1 \quad \text{or} \quad \sec \theta \geq 1,$$
$$\csc \theta \leq -1 \quad \text{or} \quad \csc \theta \geq 1.$$

In summary, the ranges of the trigonometric functions are as follows.

RANGES OF TRIGONOMETRIC FUNCTIONS

For any angle θ for which the indicated functions exist:

1. $-1 \leq \sin \theta \leq 1$ and $-1 \leq \cos \theta \leq 1$;
2. $\tan \theta$ and $\cot \theta$ may be equal to any real number;
3. $\sec \theta \leq -1$ or $\sec \theta \geq 1$ and $\csc \theta \leq -1$ or $\csc \theta \geq 1$.

(Notice that $\sec \theta$ and $\csc \theta$ are *never* between -1 and 1.)

NOTE

Using absolute value notation, we can specify the ranges of the sine, cosine, secant, and cosecant functions as follows:

$$|\sin \theta| \leq 1 \qquad |\cos \theta| \leq 1 \qquad |\sec \theta| \geq 1 \qquad |\csc \theta| \geq 1.$$

■ *Example 3*

DECIDING WHETHER CERTAIN TRIGONOMETRIC FUNCTION VALUES ARE POSSIBLE

Decide whether the following statements are *possible* or *impossible*.

(a) $\sin \theta = \sqrt{8}$

For any value of θ, $-1 \leq \sin \theta \leq 1$. Since $\sqrt{8} > 1$, there is no value of θ with $\sin \theta = \sqrt{8}$.

(b) $\tan \theta = 110.47$

Tangent can take on any value. Thus, $\tan \theta = 110.47$ is possible.

(c) $\sec \theta = .6$

Since $\sec \theta \leq -1$ or $\sec \theta \geq 1$, the statement $\sec \theta = .6$ is impossible. ■

The six trigonometric functions are defined in terms of x, y, and r, where the Pythagorean theorem shows that $r^2 = x^2 + y^2$ and $r > 0$. With these relationships, knowing the value of only one function and the quadrant in which the angle lies makes it possible to find the values of all six of the trigonometric functions. This procedure is shown in the next example.

■ *Example 4*

FINDING ALL
FUNCTION VALUES
GIVEN ONE VALUE
AND THE QUADRANT

Suppose that angle α is in quadrant II and $\sin \alpha = 2/3$. Find the values of the other five functions.

We can choose any point on the terminal side of angle α. For simplicity, since $\sin \alpha = y/r$, choose the point with $r = 3$. Then

$$\frac{y}{r} = \frac{2}{3}.$$

If $r = 3$, then y will be 2. To find x, use the result $x^2 + y^2 = r^2$.

$$x^2 + y^2 = r^2$$
$$x^2 + 2^2 = 3^2$$
$$x^2 + 4 = 9$$
$$x^2 = 5$$
$$x = \sqrt{5} \quad \text{or} \quad x = -\sqrt{5}$$

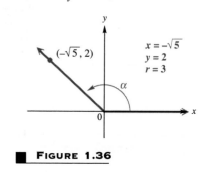

$x = -\sqrt{5}$
$y = 2$
$r = 3$

■ **FIGURE 1.36**

Since α is in quadrant II, x must be negative, as shown in Figure 1.36. Reject $\sqrt{5}$ for x, so $x = -\sqrt{5}$. This puts the point $(-\sqrt{5}, 2)$ on the terminal side of α.

Now that the values of x, y, and r are known, the values of the remaining trigonometric functions can be found.

$$\cos \alpha = \frac{x}{r} = \frac{-\sqrt{5}}{3}$$

$$\tan \alpha = \frac{y}{x} = \frac{2}{-\sqrt{5}} = \frac{-2\sqrt{5}}{5}$$

$$\cot \alpha = \frac{x}{y} = \frac{-\sqrt{5}}{2}$$

$$\sec \alpha = \frac{r}{x} = \frac{3}{-\sqrt{5}} = \frac{-3\sqrt{5}}{5}$$

$$\csc \alpha = \frac{r}{y} = \frac{3}{2} \quad ■$$

As previously mentioned, $x^2 + y^2 = r^2$. Dividing both sides by r^2 gives

$$\frac{x^2}{r^2} + \frac{y^2}{r^2} = \frac{r^2}{r^2},$$

or

$$\left(\frac{x}{r}\right)^2 + \left(\frac{y}{r}\right)^2 = 1.$$

Since $\sin \theta = y/r$ and $\cos \theta = x/r$, this result becomes

$$(\sin \theta)^2 + (\cos \theta)^2 = 1,$$

or, as it is usually written,

$$\sin^2 \theta + \cos^2 \theta = 1.$$

Starting with $x^2 + y^2 = r^2$ and dividing through by x^2 gives

$$\frac{x^2}{x^2} + \frac{y^2}{x^2} = \frac{r^2}{x^2}$$

$$1 + \left(\frac{y}{x}\right)^2 = \left(\frac{r}{x}\right)^2$$

$$1 + (\tan \theta)^2 = (\sec \theta)^2$$

or

$$\tan^2 \theta + 1 = \sec^2 \theta.$$

On the other hand, dividing through by y^2 leads to

$$1 + \cot^2 \theta = \csc^2 \theta.$$

These three identities are called the **Pythagorean identities** since the original equation that led to them, $x^2 + y^2 = r^2$, comes from the Pythagorean theorem.

PYTHAGOREAN IDENTITIES	$\sin^2 \theta + \cos^2 \theta = 1 \qquad \tan^2 \theta + 1 = \sec^2 \theta$ $1 + \cot^2 \theta = \csc^2 \theta$

As before, we have given only one form of each identity. However, algebraic transformations can be made to get equivalent identities. For example, by subtracting $\sin^2 \theta$ from both sides of $\sin^2 \theta + \cos^2 \theta = 1$ we get the equivalent identity

$$\cos^2 \theta = 1 - \sin^2 \theta.$$

You should be able to transform these identities quickly, and also recognize their equivalent forms.

Recall that $\sin \theta = y/r$ and $\cos \theta = x/r$. Consider the quotient of $\sin \theta$ and $\cos \theta$, where $\cos \theta \neq 0$.

$$\frac{\sin \theta}{\cos \theta} = \frac{y/r}{x/r} = \frac{y}{r} \div \frac{x}{r} = \frac{y}{r} \cdot \frac{r}{x} = y/x = \tan \theta$$

Similarly, it can be shown that $(\cos \theta)/(\sin \theta) = \cot \theta$, for $\sin \theta \neq 0$. Thus we have two more identities, called the **quotient identities.**

QUOTIENT IDENTITIES	$\dfrac{\sin \theta}{\cos \theta} = \tan \theta \qquad \dfrac{\cos \theta}{\sin \theta} = \cot \theta$

■ *Example 5*

FINDING OTHER
FUNCTION VALUES
GIVEN ONE VALUE
AND THE QUADRANT

Find $\sin \alpha$ and $\tan \alpha$ if $\cos \alpha = -\sqrt{3}/4$ and α is in quadrant II.
Start with $\sin^2 \alpha + \cos^2 \alpha = 1$, and replace $\cos \alpha$ with $-\sqrt{3}/4$.

$$\sin^2 \alpha + \left(-\frac{\sqrt{3}}{4}\right)^2 = 1 \qquad \text{Replace } \cos \alpha \text{ with } -\frac{\sqrt{3}}{4}.$$

$$\sin^2 \alpha + \frac{3}{16} = 1 \qquad$$

$$\sin^2 \alpha = \frac{13}{16} \qquad \text{Subtract } \frac{3}{16}.$$

$$\sin \alpha = \pm \frac{\sqrt{13}}{4} \qquad \text{Take square roots.}$$

Since α is in quadrant II, $\sin \alpha > 0$, and

$$\sin \alpha = \frac{\sqrt{13}}{4}.$$

To find $\tan \alpha$, use the quotient identity $\tan \alpha = \dfrac{\sin \alpha}{\cos \alpha}$.

$$\tan \alpha = \frac{\sin \alpha}{\cos \alpha} = \frac{\dfrac{\sqrt{13}}{4}}{\dfrac{-\sqrt{3}}{4}} = \frac{\sqrt{13}}{4} \cdot \frac{4}{-\sqrt{3}} = \frac{\sqrt{13}}{-\sqrt{3}}$$

Rationalize the denominator as follows.

$$\frac{\sqrt{13}}{-\sqrt{3}} = \frac{\sqrt{13}}{-\sqrt{3}} \cdot \frac{\sqrt{3}}{\sqrt{3}}$$

$$= \frac{\sqrt{39}}{-3}$$

$$= -\frac{\sqrt{39}}{3}$$

Therefore, $\tan \alpha = -\dfrac{\sqrt{39}}{3}$. ■

CAUTION One of the most common errors in problems like the ones in Examples 4 and 5 involves an incorrect sign choice when square roots are taken. Notice that in Example 5, we chose the positive square root for $\sin \alpha$, since α was in quadrant II, and the sine function is positive there. If the problem had specified that α was in quadrant III, then we would have had to choose the negative square root.

■ *Example 6*

FINDING sin θ AND cos θ, GIVEN tan θ AND THE QUADRANT OF θ

Find sin θ and cos θ, if tan θ = 4/3 and θ is in quadrant I.

Since θ is in quadrant I, sin θ and cos θ will both be positive. It is tempting to say that since tan θ = (sin θ)/(cos θ) and tan θ = 4/3, then sin θ = 4 and cos θ = 3. This is *incorrect,* however, since both sin θ and cos θ must be in the interval [−1, 1].

Use the Pythagorean identity $\tan^2 \theta + 1 = \sec^2 \theta$ to find sec θ, and then the reciprocal identity cos θ = 1/sec θ.

$$\tan^2 \theta + 1 = \sec^2 \theta$$

$$\left(\frac{4}{3}\right)^2 + 1 = \sec^2 \theta \qquad \tan\theta = \frac{4}{3}$$

$$\frac{16}{9} + 1 = \sec^2 \theta$$

$$\frac{25}{9} = \sec^2 \theta$$

$$\frac{5}{3} = \sec \theta \qquad \text{Choose the positive square root since } \theta \text{ is in quadrant I.}$$

$$\frac{3}{5} = \cos \theta \qquad \text{Secant and cosine are reciprocals.}$$

Since $\sin^2 \theta = 1 - \cos^2 \theta,$

$$\sin^2 \theta = 1 - \left(\frac{3}{5}\right)^2 \qquad \cos\theta = \frac{3}{5}$$

$$= 1 - \frac{9}{25}$$

$$= \frac{16}{25}$$

$$\sin \theta = \frac{4}{5}. \qquad \text{Choose the positive square root.}$$

Therefore, we have sin θ = 4/5 and cos θ = 3/5. ■

NOTE Example 6 can also be worked by drawing θ in standard position in quadrant I, finding *r* to be 5, and then using the definitions of sin θ and cos θ in terms of *x*, *y*, and *r*.

1.5 EXERCISES ■ ————————————————————————

Use the appropriate reciprocal identity to find each function value. Rationalize denominators when applicable. In Exercises 9–12, use a calculator. See Example 1.

1. $\sin \theta$, if $\csc \theta = 3$

2. $\cos \alpha$, if $\sec \alpha = -2.5$

3. $\cot \beta$, if $\tan \beta = -1/5$

4. $\sin \alpha$, if $\csc \alpha = \sqrt{15}$

5. $\csc \alpha$, if $\sin \alpha = \sqrt{2}/4$

6. $\sec \beta$, if $\cos \beta = -1/\sqrt{7}$

7. $\tan \theta$, if $\cot \theta = -\sqrt{5}/3$

8. $\cot \theta$, if $\tan \theta = \sqrt{11}/5$

9. $\sin \theta$, if $\csc \theta = 1.42716321$

10. $\cos \alpha$, if $\sec \alpha = 9.80425133$

11. $\tan \alpha$, if $\cot \alpha = .43900273$

12. $\csc \theta$, if $\sin \theta = -.37690858$

13. Can a given angle γ satisfy both $\sin \gamma > 0$ and $\csc \gamma < 0$? Explain.

14. Suppose that the following item appears on a trigonometry test:

Find $\sec \theta$, given that $\cos \theta = 3/2$.

What is wrong with this test item?

15. One form of a particular reciprocal identity is

$$\tan \theta = \frac{1}{\cot \theta}.$$

Give two other equivalent forms of this identity.

16. What is wrong with the following statement? $\tan 90° = \dfrac{1}{\cot 90°}$

Find the tangent of each of the following angles. See Example 1.

17. $\cot \gamma = 2$

18. $\cot \phi = -3$

19. $\cot \omega = \sqrt{3}/3$

20. $\cot \theta = \sqrt{6}/12$

21. $\cot \alpha = -.01$

22. $\cot \beta = .4$

Find a value of the variable in each of the following.

23. $\cos (6A + 5°) = \dfrac{1}{\sec (4A + 15°)}$

24. $\tan (3B - 4°) = \dfrac{1}{\cot (5B - 8°)}$

25. $\sin (4\theta + 2°) \csc (3\theta + 5°) = 1$

26. $\sec (2\alpha + 6°) \cos (5\alpha + 3°) = 1$

27. $\dfrac{1}{\sin (3\theta - 1°)} = \csc (2\theta + 3°)$

28. $\dfrac{1}{\tan (2k + 1°)} = \cot (4k - 3°)$

Identify the quadrant or quadrants for the angles satisfying the following conditions. See Example 2.

29. $\sin \alpha > 0$, $\cos \alpha < 0$

30. $\cos \beta > 0$, $\tan \beta > 0$

31. $\sec \theta < 0$, $\csc \theta < 0$

32. $\tan \gamma > 0$, $\cot \gamma > 0$

33. $\sin \beta < 0$, $\cos \beta > 0$

34. $\cos \beta > 0$, $\sin \beta > 0$

35. $\tan \omega < 0$, $\cot \omega < 0$

36. $\csc \theta < 0$, $\cos \theta < 0$

37. $\sin \alpha > 0$

38. $\cos \beta < 0$

39. $\tan \theta > 0$

40. $\csc \alpha < 0$

Give the signs of the six trigonometric functions for each of the following angles.

41. $74°$

42. $129°$

43. $183°$

44. $298°$

45. $302°$

46. $372°$

47. $406°$

48. $412°$

49. $-82°$

50. $-14°$

51. $-121°$

52. $-208°$

Decide whether each of the following statements is possible *or* impossible. *See Example 3.*

53. $\sin \theta = 2$ **54.** $\cos \alpha = -1.001$ **55.** $\tan \beta = .92$ **56.** $\cot \omega = -12.1$

57. $\csc \alpha = 1/2$ **58.** $\sec \alpha = 1$ **59.** $\tan \theta = 1$ **60.** $\sin \alpha = -.82$

61. $\sin \beta + 1 = .6$ **62.** $\sec \omega + 1 = 1.3$ **63.** $\csc \theta - 1 = -.2$ **64.** $\tan \alpha - 4 = 7.3$

65. $\sin \alpha = 1/2$ and $\csc \alpha = 2$ **66.** $\cos \theta = 3/4$ and $\sec \theta = 4/3$

67. $\tan \beta = 2$ and $\cot \beta = -2$ **68.** $\sec \gamma = .4$ and $\cos \gamma = 2.5$

Use identities to find the indicated function value. Use a calculator in Exercises 78–82. See Examples 4–6.

69. $\cos \theta$, if $\sin \theta = 2/3$, with θ in quadrant II   **70.** $\tan \alpha$, if $\sec \alpha = 3$, with α in quadrant IV

71. $\csc \beta$, if $\cot \beta = -1/2$, with β in quadrant IV   **72.** $\sin \alpha$, if $\cos \alpha = -1/4$, with α in quadrant II

73. $\sec \theta$, if $\tan \theta = \sqrt{7}/3$, with θ in quadrant III   **74.** $\tan \theta$, if $\cos \theta = 1/3$, with θ in quadrant IV

75. $\sin \theta$, if $\sec \theta = 2$, with θ in quadrant IV   **76.** $\cos \beta$, if $\csc \beta = -4$, with β in quadrant III

77. $\cos \alpha$, if $\sin \alpha = -\sqrt{5}/7$, with α in quadrant IV   **78.** $\sin \theta$, if $\cos \theta = .91427683$, with θ in quadrant IV

79. $\cot \alpha$, if $\csc \alpha = -3.5891420$, with α in quadrant III   **80.** $\sin \beta$, if $\cot \beta = 2.40129813$, with β in quadrant I

81. $\tan \beta$, if $\sin \beta = .49268329$, with β in quadrant II   **82.** $\csc \alpha$, if $\tan \alpha = .98244655$, with α in quadrant III

Find all the trigonometric function values for each of the following angles. Use a calculator in Exercises 93 and 94. See Examples 4–6.

83. $\cos \alpha = -3/5$, with α in quadrant III   **84.** $\tan \alpha = -15/8$, with α in quadrant II

85. $\sin \beta = 7/25$, with β in quadrant II   **86.** $\cot \gamma = 3/4$, with γ in quadrant III

87. $\csc \theta = 2$, with θ in quadrant II   **88.** $\tan \beta = \sqrt{3}$, with β in quadrant III

89. $\cot \alpha = \sqrt{3}/8$, with $\sin \alpha > 0$ **90.** $\sin \beta = \sqrt{5}/7$, with $\tan \beta > 0$

91. $\tan \theta = 3/2$, with $\csc \theta = \sqrt{13}/3$ **92.** $\csc \alpha = -\sqrt{17}/3$, with $\cot \alpha = 2\sqrt{2}/3$

93. $\sin \alpha = .164215$, with α in quadrant II   **94.** $\cot \theta = -1.49586$, with θ in quadrant IV

95. $\sin \gamma = a$, with γ in quadrant I   **96.** $\tan \omega = m$, with ω in quadrant III

97. Derive the identity $1 + \cot^2\theta = \csc^2\theta$ by dividing $x^2 + y^2 = r^2$ by y^2.

98. Using a method similar to the one given in this section showing that $(\sin \theta)/(\cos \theta) = \tan \theta$, show that

$$\frac{\cos \theta}{\sin \theta} = \cot \theta.$$

99. True or false: For all angles θ, $\sin \theta + \cos \theta = 1$. If false, give an example showing why it is false.

100. True or false: Since $\cot \theta = \dfrac{\cos \theta}{\sin \theta}$, if $\cot \theta = \dfrac{1}{2}$ with θ in quadrant I, then $\cos \theta = 1$ and $\sin \theta = 2$. If false, explain why.

CHAPTER 1 SUMMARY ■

SECTION	KEY IDEAS
1.1 Basic Concepts	**Pythagorean Theorem** If the two shorter sides (the legs) of a right triangle have lengths a and b, respectively, and if the length of the hypotenuse (the longest side, opposite the 90° angle) is c, then $$a^2 + b^2 = c^2.$$

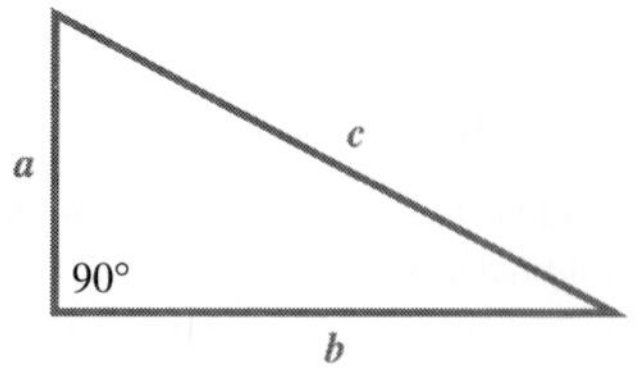

1.2 Angles

Types of Angles

Name	Angle Measure	Example
Acute angle	Between 0° and 90°	 60° 82°
Right angle	Exactly 90°	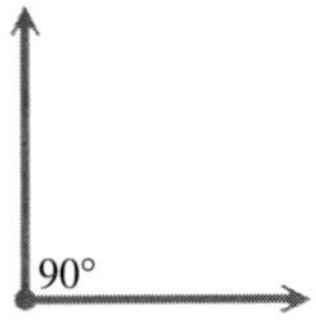 90°
Obtuse angle	Between 90° and 180°	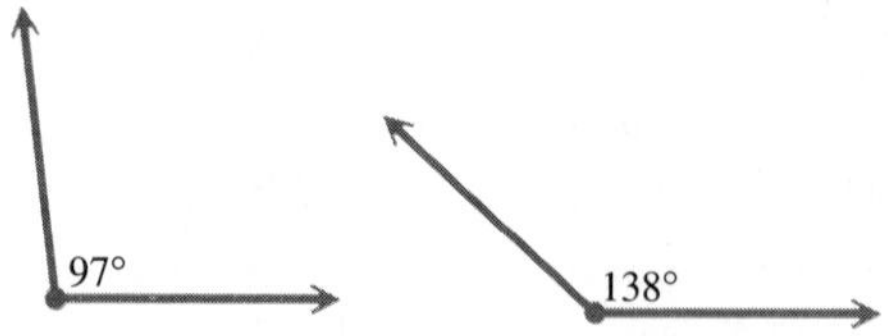 97° 138°
Straight angle	Exactly 180°	 180°

1.3 Angle Relationships and Similar Triangles

Vertical angles have equal measures.

The sum of the measures of the angles of any triangle is 180°.

When a transversal intersects parallel lines, the following angles formed have equal measure: alternate interior, alternate exterior, and corresponding. Interior angles on the same side of the transversal add to 180°.

Similar triangles have corresponding angles with the same measures, and corresponding sides proportional.

SECTION	KEY IDEAS	47

1.4 Definitions of the Trigonometric Functions

Definitions of the Trigonometric Functions

Let (x, y) be a point other than the origin on the terminal side of an angle θ in standard position. Let $r = \sqrt{x^2 + y^2}$, the distance from the origin to (x, y). Then

$$\sin \theta = \frac{y}{r} \qquad\qquad \csc \theta = \frac{r}{y} \quad (y \neq 0)$$

$$\cos \theta = \frac{x}{r} \qquad\qquad \sec \theta = \frac{r}{x} \quad (x \neq 0)$$

$$\tan \theta = \frac{y}{x} \quad (x \neq 0) \qquad \cot \theta = \frac{x}{y} \quad (y \neq 0).$$

Trigonometric Function Values for Quadrantal Angles

θ	0°	90°	180°	270°	360°
$\sin \theta$	0	1	0	-1	0
$\cos \theta$	1	0	-1	0	1
$\tan \theta$	0	undefined	0	undefined	0
$\cot \theta$	undefined	0	undefined	0	undefined
$\sec \theta$	1	undefined	-1	undefined	1
$\csc \theta$	undefined	1	undefined	-1	undefined

1.5 Using the Definitions of the Trigonometric Functions

Reciprocal Identities

$$\sin \theta = \frac{1}{\csc \theta} \qquad \csc \theta = \frac{1}{\sin \theta}$$

$$\cos \theta = \frac{1}{\sec \theta} \qquad \sec \theta = \frac{1}{\cos \theta}$$

$$\tan \theta = \frac{1}{\cot \theta} \qquad \cot \theta = \frac{1}{\tan \theta}$$

Pythagorean Identities

$$\sin^2 \theta + \cos^2 \theta = 1$$
$$\tan^2 \theta + 1 = \sec^2 \theta$$
$$1 + \cot^2 \theta = \csc^2 \theta$$

Quotient Identities

$$\frac{\sin \theta}{\cos \theta} = \tan \theta \qquad \frac{\cos \theta}{\sin \theta} = \cot \theta$$

Signs of Trigonometric Functions

$x < 0$ $y > 0$ $r > 0$	$x > 0$ $y > 0$ $r > 0$
II Sine and cosecant positive	**I** All functions positive
$x < 0$ $y < 0$ $r > 0$	$x > 0$ $y < 0$ $r > 0$
III Tangent and cotangent positive	**IV** Cosine and secant positive

CHAPTER 1 REVIEW EXERCISES ■

Write each of the following sets using interval notation.

1. $\{x \mid x \le -4\}$

2.
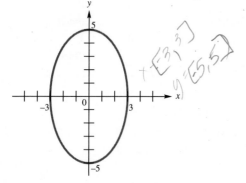

Find the distance between each of the following pairs of points.

3. $(4, -2)$ and $(1, -6)$

4. $(-6, 3)$ and $(-2, -5)$

5. Use the distance formula to determine whether the points $(-2, -2)$, $(8, 4)$, and $(2, 14)$ are the vertices of a right triangle.

6. State in your own words the vertical line test for the graph of a function.

Let $f(x) = -x^2 + 3x + 2$. Find each of the following.

7. $f(0)$ **8.** $f(-2)$ **9.** $f(a)$ **10.** $f(x + 1)$

Find the domain and range of each of the following. Identify any functions.

11. $y = 9x + 2$ **12.** $4x - 7y = 1$ **13.** $y = |x|$

14. $y = \sqrt{x}$ **15.** $x + 1 = y^2$ **16.** $x = \sqrt{y + 3}$

17.

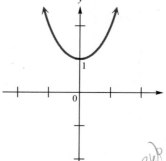

18.

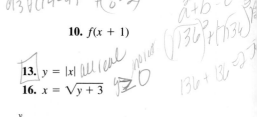

Find the angles of smallest possible positive measure coterminal with the following angles.

19. $-51°$ **20.** $-174°$ **21.** $792°$

22. Let n represent any integer, and write an expression for all angles coterminal with an angle of $270°$.

Work each of the following problems.

23. A pulley is rotating 320 times per minute. Through how many degrees does a point on the edge of the pulley move in 2/3 second?

24. The propeller of a speedboat rotates 650 times per minute. Through how many degrees will a point on the edge of the propeller rotate in 2.4 seconds?

Convert decimal degrees to degrees, minutes, seconds, and convert degrees, minutes, seconds to decimal degrees. Round to the nearest second or the nearest thousandth of a degree, as appropriate. Use a calculator as necessary.

25. $47° \, 25' \, 11''$ **26.** $119° \, 08' \, 03''$ **27.** $74.2983°$

28. $-61.5034°$ **29.** $183.0972°$ **30.** $275.1005°$

Find the measure of each marked angle.

31.

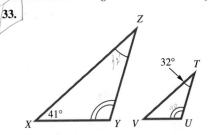

$9x+4 = 12x-14$

$18 = 3x$

$u = x$

32.

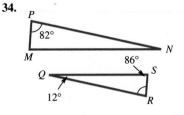

Find all unknown angle measures in each pair of similar triangles.

33.

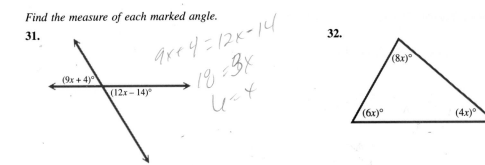

34.

35.

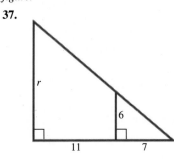

-45 $m = 30 = 4$

$\frac{75}{75} \cdot \frac{50}{60}$

$n = 40$ 60

$50.$

Find the unknown side lengths in each pair of similar triangles.

36.

Find the unknown measurement in each of the following. There are two similar triangles in each figure.

37.

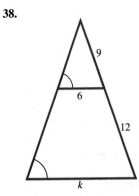

38.

39. Complete the following statement: If two triangles are similar, their corresponding sides are _____ and their corresponding angles are _____.

50

40. If a tree 20 feet tall casts a shadow of 8 feet, how long would the shadow of a 30-foot tree be at the same time?

Find the trigonometric function values of each of the following angles. If undefined, say so.

41.

42.

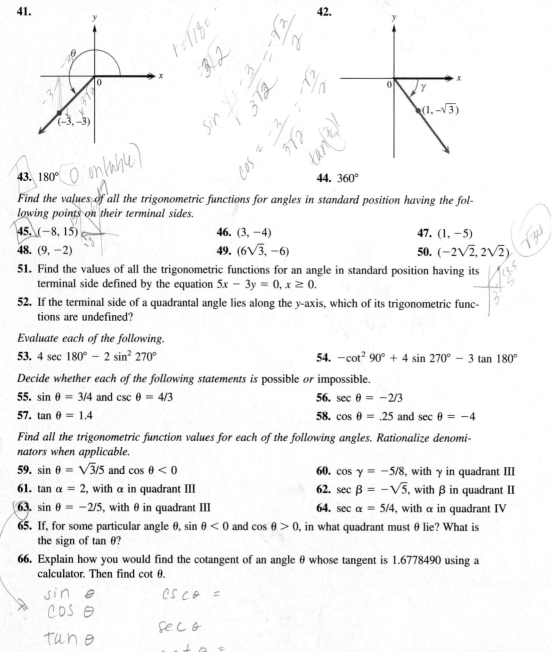

43. 180°

44. 360°

Find the values of all the trigonometric functions for angles in standard position having the following points on their terminal sides.

45. $(-8, 15)$ **46.** $(3, -4)$ **47.** $(1, -5)$

48. $(9, -2)$ **49.** $(6\sqrt{3}, -6)$ **50.** $(-2\sqrt{2}, 2\sqrt{2})$

51. Find the values of all the trigonometric functions for an angle in standard position having its terminal side defined by the equation $5x - 3y = 0$, $x \geq 0$.

52. If the terminal side of a quadrantal angle lies along the y-axis, which of its trigonometric functions are undefined?

Evaluate each of the following.

53. $4 \sec 180° - 2 \sin^2 270°$ **54.** $-\cot^2 90° + 4 \sin 270° - 3 \tan 180°$

Decide whether each of the following statements is possible *or* impossible.

55. $\sin \theta = 3/4$ and $\csc \theta = 4/3$ **56.** $\sec \theta = -2/3$

57. $\tan \theta = 1.4$ **58.** $\cos \theta = .25$ and $\sec \theta = -4$

Find all the trigonometric function values for each of the following angles. Rationalize denominators when applicable.

59. $\sin \theta = \sqrt{3}/5$ and $\cos \theta < 0$ **60.** $\cos \gamma = -5/8$, with γ in quadrant III

61. $\tan \alpha = 2$, with α in quadrant III   **62.** $\sec \beta = -\sqrt{5}$, with β in quadrant II

63. $\sin \theta = -2/5$, with θ in quadrant III   **64.** $\sec \alpha = 5/4$, with α in quadrant IV

65. If, for some particular angle θ, $\sin \theta < 0$ and $\cos \theta > 0$, in what quadrant must θ lie? What is the sign of $\tan \theta$?

66. Explain how you would find the cotangent of an angle θ whose tangent is 1.6778490 using a calculator. Then find $\cot \theta$.

ACUTE ANGLES AND RIGHT TRIANGLES

52

So far, the definitions of the trigonometric functions have been used only for angles such as 0°, 90°, 180°, or 270°. This chapter extends this work to include finding the values of trigonometric functions of other useful angles, such as 30°, 45°, and 60°. We also discuss the use of calculators for angles whose function values cannot be found directly. (While tables have historically been used for this purpose, we will not refer to tables as they have been replaced by calculators for all practical purposes.) The chapter ends with some applications of trigonometry.

2.1 ———— TRIGONOMETRIC FUNCTIONS OF ACUTE ANGLES

Figure 2.1 shows an acute angle *A* in standard position. The definitions of the trigonometric function values of angle *A* require *x*, *y*, and *r*. As drawn in Figure 2.1, *x* and *y* are the lengths of the two legs of right triangle *ABC*, and *r* is the length of the hypotenuse.

The side of length *y* is called the **side opposite** angle *A*, and the side of length *x* is called the **side adjacent** to angle *A*. The lengths of these sides can be used to replace *x* and *y* in the definition of the trigonometric functions, with *r* replaced with the length of the hypotenuse, to get the following right triangle-based definitions.

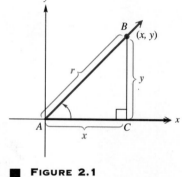

■ **FIGURE 2.1**

RIGHT TRIANGLE-BASED DEFINITIONS OF TRIGONOMETRIC FUNCTIONS	For any acute angle *A* in standard position,

$$\sin A = \frac{y}{r} = \frac{\text{side opposite}}{\text{hypotenuse}} \qquad \csc A = \frac{r}{y} = \frac{\text{hypotenuse}}{\text{side opposite}}$$

$$\cos A = \frac{x}{r} = \frac{\text{side adjacent}}{\text{hypotenuse}} \qquad \sec A = \frac{r}{x} = \frac{\text{hypotenuse}}{\text{side adjacent}}$$

$$\tan A = \frac{y}{x} = \frac{\text{side opposite}}{\text{side adjacent}} \qquad \cot A = \frac{x}{y} = \frac{\text{side adjacent}}{\text{side opposite}}.$$

■ *Example 1*
FINDING TRIGONOMETRIC FUNCTION VALUES OF AN ACUTE ANGLE IN A RIGHT TRIANGLE

Find the values of the trigonometric functions for angles *A* and *B* in the right triangle in Figure 2.2.

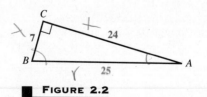

■ **FIGURE 2.2**

The length of the side opposite angle A is 7. The length of the side adjacent to angle A is 24, and the length of the hypotenuse is 25. Using the relationships given above,

$$\sin A = \frac{\text{side opposite}}{\text{hypotenuse}} = \frac{7}{25} \qquad \csc A = \frac{\text{hypotenuse}}{\text{side opposite}} = \frac{25}{7}$$

$$\cos A = \frac{\text{side adjacent}}{\text{hypotenuse}} = \frac{24}{25} \qquad \sec A = \frac{\text{hypotenuse}}{\text{side adjacent}} = \frac{25}{24}$$

$$\tan A = \frac{\text{side opposite}}{\text{side adjacent}} = \frac{7}{24} \qquad \cot A = \frac{\text{side adjacent}}{\text{side opposite}} = \frac{24}{7}.$$

The length of the side opposite angle B is 24, while the length of the side adjacent to B is 7, making

$$\sin B = \frac{24}{25} \qquad \tan B = \frac{24}{7} \qquad \sec B = \frac{25}{7}$$

$$\cos B = \frac{7}{25} \qquad \cot B = \frac{7}{24} \qquad \csc B = \frac{25}{24}. \qquad ■$$

In Example 1, you may have noticed that $\sin A = \cos B$, $\cos A = \sin B$, and so on. Such relationships are always true for the two acute angles of a right triangle. Figure 2.3 shows a right triangle with acute angles A and B and a right angle at C. (Whenever we use A, B, and C to name the angles in a right triangle, C will be the right angle.) The length of the side opposite angle A is a, and the length of the side opposite angle B is b. The length of the hypotenuse is c.

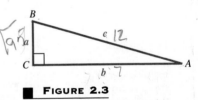

■ **FIGURE 2.3**

By the definitions given above, $\sin A = a/c$. Since $\cos B$ is also equal to a/c,

$$\sin A = \frac{a}{c} = \cos B.$$

In a similar manner,

$$\tan A = \frac{a}{b} = \cot B \qquad \sec A = \frac{c}{b} = \csc B.$$

The sum of the three angles in any triangle is 180°. Since angle C equals 90°, angles A and B must have a sum of $180° - 90° = 90°$. As mentioned in Chapter 1, angles with a sum of 90° are called complementary angles. Since angles A and B are complementary and $\sin A = \cos B$, the functions sine and cosine are called **cofunctions.** Also, tangent and cotangent are cofunctions, as are secant and cosecant. Since angles A and B are complementary, $A + B = 90°$, or

$$B = 90° - A,$$

giving $\qquad\qquad \sin A = \cos B = \cos (90° - A).$

Similar results, called the **cofunction identities,** are true for the other trigonometric functions.

COFUNCTION IDENTITIES	For any acute angle A,

$$\sin A = \cos (90° - A) \qquad \csc A = \sec (90° - A)$$
$$\cos A = \sin (90° - A) \qquad \sec A = \csc (90° - A)$$
$$\tan A = \cot (90° - A) \qquad \cot A = \tan (90° - A).$$

(These identities will be extended to *any* angle A, and not just acute angles, in Chapter 5.) It would be wise to learn all the identities presented in this book.

■ *Example 2*
WRITING FUNCTIONS IN TERMS OF COFUNCTIONS

Write each of the following in terms of cofunctions.

(a) $\cos 52°$

$$\cos 52 = \sin (90 - 52)$$
$$\sin 38$$

Since $\cos A = \sin (90° - A)$,

$$\cos 52° = \sin (90° - 52°) = \sin 38°.$$

(b) $\tan 71° = \cot 19°$ $\tan 71° = \cot (90 - 71) = \cot 19$

(c) $\sec 24° = \csc 66°$ ■

■ *Example 3*
SOLVING EQUATIONS BY USING THE COFUNCTION IDENTITIES

Find a value of θ satisfying each of the following. Assume that all angles involved are acute angles.

(a) $\cos (\theta + 4°) = \sin (3\theta + 2°)$

Since sine and cosine are cofunctions, this equation is true if the sum of the angles is 90°, or

$$(\theta + 4°) + (3\theta + 2°) = 90°$$
$$4\theta + 6° = 90°$$
$$4\theta = 84°$$
$$\theta = 21°.$$

(b) $\tan (2\theta - 18°) = \cot (\theta + 18°)$

$$(2\theta - 18°) + (\theta + 18°) = 90°$$
$$3\theta = 90°$$
$$\theta = 30° ■$$

Figure 2.4 shows three right triangles. From left to right, the length of each hypotenuse is the same, but angle A increases in measure. As angle A increases in measure from 0° to 90°, the length of the side opposite angle A also increases.

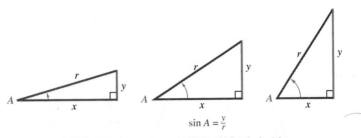

$$\sin A = \frac{y}{r}$$

As A increases, y increases. Since r is fixed, $\sin A$ increases.

■ **FIGURE 2.4**

Since

$$\sin A = \frac{\text{side opposite}}{\text{hypotenuse}},$$

as angle A increases, the numerator of this fraction also increases, while the denominator is fixed. This means that $\sin A$ *increases* as A increases from $0°$ to $90°$.

As angle A increases from $0°$ to $90°$, the length of the side adjacent to A decreases. Since r is fixed, the ratio x/r will decrease. This ratio gives $\cos A$, showing that the values of cosine *decrease* as the angle measure changes from $0°$ to $90°$. Finally, increasing A from $0°$ to $90°$ causes y to increase and x to decrease, making the values of $y/x = \tan A$ increase.

A similar discussion shows that as A increases from $0°$ to $90°$, the values of $\sec A$ increase, while the values of $\cot A$ and $\csc A$ decrease.

■ *Example 4*

COMPARING
FUNCTION VALUES
OF ACUTE ANGLES

Tell whether each of the following is *true* or *false*.

(a) $\sin 21° > \sin 18°$

In the interval from $0°$ to $90°$, as the angle increases, so does the sine of the angle, which makes $\sin 21° > \sin 18°$ a true statement.

(b) $\cos 49° \le \cos 56°$

As the angle increases, the cosine of the angle decreases. The given statement $\cos 49° \le \cos 56°$ is false. ■

Certain special angles, such as $30°$, $45°$, and $60°$, occur so often in trigonometry and in more advanced mathematics that they deserve special study. We can find the exact trigonometric function values of these angles by using properties of geometry and the Pythagorean theorem.

To find the trigonometric function values for $30°$ and $60°$, we start with an equilateral triangle, a triangle with all sides of equal length. Each angle of such a triangle has a measure of $60°$. While the results we will obtain are independent of the length, for convenience, we choose the length of each side to be 2 units. See Figure 2.5(a).

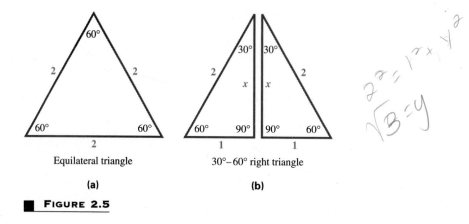

Equilateral triangle

30°–60° right triangle

(a)

(b)

■ **FIGURE 2.5**

Bisecting one angle of this equilateral triangle leads to two right triangles, each of which has angles of 30°, 60°, and 90°, as shown in Figure 2.5(b). Since the hypotenuse of one of these right triangles has a length of 2, the shortest side will have a length of 1. (Why?) If x represents the length of the medium side, then, by the Pythagorean theorem,

$$2^2 = 1^2 + x^2$$
$$4 = 1 + x^2$$
$$3 = x^2$$
$$\sqrt{3} = x.$$

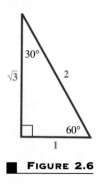

■ **FIGURE 2.6**

Figure 2.6 summarizes our results, showing a 30°–60° right triangle.

As shown in the figure, the side opposite the 30° angle has length 1; that is, for the 30° angle,

$$\text{hypotenuse} = 2, \qquad \text{side opposite} = 1, \qquad \text{side adjacent} = \sqrt{3}.$$

Using the definitions of the trigonometric functions,

$$\sin 30° = \frac{\text{side opposite}}{\text{hypotenuse}} = \frac{1}{2} \qquad\qquad \csc 30° = \frac{2}{1} = 2$$

$$\cos 30° = \frac{\text{side adjacent}}{\text{hypotenuse}} = \frac{\sqrt{3}}{2} \qquad\qquad \sec 30° = \frac{2}{\sqrt{3}} = \frac{2\sqrt{3}}{3}$$

$$\tan 30° = \frac{\text{side opposite}}{\text{side adjacent}} = \frac{1}{\sqrt{3}} = \frac{\sqrt{3}}{3} \qquad \cot 30° = \frac{\sqrt{3}}{1} = \sqrt{3}.$$

The denominator was rationalized for tan 30° and sec 30°.

In a similar manner,

$$\sin 60° = \frac{\sqrt{3}}{2} \qquad \tan 60° = \sqrt{3} \qquad \sec 60° = 2$$

$$\cos 60° = \frac{1}{2} \qquad \cot 60° = \frac{\sqrt{3}}{3} \qquad \csc 60° = \frac{2\sqrt{3}}{3}.$$

The values of the trigonometric functions for 45° can be found by starting with a 45°–45° right triangle, as shown in Figure 2.7. This triangle is isosceles, and, for convenience, we choose the lengths of the equal sides to be 1 unit. (As before, the results are independent of the length of the equal sides of the right triangle.) Since the shorter sides each have length 1, if r represents the length of the hypotenuse, then

$$1^2 + 1^2 = r^2$$
$$2 = r^2$$
$$\sqrt{2} = r.$$

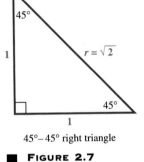

45°–45° right triangle

■ **FIGURE 2.7**

Using the measures indicated on the 45°–45° right triangle in Figure 2.7, we find

$$\sin 45° = \frac{1}{\sqrt{2}} = \frac{\sqrt{2}}{2} \qquad \tan 45° = \frac{1}{1} = 1 \qquad \sec 45° = \frac{\sqrt{2}}{1} = \sqrt{2}$$

$$\cos 45° = \frac{1}{\sqrt{2}} = \frac{\sqrt{2}}{2} \qquad \cot 45° = \frac{1}{1} = 1 \qquad \csc 45° = \frac{\sqrt{2}}{1} = \sqrt{2}.$$

The importance of these exact trigonometric function values of 30°, 60°, and 45° angles cannot be overemphasized. It is essential to learn them. They are summarized in the chart that follows.

FUNCTION VALUES OF SPECIAL ANGLES	θ	$\sin \theta$	$\cos \theta$	$\tan \theta$	$\cot \theta$	$\sec \theta$	$\csc \theta$
	30°	$\frac{1}{2}$	$\frac{\sqrt{3}}{2}$	$\frac{\sqrt{3}}{3}$	$\sqrt{3}$	$\frac{2\sqrt{3}}{3}$	2
	45°	$\frac{\sqrt{2}}{2}$	$\frac{\sqrt{2}}{2}$	1	1	$\sqrt{2}$	$\sqrt{2}$
	60°	$\frac{\sqrt{3}}{2}$	$\frac{1}{2}$	$\sqrt{3}$	$\frac{\sqrt{3}}{3}$	2	$\frac{2\sqrt{3}}{3}$

NOTE You should be able to reproduce this chart quickly. It is not difficult to do if you learn the values of sin 30°, cos 30°, and sin 45°. Then complete the rest of the chart using the reciprocal identities, the cofunction identities, and the quotient identities.

58

In Exercises 61 and 62, we generalize the relationships among the sides of a 30°–60° right triangle and a 45°–45° right triangle.

If you have a calculator that finds trigonometric function values at the touch of a key, you may wonder why we spend so much time in finding values for special angles. We do this because a calculator gives only *approximate* values in most cases, while we often need *exact* values. For example, tan 30° can be found on a calculator by first setting the machine in the *degree mode,* then entering 30 and pressing the [tan] key.

Enter: 30 [tan] **Display:** [.57735027],

so that

$$\tan 30° \approx .57735027$$

(the symbol ≈ means "is approximately equal to"). Earlier, however, we found the exact value:

$$\tan 30° = \frac{\sqrt{3}}{3}.$$

2.1 EXERCISES ■

In each of the following exercises, find the values of the six trigonometric functions for angle A. Leave answers as fractions. See Example 1.

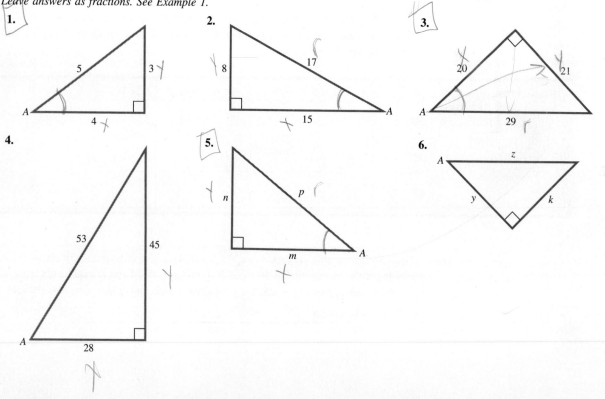

1.

5   3

A   4

2.

8   17

15   A

3.

20   21

A   29

4.

53   45

A   28

5.

n   p

m   A

6.

A   z

y   k

acute angles = 0° - 90° (handwritten)

Use a calculator to find the values of the six trigonometric functions for each angle A. Round answers to four decimal places. See Example 1.

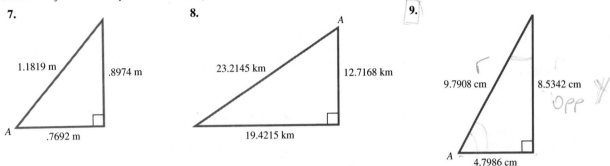

7. 1.1819 m   .8974 m   A   .7692 m

8. 23.2145 km   12.7168 km   A   19.4215 km

9. 9.7908 cm   8.5342 cm   *opp* (handwritten)   A   4.7986 cm   *adj.* (handwritten)

Suppose that ABC is a right triangle with sides of lengths a, b, and c with right angle at C (see Figure 2.3). Find the unknown side length using the Pythagorean theorem, and then find the values of the six trigonometric functions for angle B. Rationalize denominators when applicable.

10. $a = 3, b = 5$ **11.** $b = 7, c = 12$ **12.** $a = 6, b = 7$

13. Write a summary of the relationships between cofunctions of complementary angles.

Write each of the following in terms of the cofunction. Assume that all angles in which an unknown appears are acute angles. See Example 2.

14. cot 73°   **15.** tan 50°   **16.** sec 39°   **17.** csc 47° *Sin 43°* (handwritten)

18. cos 43°   **19.** sin 38° 29'   **20.** tan 25° 43'   **21.** sin γ

22. cot α   **23.** cos (α + 20°)   **24.** cot (β − 10°)

Find a solution for each of the following equations. Assume that all angles in which an unknown appears are acute angles. See Example 3.

25. cos θ = sin 2θ   **26.** tan α = cot (α + 10°)

27. sec (β + 10°) = csc (2β + 20°)   **28.** sin (2γ + 10°) = cos (3γ − 20°)

29. cot (5θ + 2°) = tan (2θ + 4°)   **30.** tan (3B + 4°) = cot (5B − 10°)

31. $\sec\left(\dfrac{\beta}{2} + 5°\right) = \csc\left(\dfrac{\beta}{2} + 15°\right)$ **32.** $\sin\left(\dfrac{3A}{2} - 5°\right) = \cos\left(\dfrac{2}{3}A + 30°\right)$

Tell whether each of the following is true or false. See Example 4.

33. tan 28° ≤ tan 40°   **34.** sin 50° > sin 40°

35. sin 46° < cos 46°   **36.** cos 28° < sin 28°
     (Hint: cos 46° = sin 44°)

37. tan 41° < cot 41°   *tan 49° > tan 41°*   *tan 49° = cot 41°* (handwritten)   **38.** cot 30° < tan 40°

39. sin 60° ≤ cos 30°   *cot 41° > tan 41°* (handwritten)   **40.** sec 80° ≥ csc 10°

Refer to the discussion in this section to give the exact trigonometric function values in each of the following. Do not use a calculator.

41. tan 30°   **42.** cot 30°   **43.** sin 30°   **44.** cos 30°   **45.** sec 30°   **46.** csc 30°

47. csc 45°   **48.** sec 45°   **49.** cos 45°   **50.** sin 45°   **51.** cot 45°   **52.** tan 45°

53. sin 60°   **54.** cos 60°   **55.** tan 60°   **56.** cot 60°   **57.** sec 60°   **58.** csc 60°

59. Which pair of trigonometric functions are both reciprocals and cofunctions?

60. Find the angle θ in the interval $[0°, 90°)$ for which $\sin \theta = \cos \theta$.

61. Construct an equilateral triangle with each side having length $2k$.
(a) What is the measure of each angle?
(b) Label one angle A. Drop a perpendicular from A to the side opposite A. Two 30° angles are formed at A, and two right triangles are formed. What is the length of each side opposite each 30° angle?
(c) What is the length of the perpendicular constructed in part (b)?
(d) From the results of parts (a)–(c), complete the following statement: In a 30°–60° right triangle, the hypotenuse is always _____ times as long as the shorter leg, and the longer leg has a length that is _____ times as long as that of the shorter leg. Also, the shorter leg is opposite the _____ angle, and the longer leg is opposite the _____ angle.

62. Construct a square with each side of length k.
(a) Draw a diagonal of the square. What is the measure of each angle formed by a side of the square and this diagonal?
(b) What is the length of the diagonal?
(c) From the results of parts (a) and (b), complete the following statement: In a 45°–45° right triangle, the hypotenuse has a length that is _____ times as long as either leg.

Use the results of Exercises 61 and 62 to find the exact value of each labeled part in each of the following figures.

63.

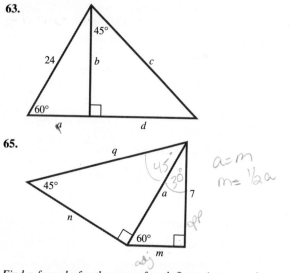

64.

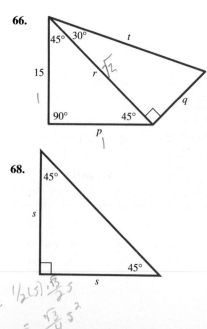

65.

66.

Find a formula for the area of each figure in terms of s.

67.

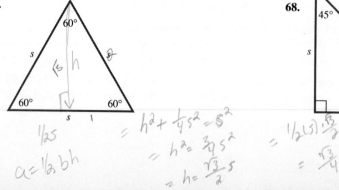

68.

69. The table shown here illustrates an interesting pattern for the values of the sine of 0°, 30°, 45°, 60°, and 90°. (There is no mathematical justification for this pattern.)

θ	$\sin \theta$
0°	$\sqrt{0}/2 = 0$
30°	$\sqrt{1}/2 = 1/2$
45°	$\sqrt{2}/2$
60°	$\sqrt{3}/2$
90°	$\sqrt{4}/2 = 1$

Explain why this pattern cannot possibly continue past 90°. (*Hint:* What is the maximum value of the sine ratio?)

70. Construct a table similar to the one in Exercise 69 for values of the cosine of those angles.

2.2 ———— REFERENCE ANGLES; COTERMINAL ANGLES

Associated with every non-quadrantal angle in standard position is a positive acute angle called its reference angle. A **reference angle** for an angle θ, written θ', is the positive acute angle made by the terminal side of angle θ and the x-axis. Figure 2.8 shows several angles θ (each less than one complete counterclockwise revolution) in quadrants II, III, and IV, respectively, with the reference angle θ' also shown. In quadrant I, θ and θ' are the same. If an angle θ is negative or has measure greater than 360°, its reference angle is found by first finding its coterminal angle that is between 0° and 360°, and then using the diagrams in Figure 2.8.

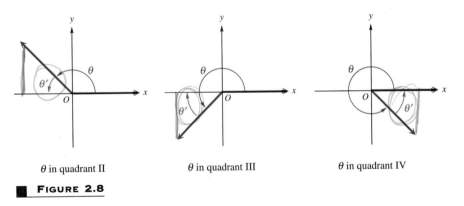

θ in quadrant II θ in quadrant III θ in quadrant IV

∎ **FIGURE 2.8**

CAUTION | A very common error is to find the reference angle by using the terminal side of θ and the y-axis. *The reference angle is always found with reference to the x-axis.*

■ *Example 1*
FINDING
REFERENCE ANGLES

Find the reference angles for the following three angles.

(a) 218°

As shown in Figure 2.9, the positive acute angle made by the terminal side of this angle and the x-axis is $218° - 180° = 38°$. For $\theta = 218°$, the reference angle $\theta' = 38°$.

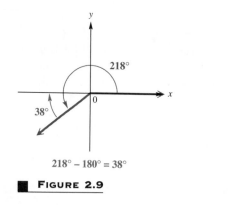

$$218° - 180° = 38°$$

■ **FIGURE 2.9**

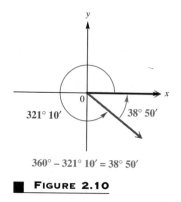

$$360° - 321° \, 10' = 38° \, 50'$$

■ **FIGURE 2.10**

(b) 321° 10′

As shown in Figure 2.10, the positive acute angle made by the terminal side of this angle and the x-axis is $360° - 321° \, 10'$. Write $360°$ as $359° \, 60'$ so that the reference angle is

$$359° \, 60' - 321° \, 10' = 38° \, 50'.$$

Note that an angle of $-38° \, 50'$, which has the same terminal ray as $321° \, 10'$, also has a reference angle of $38° \, 50'$.

(c) 1387°

First find a coterminal angle between $0°$ and $360°$. Divide $1387°$ by $360°$ to get a quotient of about 3.9. Begin by subtracting $360°$ three times (because of the 3 in 3.9):

$$1387° - 3 \cdot 360° = 307°.$$

The reference angle for $307°$ (and thus for $1387°$) is $360° - 307° = 53°$. ■

The preceding example suggests the following table for finding the reference angle θ' for any angle θ between $0°$ and $360°$.

REFERENCE ANGLES FOR θ IN (0°, 360°)	θ in Quadrant	θ′ Is	Example
	I	θ	
	II	180° − θ	
	III	θ − 180°	
	IV	360° − θ	

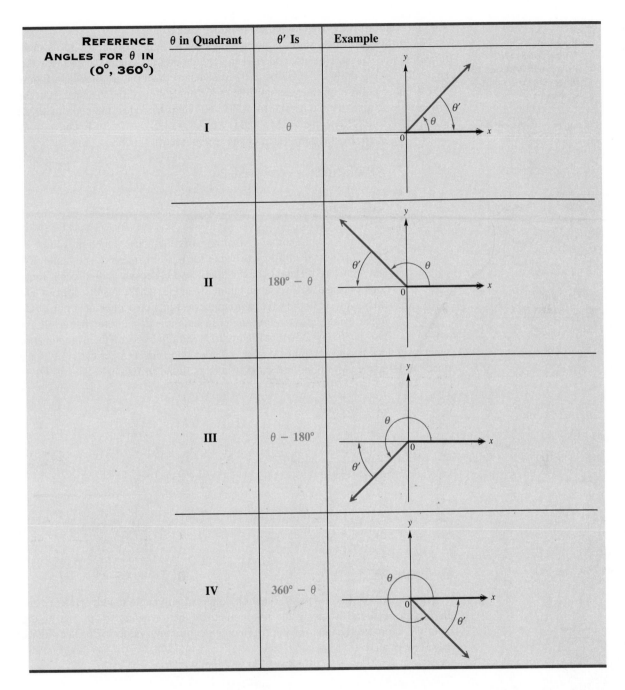

We can now find exact trigonometric function values of angles with reference angles of 30°, 60°, or 45°. In Example 2 we show how to use these function values to find the trigonometric function values for 210°.

64

■ *Example 2*

FINDING TRIGONOMETRIC FUNCTION VALUES OF 210° (USING x, y, AND r)

Find the values of the trigonometric functions for 210°.

Even though a 210° angle is not an angle of a right triangle, the ideas mentioned earlier can still be used to find the trigonometric function values for this angle. To do so, draw an angle of 210° in standard position, as shown in Figure 2.11. Choose point P on the terminal side of the angle so that the distance from the origin O to P is 2. By the results from 30°–60° right triangles, the coordinates of point P become $(-\sqrt{3}, -1)$, with $x = -\sqrt{3}$, $y = -1$, and $r = 2$. Then, by the definitions of the trigonometric functions,

$$\sin 210° = -\frac{1}{2} \qquad \tan 210° = \frac{\sqrt{3}}{3} \qquad \sec 210° = -\frac{2\sqrt{3}}{3}$$

$$\cos 210° = -\frac{\sqrt{3}}{2} \qquad \cot 210° = \sqrt{3} \qquad \csc 210° = -2. \quad ■$$

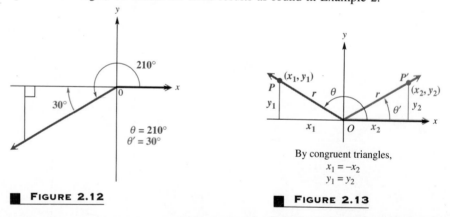

■ FIGURE 2.11

Notice in Example 2 that the trigonometric function values of 210° correspond in absolute value to those of its reference angle 30°. The signs are different for the sine, cosine, secant, and cosecant functions because 210° is a quadrant III angle. These results suggest another way of finding the trigonometric function values of a non-acute angle, using the reference angle. In Example 2, the reference angle for 210° is 30°, as shown in Figure 2.12. Simply by using the trigonometric function values of the reference angle, 30°, and choosing the correct signs for a quadrant III angle, we obtain the same results as found in Example 2.

210°

30°

$\theta = 210°$
$\theta' = 30°$

■ FIGURE 2.12

(x_1, y_1)

P

r θ

y_1

x_1 O x_2

P'

(x_2, y_2)

r

θ' y_2

By congruent triangles,
$x_1 = -x_2$
$y_1 = y_2$

■ FIGURE 2.13

In general, the trigonometric function values of any angle θ with reference angle θ' can be found by using the trigonometric function values of θ' and giving them the appropriate signs for the trigonometric function values of θ. See Figure 2.13. Notice that

$$\sin \theta = \frac{y_1}{r} = \frac{y_2}{r} = \sin \theta',$$

$$\cos \theta = \frac{x_1}{r} = \frac{-x_2}{r} = -\cos \theta',$$

$$\tan \theta = \frac{y_1}{x_1} = \frac{y_2}{-x_2} = -\tan \theta',$$

and so on.

Based on this work, the values of the trigonometric functions for any non-quadrantal angle θ can be found by finding the function values for an angle between $0°$ and $90°$. To do this, perform the following steps.

FINDING TRIGONOMETRIC FUNCTION VALUES FOR ANY NON-QUADRANTAL ANGLE

1. If $\theta \geq 360°$, or if $\theta < 0°$, find a coterminal angle by adding or subtracting $360°$ as many times as needed to get an angle of at least $0°$ but less than $360°$.
2. Find the reference angle θ'.
3. Find the necessary values of the trigonometric functions for the reference angle θ'.
4. Find the correct signs for the values found in Step 3. (Use the table of signs in Section 1.5). This result gives the value of the trigonometric functions for angle θ.

▪ *Example 3*

FINDING TRIGONOMETRIC FUNCTION VALUES USING REFERENCE ANGLES

Use reference angles to find the exact value of each of the following.

(a) $\cos(-240°)$

The reference angle is $60°$, as shown in Figure 2.14. Since the cosine is negative in quadrant II,

$$\cos(-240°) = -\cos 60° = -\frac{1}{2}.$$

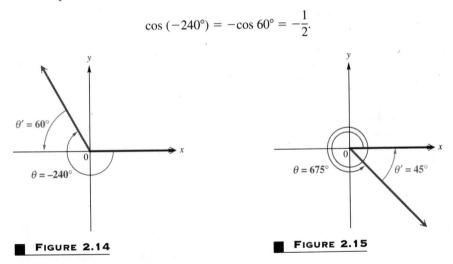

▪ **FIGURE 2.14**   ▪ **FIGURE 2.15**

(b) $\tan 675°$

Begin by subtracting $360°$ to get a coterminal angle between $0°$ and $360°$.

$$675° - 360° = 315°$$

As shown in Figure 2.15, the reference angle is $360° - 315° = 45°$. An angle of $315°$ is in quadrant IV, so the tangent will be negative, and

$$\tan 675° = \tan 315° = -\tan 45° = -1. \quad ▪$$

The ideas discussed in this section can be reversed to find the measures of certain angles, given a trigonometric function value and an interval in which the angle must lie. We are most often interested in the interval [0°, 360°).

■ *Example 4*

FINDING ANGLE MEASURES GIVEN AN INTERVAL AND A FUNCTION VALUE

Find all values of θ, if θ is in the interval [0°, 360°) and cos θ = $-\sqrt{2}/2$.

Since cosine here is negative, θ must lie in either quadrant II or III. Since the absolute value of cos θ is $\sqrt{2}/2$, the reference angle θ' must be 45°. The two possible angles θ are sketched in Figure 2.16.

The quadrant II angle θ must equal 180° − 45° = 135°, and the quadrant III angle θ must equal 180° + 45° = 225°. ■

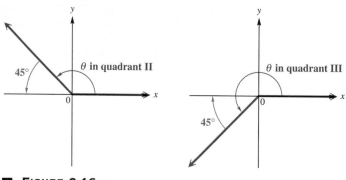

■ **FIGURE 2.16**

Exact values can be used in evaluating expressions, as shown in the next example.

■ *Example 5*

EVALUATING AN EXPRESSION WITH FUNCTION VALUES OF SPECIAL ANGLES

Evaluate cos 120° + 2 sin² 60° − tan² 30°.

Since cos 120° = −1/2, sin 60° = $\sqrt{3}/2$, and tan 30° = $\sqrt{3}/3$,

$$\cos 120° + 2\sin^2 60° - \tan^2 30° = -\frac{1}{2} + 2\left(\frac{\sqrt{3}}{2}\right)^2 - \left(\frac{\sqrt{3}}{3}\right)^2$$

$$= -\frac{1}{2} + 2\left(\frac{3}{4}\right) - \frac{3}{9}$$

$$= \frac{2}{3}. \quad ■$$

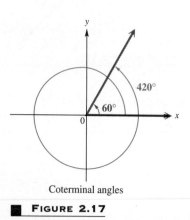

Coterminal angles

■ **FIGURE 2.17**

Recall from Chapter 1 that coterminal angles have the same initial side and the same terminal side. For example, Figure 2.17 shows an angle of 60°. Angles of 360° + 60° = 420°, or 720° + 60° = 780°, or −360° + 60° = −300°, and so on, are coterminal with this angle. Because the trigonometric functions are defined by a point on the terminal side of the angle (and not by the amount of rotation),

$$\sin 60° = \sin 420° = \sin 780° = \sin (-300°),$$

and so on. Similar statements are true for the other five functions. Rotating the terminal side of an angle through 360° results in a new angle having its terminal side coterminal with that of the original angle. The values of the trigonometric functions of both angles are the same, leading to the following **identities for coterminal angles.**

IDENTITIES FOR COTERMINAL ANGLES	For any angle θ, and any integer n,
	$\sin \theta = \sin (\theta + n \cdot 360°)$ $\cot \theta = \cot (\theta + n \cdot 360°)$
	$\cos \theta = \cos (\theta + n \cdot 360°)$ $\sec \theta = \sec (\theta + n \cdot 360°)$
	$\tan \theta = \tan (\theta + n \cdot 360°)$ $\csc \theta = \csc (\theta + n \cdot 360°)$.

▪ *Example 6*
USING THE IDENTITIES FOR COTERMINAL ANGLES

Evaluate each of the following by first expressing the function in terms of an angle between 0° and 360°.

(a) cos 780°

Add or subtract 360° as many times as necessary so that the final angle is between 0° and 360°. Subtracting 720°, which is 2 · 360°, gives

$$\cos 780° = \cos (60° + 2 \cdot 360°)$$
$$= \cos 60°$$
$$= 1/2.$$

(b) tan (−210°)

Since −210° = 150° − 1 · 360°,

$$\tan (-210°) = \tan (150° - 1 \cdot 360°)$$
$$= \tan 150°$$
$$= -\sqrt{3}/3 \qquad\qquad \tan 150° = -\tan 30° \ ▪$$

2.2 EXERCISES ▪

Find the reference angle for each of the following. See Example 1.

1. 120°   **2.** 135°   **3.** 225°   **4.** 240°   **5.** 300°   **6.** 315°   **7.** −135°   **8.** −210°

9. 98°   **10.** 212°   **11.** 285° 30′   **12.** 314° 50′   **13.** 480°   **14.** 750°   **15.** −600°   **16.** −780°

17. Can 120° be the measure of a reference angle? Explain. *Its an acute angle*

18. Name the angle satisfying the given conditions and having a reference angle of 30°.
   (a) positive measure, less than one complete revolution, in quadrant II
   (b) negative measure, less than one complete clockwise revolution, in quadrant IV
   (c) between one and two counterclockwise revolutions, in quadrant III

Find an angle θ, where θ is in the interval [0°, 360°), that is coterminal with each of the following. See Example 1(c).

19. 615°   **20.** 708°   **21.** 458° 20′   **22.** 506° 40′   **23.** −321° 50′   **24.** −243° 30′

68

Note: *The remaining exercises in this set are* **not** *to be worked with a calculator.*

Use the methods of this section to find the exact values *of the six trigonometric functions for each of the following angles. Rationalize denominators when applicable. See Examples 2, 3, and 6.*

25. 120°   **26.** 135°   **27.** 150°   **28.** 225°   **29.** 240°

30. 300°   **31.** 330°   **32.** 390°   **33.** 420°   **34.** 495°

35. 510°   **36.** 570°   **37.** 750°   **38.** 1320°   **39.** 1500°

40. 2670°   **41.** −390°   **42.** −510°   **43.** −1020°   **44.** −1290°

Complete the following table with exact trigonometric function values using the methods of this section. See Examples 2 and 3.

θ	$\sin\theta$	$\cos\theta$	$\tan\theta$	$\cot\theta$	$\sec\theta$	$\csc\theta$
45. 30°	1/2	$\sqrt{3}/2$	___	___	$2\sqrt{3}/3$	2
46. 45°	___	___	1	1	___	___
47. 60°	___	1/2	$\sqrt{3}$	___	2	___
48. 120°	$\sqrt{3}/2$	___	$-\sqrt{3}$	___	___	$2\sqrt{3}/3$
49. 135°	$\sqrt{2}/2$	$-\sqrt{2}/2$	___	___	$-\sqrt{2}$	$\sqrt{2}$
50. 150°	___	$-\sqrt{3}/2$	$-\sqrt{3}/3$	___	___	2
51. 210°	$-1/2$	___	$\sqrt{3}/3$	$\sqrt{3}$	___	-2
52. 240°	$-\sqrt{3}/2$	$-1/2$	___	___	-2	$-2\sqrt{3}/3$

Find all values of θ, *if* θ *is in the interval* [0°, 360°) *and has the given function value. See Example 4.*

53. $\sin\theta = \dfrac{1}{2}$ **54.** $\cos\theta = \dfrac{\sqrt{3}}{2}$ **55.** $\tan\theta = \sqrt{3}$ **56.** $\sec\theta = \sqrt{2}$

57. $\cos\theta = -\dfrac{1}{2}$ **58.** $\cot\theta = -\dfrac{\sqrt{3}}{3}$ **59.** $\sin\theta = -\dfrac{\sqrt{3}}{2}$ **60.** $\cos\theta = -\dfrac{\sqrt{2}}{2}$

61. $\tan\theta = -1$ **62.** $\cot\theta = -\sqrt{3}$ **63.** $\cos\theta = 0$ **64.** $\sin\theta = 1$

65. $\cot\theta$ is undefined   **66.** $\csc\theta$ is undefined

67. Does there exist an angle θ with the function values $\cos\theta = .6$ and $\sin\theta = -.8$?

68. Does there exist an angle θ with the function values $\cos\theta = 2/3$ and $\sin\theta = 3/4$?

Suppose that θ *is in the interval* (90°, 180°). *Find the sign of each of the following.*

69. $\sin\dfrac{\theta}{2}$ **70.** $\cos\dfrac{\theta}{2}$ **71.** $\cot(\theta + 180°)$

72. $\sec(\theta + 180°)$ **73.** $\cos(-\theta)$ **74.** $\sin(-\theta)$

Evaluate each of the following. See Example 5.

75. $\sin^2 120° + \cos^2 120°$

76. $\sin^2 225° + \cos^2 225°$

77. $2\tan^2 120° + 3\sin^2 150° - \cos^2 180°$

78. $\cot^2 135° - \sin 30° + 4\tan 45°$

79. $\sin^2 225° - \cos^2 270° + \tan 60°$

80. $\cot^2 90° - \sec^2 180° + \csc^2 135°$

81. $\cos^2 60° + \sec^2 150° - \csc^2 210°$

82. $\cot^2 135° + \tan^4 60° - \sin^4 180°$

Tell whether each of the following is true *or* false. *If false, tell why.*

83. $\sin 30° + \sin 60° = \sin (30° + 60°)$

84. $\sin(30° + 60°) = \sin 30° \cdot \cos 60° + \sin 60° \cdot \cos 30°$

85. $\cos 60° = 2 \cos^2 30° - 1$

86. $\cos 60° = 2 \cos 30°$

87. $\sin 120° = \sin 150° - \sin 30°$

88. $\sin 210° = \sin 180° + \sin 30°$

89. $\sin 120° = \sin 180° \cdot \cos 60° - \sin 60° \cdot \cos 180°$

90. $\cos 300° = \cos 240° \cdot \cos 60° - \sin 240° \cdot \sin 60°$

2.3 ———————— **FINDING TRIGONOMETRIC FUNCTION VALUES USING A CALCULATOR**

To say that the trigonometry student of today has an easier task than his or her counterpart of twenty-five years ago is an understatement. The tables of trigonometric function values that yesterday's students were required to use have been replaced by keys of scientific calculators. When values obtained from tables did not provide enough accuracy, a process called linear interpolation (although an important concept in its own right) was used to increase accuracy, but also used up valuable time. Today's calculators give more accuracy than we usually ever need. When involved calculations were required, computation with logarithms eased the work somewhat. However, learning to calculate with logarithms and read tables of logarithms usually required an additional textbook chapter, and still did not provide the accuracy that calculators afford.

With the technological advances of this era in mind, the examples and exercises in this text are written with the assumption that all students have access to scientific calculators. However, since calculators differ among makes and models, students should always consult their owner's manuals for specific information if questions arise concerning their use. Some newer models of calculators (in particular, graphing calculators) require a different order of keystrokes than those described in the examples in this section.

CAUTION Thus far in this book, we have studied only one type of measure for angles, degree measure; another type of measure, radians, will be studied in Chapter 3. When evaluating trigonometric functions of angles given in degrees, it is a common error to use the incorrect mode; remember that the calculator must be set in the *degree mode*. One way to avoid this problem is to get in the habit of always starting work as follows.

Enter: 90 $\boxed{\text{sin}}$

If the displayed answer is 1, the calculator is set for degree measure; otherwise it is not.

70

■ *Example 1*
FINDING FUNCTION
VALUES WITH A
CALCULATOR

Use a calculator to find approximate values of the following trigonometric functions.

(a) sin 49° 12′

Convert 49° 12′ to decimal degrees, as explained in Chapter 1.

$$49° \; 12′ = 49\frac{12°}{60} = 49.2°$$

(As mentioned earlier, some calculators will do this conversion automatically.) Then press the $\boxed{\text{sin}}$ key.

Enter: 49.2 $\boxed{\text{sin}}$ **Display:** $\boxed{.75699506}$

To eight decimal places,

$$\text{sin } 49° \; 12′ = .75699506.$$

(b) $\sec 97° \; 58′ \; 37″ = \sec\left(97 + \dfrac{58}{60} + \dfrac{37}{3600}\right)^{°}$

$$= \sec 97.976944°$$

Calculators do not have secant keys. However,

$$\sec \theta = \frac{1}{\cos \theta}$$

for all angles θ when cos θ is not 0. So, find sec 97.976944° by pressing the cosine key and then taking the reciprocal, by pressing the $\boxed{1/x}$ key.

Enter: 97.976944 $\boxed{\text{cos}}$ $\boxed{1/x}$ **Display:** $\boxed{-7.2059291}$

(c) cot 51.4283°

This angle is already in decimal degrees. Use the identity cot θ = 1/tan θ.

Enter: 51.4283 $\boxed{\text{tan}}$ $\boxed{1/x}$ **Display:** $\boxed{.79748114}$

(d) sin (−246°)

Enter 246, and then use the key that changes the sign of the displayed number. (This key is often labeled $\boxed{+/-}$.)

Enter: 246 $\boxed{+/-}$ $\boxed{\text{sin}}$ **Display:** $\boxed{.91354546}$

(e) sin 130° 48′

130° 48′ is equal to 130.8 in decimal degrees. Enter 130.8 and press the $\boxed{\text{sin}}$ key.

Enter: 130.8 $\boxed{\text{sin}}$ **Display:** $\boxed{.75699506}$ ■

Notice that the values found in parts (a) and (e) of Example 1 are the same. The reason for this is that 49° 12′ is the reference angle for 130° 48′ and the sine function is positive for a quadrant II angle.

FINDING ANGLES So far in this section we have used a calculator to find trigonometric function values of angles. This process can be reversed. An angle can be found from its trigonometric function value as shown in the next example.

■ *Example 2*
FINDING ANGLES
WITH A
CALCULATOR

Use a calculator to find a value of θ in the interval [0°, 90°) satisfying each of the following. Leave answers in decimal degrees.

(a) sin θ = .81815000

Find θ using a key labeled ⌗arc⌗ or ⌗INV⌗ together with the ⌗sin⌗ key. Some calculators require a different sequence of keystrokes, or may use a key labeled ⌗sin⁻¹⌗. Again, make sure the calculator is set for degree measure.

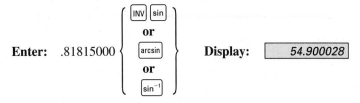

Enter: .81815000 { INV sin **or** arcsin **or** sin⁻¹ } **Display:** 54.900028

(The *arc* and *sin*⁻¹ notations are explained in Chapter 6.) Finally,

$$\theta = 54.900028°.$$

The second function key may be required in addition to one of the combinations shown above.

(b) sec θ = 1.0545829
Use the identity cos θ = 1/sec θ.

Enter: 1.0545829 ⌗1/x⌗   Use appropriate keys involving cos   **Display:** 18.514704

Therefore,   θ = 18.514704°. ■

CAUTION | In Examples 1(b) and 2(b) note that the ⌗1/x⌗ key is used *before* the ⌗cos⌗ key when finding the angle, but *after* the ⌗cos⌗ key when finding the trigonometric function given the angle.

■ *Example 3*
FINDING ANGLES
WITH A
CALCULATOR

Find two angles in the interval [0°, 360°) that satisfy cos θ = .81411552.

Using the keystrokes described in Example 2, we find that one value of θ is 35.5°. However, we must find another value of θ in [0°, 360°) that satisfies this function value. This other value of θ will have its reference angle θ′ equal to 35.5°, and must be in quadrant IV, since cosine values are positive in quadrants I and IV. Recall the table in Section 2.2; we find that the other value of θ is

$$\theta = 360° - \theta' = 360° - 35.5° = 324.5°. \quad ■$$

2.3 EXERCISES ■

Use a calculator to find approximations for each of the following. See Example 1.

1. sin 38° 40′	**2.** tan 29° 30′	**3.** cot 41° 20′	**4.** cos 27° 10′
5. sin 58° 30′	**6.** cos 46° 10′	**7.** tan 17° 20′	**8.** sin 39° 40′
9. cot 128° 30′	**10.** tan 153° 20′	**11.** sin 179° 20′	**12.** cos 124° 10′
13. sin 204° 20′	**14.** sec 218° 50′	**15.** cos 251° 10′	**16.** cot 298° 30′
17. sin 274° 30′	**18.** cos 304° 50′	**19.** csc 421° 10′	**20.** cot 512° 20′
21. sec (−108° 20′)	**22.** csc (−29° 30′)	**23.** tan (−197° 50′)	**24.** cos (−299° 40′)
25. sin 59.642°	**26.** cos 38.1219°	**27.** tan (−80.612°)	**28.** sec (−19.702°)
29. cos 258.409°	**30.** sin 315.768°	**31.** cot 109.713°	**32.** csc 278.143°
33. cos 74° 11′	**34.** tan 58° 46′	**35.** cot 125° 52′ 10″	**36.** csc 211° 40′ 38″
37. sin (−15° 28′ 17″)	**38.** cos (−29° 03′ 05″)	**39.** sec (−121° 53′ 42″)	**40.** cot (−276° 18′ 34″)

41. A student, wishing to use a calculator to verify the value of sin 30°, enters the information correctly but gets a display of −.98803162. He knows that the display should be .5, and he also knows that his calculator is in good working order. What do you think is the problem?

42. A certain make of calculator does not allow the input of angles outside of a particular interval when finding trigonometric function values. For example, trying to find cos 2000° using the methods of this section would give an error message, despite the fact that cos 2000° can be evaluated. Explain how you would find cos 2000° using this calculator.

Use a calculator to find a value of θ in [0°, 90°) satisfying each of the following. Leave your answer in decimal degrees. See Example 2.

43. sin θ = .84802194	**44.** tan θ = 1.4739716	**45.** sec θ = 1.1606249	**46.** cot θ = 1.2575516
47. sin θ = .72144101	**48.** sec θ = 2.7496222	**49.** tan θ = 6.4358841	**50.** sin θ = .27843196

Use a calculator to find two angles in the interval [0°, 360°) that satisfy each of the following. Leave your answer in decimal degrees. See Example 3.

51. cos θ = .68716510	**52.** cos θ = .96476120	**53.** sin θ = .41298643
54. sin θ = .63898531	**55.** tan θ = .87692035	**56.** tan θ = 1.2841996

Use a calculator to find each of the following. (As shown later in Chapter 5, all these answers should be integers.)

57. cos 100° cos 80° − sin 100° sin 80°

58. sin 35° cos 55° + cos 35° sin 55°

59. sin 28° 14′ cos 61° 46′ + cos 28° 14′ sin 61° 46′

60. cos 75° 29′ cos 14° 31′ − sin 75° 29′ sin 14° 31′

When a light ray travels from one medium, such as air, to another medium, such as water or glass, the speed of the light changes, and the direction that the ray is traveling changes. (This is why a fish under water is in a different position than it appears to be.) These changes are given by Snell's law

$$\frac{c_1}{c_2} = \frac{\sin \theta_1}{\sin \theta_2},$$

where c_1 is the speed of light in the first medium, c_2 is the speed of light in the second medium, and θ_1 and θ_2 are the angles shown in the figure. In the following exercises, assume that $c_1 = 3 \times 10^8$ m per sec. Find the speed of light in the second medium.

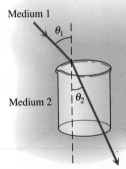

Medium 1

If this medium is less dense, light travels at a faster speed, c_1.

Medium 2

If this medium is more dense, light travels at a slower speed, c_2.

61. $\theta_1 = 46°$, $\theta_2 = 31°$

62. $\theta_1 = 39°$, $\theta_2 = 28°$

Find θ_2 for the following values of θ_1 and c_2. Round to the nearest degree.

63. $\theta_1 = 40°$, $c_2 = 1.5 \times 10^8$ m per sec

64. $\theta_1 = 62°$, $c_2 = 2.6 \times 10^8$ m per sec

The figure below shows a fish's view of the world above the surface of the water. *
Suppose that a light ray comes from the horizon, enters the water, and strikes the fish's eye.

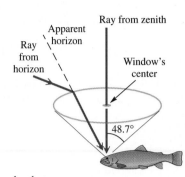

65. Let us assume that this ray gives a value of 90° for angle θ_1 in the formula for Snell's law. (In a practical situation this angle would probably be a little less than 90°.) The speed of light in water is about 2.254×10^8 m per sec. Find angle θ_2.

   (Your result should have been about 48.7°. This means that a fish sees the world above the water as a cone, making an angle of 48.7° with the vertical.)

66. Suppose that an object is located at a true angle of 29.6° above the horizon. Find the apparent angle above the horizon to a fish.

Use a calculator to decide whether the following statements are true or false. It may be that a true statement will lead to results that differ in the last decimal place, due to rounding error.

67. $\sin 10° + \sin 10° = \sin 20°$

68. $\cos 40° = 2 \cos 20°$

69. $\sin 50° = 2 \sin 25° \cdot \cos 25°$

70. $\cos 70° = 2 \cos^2 35° - 1$

71. $\cos 40° = 1 - 2 \sin^2 80°$

72. $2 \cos 38° 22' = \cos 76° 44'$

73. $\sin 39° 48' + \cos 39° 48' = 1$

74. $\dfrac{1}{2} \sin 40° = \sin \dfrac{1}{2} (40°)$

TRIGONOMETRIC FUNCTION VALUES:
EXACT AND APPROXIMATE ■ ————————————————

In Sections 2.2 and 2.3, we have seen that trigonometric function values of some angles can be determined exactly, while for others we must rely on calculator approximations. Students should be aware of the distinction between exact and approximate values. For example, we know that the exact value of $\sin 60°$ is $\sqrt{3}/2$, while a calculator will give only an approximate value for $\sin 60°$, .86602540.

   For each of the functions below, determine whether the reference angle is 30°, 45°, 60°, or some other acute angle. If it is one of these three special angles, give the exact function value. (Don't forget to determine the sign by using the quadrant of the angle.) If the reference angle is some other acute angle, find a calculator approximation.

1. $\sin 240°$

2. $\sin 225°$

3. $\cos 130°$

4. $\cos 290°$

5. $\tan (-60°)$

6. $\tan (-300°)$

7. $\sin (-60°)$

8. $\sin (-225°)$

9. $\sin 190°$

10. $\sin 170°$

11. $\tan 210°$

12. $\tan 330°$

13. $\sec 150°$

14. $\sec 330°$

15. $\csc 140°$

16. $\csc 280°$

17. $\tan 495°$

18. $\cot (-225°)$

19. $\cot (-80°)$

20. $\tan 170°$

2.4 ———————— **SOLVING RIGHT TRIANGLES**

One of the main applications of trigonometry is solving triangles. **To solve a triangle** means to find the measures of all the angles and sides of the triangle. This section and the next discuss methods of solving right triangles. (Methods for solving other triangles are presented in Chapter 7.)

Before we solve triangles, a short discussion concerning accuracy and significant digits is appropriate. Suppose that we glance quickly at a room and guess that it is 15 feet by 18 feet. To calculate the length of a diagonal of the room, the Pythagorean theorem can be used.

$$d^2 = 15^2 + 18^2$$
$$d^2 = 549$$
$$d = \sqrt{549}$$

On a calculator, $\sqrt{549} = 23.430749.$

Should this answer be given as the length of the diagonal of the room? Of course not. The number 23.430749 contains 6 decimal places, while the original data of 15 feet and 18 feet are only accurate to the nearest foot. Since the results of a problem can be no more accurate than the least accurate number in any calculation, we really should say that the diagonal of the 15-by-18-foot room is 23 feet.

If a wall is measured to the nearest foot and is found to be 18 feet long, actually this means that the wall has a length between 17.5 feet and 18.5 feet. If the wall is measured more accurately and found to be 18.3 feet long, then its length is really between 18.25 feet and 18.35 feet. A measurement of 18.00 feet would indicate that the length of the wall is between 17.995 feet and 18.005 feet. The measurement 18 feet is said to have two **significant digits** of accuracy; 18.0 has three significant digits, and 18.00 has four.

A significant digit is a digit obtained by actual measurement. A number that represents the result of counting, or a number that results from theoretical work and is not the result of a measurement, is an **exact number.** There are fifty states in the United States, so 50 is an exact number. The number of states is not 49 3/4 or 50 1/4; nor is the number 50 used here to represent "some number between 45 and 55." In the formula for the perimeter of a rectangle, $P = 2L + 2W$, the 2's are obtained from the definition of perimeter, and are exact numbers.

Most values of trigonometric functions are approximations, and virtually all measurements are approximations. To perform calculations on such approximate numbers, follow the rules given below.

CALCULATION WITH SIGNIFICANT DIGITS

For *adding* and *subtracting,* round the answer so that the last digit you keep is in the right-most column in which all the numbers have significant digits.

For *multiplying* or *dividing,* round the answer to the least number of significant digits found in any of the given numbers.

For *powers* and *roots,* round the answer so that it has the same number of significant digits as the number whose power or root you are finding.

When solving triangles, use the following table for deciding on significant digits in angle measure.

SIGNIFICANT DIGITS FOR ANGLES	Number of Significant Digits	Angle Measure to Nearest:
	2	Degree
	3	Ten minutes, or nearest tenth of a degree
	4	Minute, or nearest hundredth of a degree
	5	Tenth of a minute, or nearest thousandth of a degree

For example, an angle measuring 52° 30′ has three significant digits (assuming that 30′ is measured to the nearest ten minutes).

In using trigonometry to solve triangles, it is convenient to use a to represent the length of the side opposite angle A, b for the length of the side opposite angle B, and so on. As mentioned earlier, in a right triangle the letter c is reserved for the hypotenuse. Figure 2.18 shows the labeling of a typical right triangle.

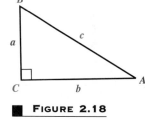

■ **FIGURE 2.18**

■ *Example 1*

SOLVING A RIGHT TRIANGLE GIVEN AN ANGLE AND A SIDE

Solve right triangle ABC, with $A = 34° 30′$ and $c = 12.7$ in. See Figure 2.19.

To solve the triangle, find the measures of the remaining sides and angles. The value of a can be found with a trigonometric function involving the known values of angle A and side c. Since the sine of angle A is given by the quotient of the side opposite A and the hypotenuse, use $\sin A$.

$$\sin A = \frac{a}{c}$$

Substituting known values gives

$$\sin 34° 30′ = \frac{a}{12.7},$$

or, upon multiplying both sides by 12.7,

$a = 12.7 \sin 34° 30′$

$a = 12.7(.56640624)$ Use a calculator.

$a = 7.19$ in.

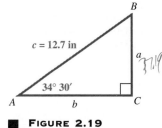

■ **FIGURE 2.19**

The value of b could be found with the Pythagorean theorem. It is better, however, to use the information given in the problem rather than a result just calculated. If a mistake were to be made in finding *a*, then *b* also would be incorrect. Also, rounding more than once may cause the result to be less accurate. Using cos *A* gives

$$\cos A = \frac{\text{side adjacent}}{\text{hypotenuse}} = \frac{b}{c}$$

$$\cos 34° 30' = \frac{b}{12.7}$$

$$b = 12.7 \cos 34° 30'$$

$$b = 10.5 \text{ in.}$$

Once *b* has been found, the Pythagorean theorem could be used as a check. All that remains to solve triangle *ABC* is to find the measure of angle *B*. Since *A* + *B* = 90° and *A* = 34° 30',

$$A + B = 90°$$

$$B = 90° - A$$

$$B = 89° 60' - 34° 30'$$

$$B = 55° 30'. \quad ■$$

NOTE   In Example 1 we could have started by finding the measure of angle *B* and then used the trigonometric function values of *B* to find the unknown sides. The process of solving a right triangle (like many problems in mathematics) can usually be done in several ways, each resulting in the correct answer. However, in order to retain as much accuracy as can be expected, always use given information as much as possible, and avoid rounding off in intermediate steps.

■ *Example 2*

SOLVING A RIGHT TRIANGLE GIVEN TWO SIDES

Solve right triangle *ABC* if *a* = 29.43 cm and *c* = 53.58 cm.

Draw a sketch showing the given information, as in Figure 2.20. One way to begin is to find angle *A* by using the sine.

$$\sin A = \frac{\text{side opposite}}{\text{hypotenuse}}$$

$$\sin A = \frac{29.43}{53.58}$$

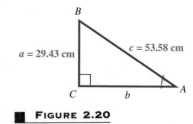

■ **FIGURE 2.20**

Using [INV] [sin] or [sin⁻¹] on a calculator, we find that *A* = 33.32°. The measure of *B* is 90° − 33.32° = 56.68°.

We now find *b* from the Pythagorean theorem, $a^2 + b^2 = c^2$, or $b^2 = c^2 - a^2$. Since *c* = 53.58 and *a* = 29.43,

$$b^2 = 53.58^2 - 29.43^2$$

giving $b = 44.77 \text{ cm.} \quad ■$

PROBLEM SOLVING

The process of solving right triangles is easily adapted to solving applied problems. A crucial step in such applications involves sketching the triangle and labelling the given parts correctly. Then we can use the methods described in the earlier examples to find the unknown value or values. ■

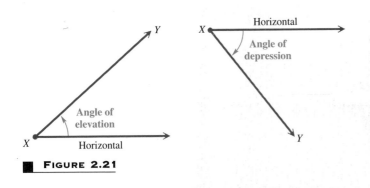

FIGURE 2.21

Many applications of right triangles involve the angle of elevation or the angle of depression. The **angle of elevation** from point X to point Y (above X) is the angle made by ray XY and a horizontal ray with endpoint at X. The angle of elevation is always measured from the horizontal. See Figure 2.21. The **angle of depression** from **point** X to point Y (below X) is the angle made by ray XY and a horizontal ray with endpoint X. Again, see Figure 2.21.

CAUTION Errors are often made in interpreting the angle of depression. Remember that both the angle of elevation *and* the angle of depression are measured *from* the horizontal *to* the line of sight.

■ *Example 3*
FINDING A LENGTH WHEN THE ANGLE OF ELEVATION IS KNOWN

Donna Garbarino knows that when she stands 123 feet from the base of a flagpole, the angle of elevation to the top is 26° 40′. If her eyes are 5.30 feet above the ground, find the height of the flagpole.

The length of the side adjacent to Donna is known and the length of the side opposite her is to be found. See Figure 2.22. The ratio that involves these two values is the tangent.

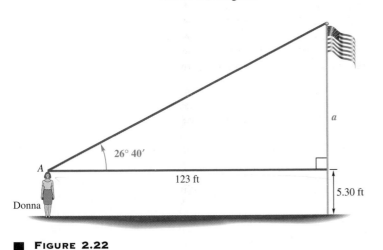

FIGURE 2.22

$$\tan A = \frac{\text{side opposite}}{\text{side adjacent}}$$

$$\tan 26° 40′ = \frac{a}{123}$$

$$a = 123 \tan 26° 40′$$

$$a = 61.8 \text{ feet}$$

Since Donna's eyes are 5.30 feet above the ground, the height of the flagpole is

$$61.8 + 5.30 = 67.1 \text{ feet.} \quad ■$$

■ *Example 4*

FINDING THE ANGLE
OF ELEVATION
WHEN LENGTHS
ARE KNOWN

The length of the shadow of a building 34.09 meters tall is 37.62 meters. Find the angle of elevation of the sun.

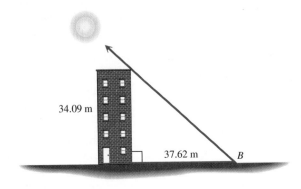

34.09 m

37.62 m   *B*

■ FIGURE 2.23

As shown in Figure 2.23, the angle of elevation of the sun is angle *B*. Since the side opposite *B* and the side adjacent to *B* are known, use the tangent ratio to find *B*.

$$\tan B = \frac{34.09}{37.62}$$

$$B = 42.18°$$ Use [INV] [tan] on a calculator.

The angle of elevation of the sun is 42.18°.   ■

2.4 EXERCISES ■

Refer to the discussion of accuracy and significant digits in this section to work the following problems.

1. When Mt. Everest was first surveyed, the surveyors obtained a height of 29,000 ft to the nearest foot. State the range represented by this number. (The surveyors thought that no one would believe a measurement of 29,000 ft, so they reported it as 29,002.)

2. New Orleans Saints kicker Tom Dempsey holds the National Football League record for the longest field goal. On November 8, 1970, he kicked a 63-yd field goal against the Detroit Lions to win the game. What range does the number 63 represent here?

3. According to the *Guinness Book of World Records,* the widest long-span bridge is the 1650-ft-long Sydney Harbour Bridge in Australia, which is 160 ft wide. State the ranges represented by these two numbers if they represent accuracy to the nearest foot.

4. At Denny's, a chain of restaurants, the Low-Cal Special is said to have "approximately 472 calories." What is the range of calories represented by this number? By claiming "approximately 472 calories," they are probably claiming more accuracy than is possible. In your opinion, what might be a better claim?

Find the error in each statement.

5. I have 2 bushel baskets, each containing 65 apples. I know that $2 \times 65 = 130$, but 2 has only one significant figure, so I must write the answer as 1×10^2, or 100. I therefore have 100 apples.

6. The formula for the circumference of a circle is $C = 2\pi r$. My circle has a radius of 54.98 cm, and my calculator has a $\boxed{\pi}$ key, giving fifteen digits of accuracy. Pressing the right buttons gives 345.44953. Because 2 has only one significant digit, however, the answer must be given as 3×10^2, or 300 cm. (What is the correct answer?)

In the remaining exercises in this set, use a calculator as necessary.

Solve each right triangle. See Examples 1 and 2.

7. **8.** **9.**

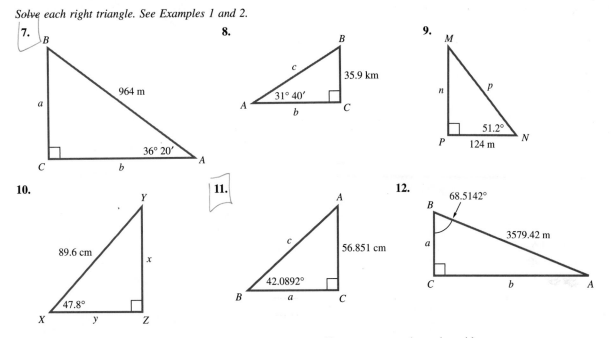

10. **11.** **12.**

13. Can a right triangle be solved if we are given the measures of its two acute angles and no side lengths? Explain.

14. If we are given an acute angle and a side in a right triangle, what unknown part of the triangle requires the least work to find?

Solve each of the following right triangles. In each case, $C = 90°$. If the angle information is given in degrees and minutes, give the answers in the same way. If given in decimal degrees, do likewise in your answers. When two sides are given, give answers in degrees and minutes.

15. $A = 28° \, 00'$, $c = 17.4$ ft
16. $B = 46° \, 00'$, $c = 29.7$ m
17. $B = 73° \, 00'$, $b = 128$ in
18. $A = 61° \, 00'$, $b = 39.2$ cm
19. $a = 76.4$ yd, $b = 39.3$ yd
20. $a = 958$ m, $b = 489$ m
21. $a = 18.9$ cm, $c = 46.3$ cm
22. $b = 219$ m, $c = 647$ m
23. $A = 53° \, 24'$, $c = 387.1$ ft
24. $A = 13° \, 47'$, $c = 1285$ m
25. $B = 39° \, 09'$, $c = .6231$ m
26. $B = 82° \, 51'$, $c = 4.825$ cm
27. $c = 7.813$ m, $b = 2.467$ m
28. $c = 44.91$ mm, $a = 32.71$ mm
29. $B = 42.432°$, $a = 157.49$ m
30. $A = 36.704°$, $c = 1461.3$ cm
31. $A = 57.209°$, $c = 186.49$ cm
32. $B = 12.099°$, $b = 7.0463$ m
33. $b = 173.921$ m, $c = 208.543$ m
34. $a = 864.003$ cm, $c = 1092.84$ cm

35. Use the ideas found in Section 1.3 involving a transversal intersecting parallel lines to explain why the angle of depression *DAB* has the same measure as the angle of elevation *ABC* in the accompanying figure.

36. Why is angle *CAB not* an angle of depression in the accompanying figure?

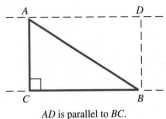

AD is parallel to BC.

Solve each of the following. See Example 3.

37. A 13.5-m fire-truck ladder is leaning against a wall. Find the distance the ladder goes up the wall if it makes an angle of 43° 50′ with the ground.

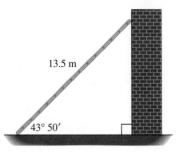

13.5 m

43° 50′

38. A swimming pool is 40.0 ft long and 4.00 ft deep at one end. If it is 8.00 ft deep at the other end, find the total distance along the bottom.

40.0 ft

4.00 ft

8.00 ft

39. A guy wire 77.4 m long is attached to the top of an antenna mast that is 71.3 m high. Find the angle that the wire makes with the ground.

40. Find the length of a guy wire that makes an angle of 45° 30′ with the ground if the wire is attached to the top of a tower 63.0 m high.

41. To measure the height of a flagpole, Kitty Pellissier finds that the angle of elevation from a point 24.73 ft from the base to the top is 38° 12′. Find the height of the flagpole.

42. A rectangular piece of land is 528.2 ft by 630.7 ft. Find an acute angle made by the diagonal of the rectangle.

43. To find the distance *RS* across a lake, a surveyor lays off *RT* = 53.1 m, with angle *T* = 32° 10′, and angle *S* = 57° 50′. Find length *RS*.

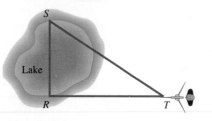

S

Lake

R   T

44. A surveyor must find the distance *QM* across a depressed freeway. She lays off *QN* = 769 ft along one side of the freeway, with angle *N* = 21° 50′, and with angle *M* = 68° 10′. Find *QM*.

45. The length of the base of an isosceles triangle is 42.36 in. Each base angle is 38.12°. Find the length of each of the two equal sides of the triangle. (*Hint:* Divide the triangle into two right triangles.)

46. Find the altitude of an isosceles triangle having a base of 184.2 cm if the angle opposite the base is 68° 44′.

Work the following problems involving angles of elevation or depression. See Examples 3 and 4.

47. Suppose that the angle of elevation of the sun is 23.4°. Find the length of the shadow cast by Cindy Newman, who is 5.75 ft tall.

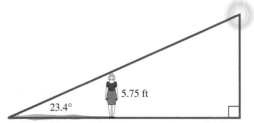

5.75 ft

23.4°

48. The shadow of a vertical tower is 40.6 m long when the angle of elevation of the sun is 34.6°. Find the height of the tower.

$$\tan = \frac{opp}{adj} = \frac{h}{40.6}$$

$$h = 40.6 \cdot \tan 34.6°$$

$$h = 28.0 m$$

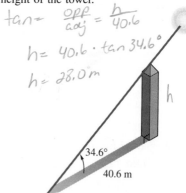

h

34.6°

40.6 m

49. Find the angle of elevation of the sun if a 48.6-ft flagpole casts a shadow 63.1 ft long.

50. The angle of depression from the top of a building to a point on the ground is 32° 30′. How far is the point on the ground from the top of the building if the building is 252 m high?

51. An airplane is flying 10,500 feet above the level ground. The angle of depression from the plane to the base of a tree is 13° 50′. How far horizontally must the plane fly to be directly over the tree?

10,500 ft

52. The angle of elevation from the top of a small building to the top of a nearby taller building is 46° 40′, while the angle of depression to the bottom is 14° 10′. If the smaller building is 28.0 m high, find the height of the taller building.

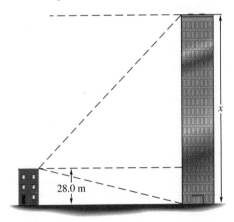

x

28.0 m

53. A television camera is to be mounted on a bank wall so as to have a good view of the head teller (see the figure). Find the angle of depression that the lens should make with the horizontal.

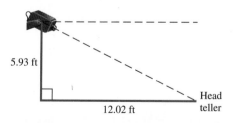

5.93 ft

12.02 ft

Head teller

54. A company safety committee has recommended that a floodlight be mounted in a parking lot so as to illuminate the employee exit (see the following figure). Find the angle of depression of the light.

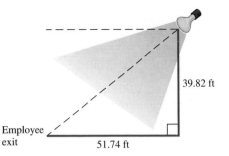

39.82 ft

Employee exit

51.74 ft

55. Atoms in metals can be arranged in patterns called *unit cells.* One such unit cell, called a *primitive cell,* is a cube with an atom at each corner. A right triangle can be formed from one edge of the cell, a face diagonal, and a cube diagonal, as shown in the figure. If each cell edge is 3×10^{-8} cm and the face diagonal is 4.24×10^{-8} cm, what is the angle between the cell edge and a cube diagonal?

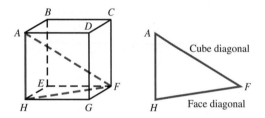

56. To determine the diameter of the sun, an astronomer might sight with a transit (a device used by surveyors for measuring angles) first to one edge of the sun and then to the other, finding that the included angle equals 1° 4′. Assuming that the distance from the earth to the sun is 92,919,800 mi, calculate the diameter of the sun. (See the figure.)

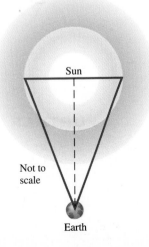

57. Very accurate measurements have shown that the distance between California's Owens Valley Radio Observatory and the Haystack Observatory in Massachusetts is 2441.2938 mi. Suppose that the two observatories focus on a distant star and find that angles E and E' in the figure are both 89.99999°. Find the distance to the star from Haystack. (Assume that the earth is flat.)

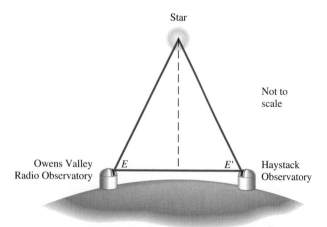

58. The figure shows a magnified view of the threads of a bolt. Find x if d is 2.894 mm.

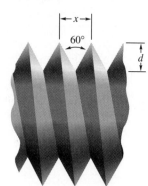

59. You have a summer job putting up outhouses in county parks. You must precut the rafter ends so that they will be vertical when in place. The front wall is 8 ft high, the back wall is 6 1/2 ft high and the distance between walls is 8 ft. At what angle should you cut the rafters? (Round to the nearest degree.)*

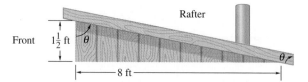

60. A degree may seem like a very small unit, but an error of one degree in measuring an angle may be very significant. For example, suppose that a laser beam is directed toward the visible center of the moon and that it misses its assigned target by 30 sec. How far is it (in mi) from its assigned target? (Take the distance from the surface of the earth to that of the moon to be 234,000 mi.)*

*From *A Sourcebook of Applications of School Mathematics* by Donald Bushaw et al. Copyright © 1980 by The Mathematical Association of America. Reprinted by permission. The material was prepared with the support of National Science Foundation Grant No. SED72-01123 A05. However, any opinions, findings, conclusions, or recommendations expressed herein are those of the authors and do not necessarily reflect the views of NSF.

2.5 — FURTHER APPLICATIONS OF RIGHT TRIANGLES

Other applications of right triangles involve **bearing,** an important idea in navigation. There are two common ways to express bearing. *When a single angle is given, such as 164°, it is understood that the bearing is measured in a clockwise direction from due north.* Several sample bearings using this first type of system are shown in Figure 2.24.

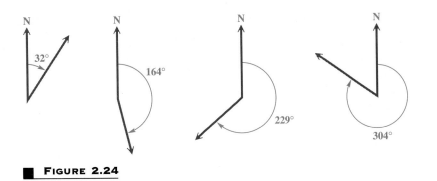

■ FIGURE 2.24

NOTE | In the following examples and exercises, the problems all result in right triangles, so the methods of the previous section apply. Chapter 7 will include problems involving bearing that result in triangles that are *not* right triangles and require other methods to solve.

■ *Example 1*

SOLVING A PROBLEM INVOLVING BEARING (FIRST TYPE)

Radar stations *A* and *B* are on an east-west line, 3.7 kilometers apart. Station *A* detects a plane at *C*, on a bearing of 61°. Station *B* simultaneously detects the same plane, on a bearing of 331°. Find the distance from *A* to *C*.

Draw a sketch showing the given information, as in Figure 2.25. Since a line drawn due north is perpendicular to an east-west line, right angles are formed at *A* and *B*, so that angles *CAB* and *CBA* can be found. Angle *C* is a right angle because angles *CAB* and *CBA* are complementary. (If *C* were not a right angle, the methods of Chapter 7 would be needed.) Find distance *b* by using the cosine function.

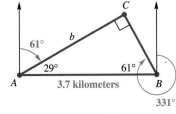

■ FIGURE 2.25

$$\cos 29° = \frac{b}{3.7}$$

$$3.7 \cos 29° = b$$

$$b = 3.2 \text{ kilometers}$$

Use a calculator and round to the nearest tenth. ■

PROBLEM SOLVING

It would be foolish to attempt to solve the problem in Example 1 without drawing a sketch. The importance of a correctly labeled sketch in applications such as this cannot be overemphasized, as some of the necessary information is often not given in the problem, and can only be determined from the sketch. ■

The second common system for expressing bearing starts with a north-south line and uses an acute angle to show the direction, either east or west, from this line. Figure 2.26 shows several sample bearings using this system. Either N or S always comes first, followed by an acute angle, and then E or W.

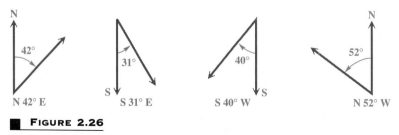

■ **FIGURE 2.26**

■ *Example 2*

SOLVING A
PROBLEM INVOLVING
BEARING (SECOND
TYPE)

The bearing from A to C is S 52° E. The bearing from A to B is N 84° E. The bearing from B to C is S 38° W. A plane flying at 250 miles per hour takes 2.4 hours to go from A to B. Find the distance from A to C.

Make a sketch of the situation. First draw the two bearings from point A. Choose a point B on the bearing N 84° E from A and draw the bearing to C. Point C will be located where the bearing lines from A and B intersect as shown in Figure 2.27.

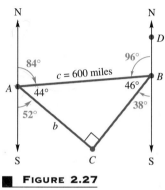

■ **FIGURE 2.27**

Since the bearing from A to B is N 84° E, angle ABD is 180° − 84° = 96°. Thus, angle ABC is 46°. Also, angle BAC is 180° − (84° + 52°) = 44°. Angle C is 180° − (44° + 46°) = 90°. From the statement of the problem, a plane flying at 250 miles per hour takes 2.4 hours to go from A to B. The distance from A to B is the product of rate and time, or

$$c = \text{rate} \times \text{time} = 250(2.4) = 600 \text{ miles.}$$

To find b, the distance from A to C, use the sine. (The cosine could also have been used.)

$$\sin 46° = \frac{b}{c}$$

$$\sin 46° = \frac{b}{600}$$

$$600 \sin 46° = b$$

$$b = 430 \text{ miles} \qquad \text{Use a calculator.} \quad ■$$

The next example uses the idea of the angle of elevation first discussed in the previous section.

■ *Example 3*

SOLVING A
PROBLEM INVOLVING
ANGLE OF
ELEVATION

Francisco needs to know the height of a tree. From a given point on the ground he finds that the angle of elevation to the top of the tree is 36° 40′. He then moves back 50 feet. From the second point, the angle of elevation to the top of the tree is 22° 10′. See Figure 2.28. Find the height of the tree.

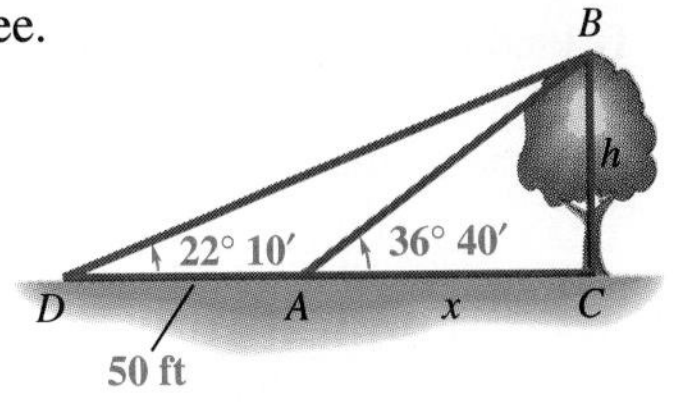

The figure shows two unknowns: x, the distance from the center of the trunk of the tree to the point where the first observation was made, and h, the height of the tree. Since nothing is given about the length of the hypotenuse of either triangle ABC or triangle BCD, use a ratio that does not involve the hypotenuse—the tangent.

■ **FIGURE 2.28**

In triangle ABC, $\quad \tan 36° 40′ = \dfrac{h}{x} \quad$ or $\quad h = x \tan 36° 40′$.

In triangle BCD, $\quad \tan 22° 10′ = \dfrac{h}{50 + x} \quad$ or $\quad h = (50 + x) \tan 22° 10′$.

Since each of these expressions equals h, these expressions must be equal. Thus,

$$x \tan 36° 40′ = (50 + x) \tan 22° 10′.$$

Now use algebra to solve for x.

$$x \tan 36° 40′ = 50 \tan 22° 10′ + x \tan 22° 10′ \qquad \text{Distributive property}$$

$$x \tan 36° 40′ - x \tan 22° 10′ = 50 \tan 22° 10′ \qquad \text{Get } x \text{ terms on one side.}$$

$$x(\tan 36° 40′ - \tan 22° 10′) = 50 \tan 22° 10′ \qquad \text{Factor out } x \text{ on the left.}$$

$$x = \frac{50 \tan 22° 10′}{\tan 36° 40′ - \tan 22° 10′} \qquad \text{Divide by the coefficient of } x.$$

We saw above that $h = x \tan 36° 40′$. Substituting for x,

$$h = \left(\frac{50 \tan 22° 10′}{\tan 36° 40′ - \tan 22° 10°}\right)(\tan 36° 40′).$$

From a calculator,

$$\tan 36° 40′ = .74447242$$
$$\tan 22° 10′ = .40741394,$$

so

$$\tan 36° 40′ - \tan 22° 10′ = .74447242 - .40741394 = .33705848,$$

and

$$h = \left(\frac{50(.40741394)}{.33705848}\right)(.74447242) = 45 \text{ (rounded)}.$$

The height of the tree is approximately 45 feet. ■

NOTE
In practice we usually do not write down the intermediate calculator approximation steps. However, we have done this in Example 3 so that the reader may follow the steps more easily.

2.5 EXERCISES ■

Solve these problems involving right triangles.

1. Find the angle of the sun above the horizon when a person 5.25 ft tall casts a shadow 8.65 ft long. (Find θ in the figure.)

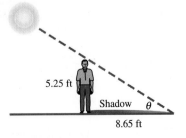

2. The angle of the sun above the horizon is 27.5°. Find the length of the shadow of a person 4.75 ft tall. (Find *x* in the figure.)

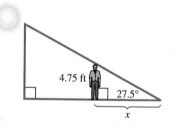

3. The solar panel shown in the figure must be tilted so that angle θ is 94° when the angle of elevation of the sun is 38°. (The sun's rays are assumed to be parallel, and they make an angle of 38° with the horizontal.) The panel is 4.5 ft long. Find *h*.

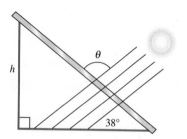

4. Find the minimum height *h* above the surface of the earth so that a pilot at point *A* in the figure can see an object on the horizon at *C*, 125 mi away. Assume that the radius of the earth is 4.00×10^3 mi.

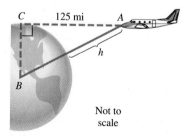

Not to scale

5. In one area, the lowest angle of elevation of the sun in winter is 23° 20′. Find the minimum distance *x* that a plant needing full sun can be placed from a fence 4.65 ft high (see the figure).

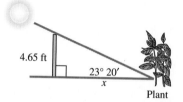

6. A tunnel is to be dug from *A* to *B* (see the figure). Both *A* and *B* are visible from *C*. If *AC* is 1.4923 mi and *BC* is 1.0837 mi, and if *C* is 90°, find the measures of angles *A* and *B*.

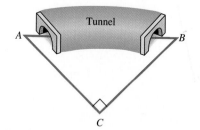

7. A piece of land has the shape as shown in the figure. Find *x*.

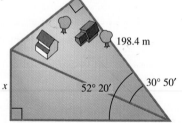

8. Find the value of *x* in the figure.

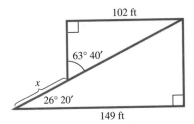

102 ft

63° 40'

x

26° 20'

149 ft

***9.** The leaning tower of Pisa is approximately 179 ft in height and is approximately 16.5 ft out of plumb (that is, tilted away from the vertical). Find the angle at which it deviates from the vertical.

***10.** A regular pentagon (five-sided polygon with equal lengths of sides and equal angles) is inscribed in a circle of radius 7. Find the length of a side of the pentagon. Give your answer to the nearest thousandth.

11. Find *h* as indicated in the figure.

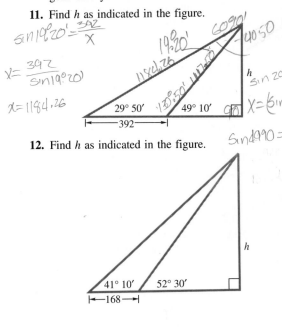

$\sin 19°20' = \frac{392}{X}$

$X = \frac{392}{\sin 19°20'}$

$X = 1184.26$

19°20'

60°90'

40 50

11°40.26

30°50' 117°20

29° 50'   49° 10'   90

392

h $\sin 20°50 = \frac{X}{1184.26}$

$X = \sin 20°50' \cdot 1184.26$

$\sin 40°90 = \frac{h}{1117.50}$

12. Find *h* as indicated in the figure.

h

41° 10'   52° 30'

168

Solve each of the following. See Examples 1–3. Drawing a sketch for these problems where one is not given may be helpful.

13. The angle of elevation from a point on the ground to the top of a pyramid is 35° 30'. The angle of elevation from a point 135 ft farther back to the top of the pyramid is 21° 10'. Find the height of the pyramid.

14. Debbie Glockner, a lighthouse keeper, is watching a whale approach the lighthouse directly. When she first begins watching the whale, the angle of depression of the whale is 15° 50'. Just as the whale turns away from the lighthouse, the angle of depression is 35° 40'. If the height of the lighthouse is 68.7 m, find the distance traveled by the whale as it approaches the lighthouse.

15. A scanner antenna is on top of the center of a house. The angle of elevation from a point 28.0 m from the center of the house to the top of the antenna is 27° 10', and the angle of elevation to the bottom of the antenna is 18° 10'. Find the height of the antenna.

16. The angle of elevation from Lone Pine to the top of Mt. Whitney is 10° 50'. Van Dong Le, traveling 7.00 km from Lone Pine along a straight, level road toward Mt. Whitney, finds the angle of elevation to be 22° 40'. Find the height of the top of Mt. Whitney above the level of the road.

17. A plane is found by radar to be flying 6000 m above the ground. The angle of elevation from the radar to the plane is 80°. Fifteen seconds later, the plane is directly over the station. (See the figure.) Find the speed of the plane, assuming that it is flying parallel to the ground.

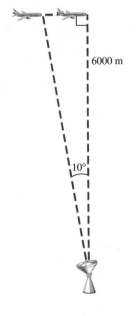

6000 m

10°

---

*Exercises 9 and 10 are excerpts from *Plane Trigonometry,* Revised Edition by Frank A. Rickey and J. P. Cole, copyright © 1964 by Holt, Rinehart and Winston, Inc., reprinted by permission of the publisher.

18. A chisel is to be made from a steel rod with a diameter of 4.6 cm. If the angle at the tip is 64°, how long will the tip be? (See the figure.)

19. A plane flies 1.5 hr at 110 mph on a bearing of 40°. It then turns and flies 1.3 hr at the same speed on a bearing of 130°. How far is the plane from its starting point?

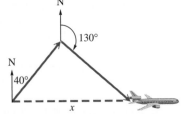

20. A ship travels 50 km on a bearing of 27°, and then travels on a bearing of 117° for 140 km. Find the distance traveled from the starting point to the ending point.

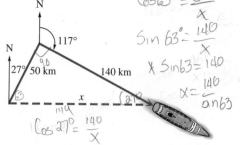

$$\cos 63° = \frac{50°}{x}$$

$$\sin 63° = \frac{140}{x}$$

$$x \sin 63 = 140$$

$$x = \frac{140}{\sin 63}$$

$$\cos 27° = \frac{140}{x}$$

21. Two ships leave a port at the same time. The first ship sails on a bearing of 40° at 18 knots (nautical miles per hour) and the second at a bearing of 130° at 26 knots. How far apart are they after 1.5 hours?

22. Two lighthouses are located on a north-south line. From lighthouse A the bearing of a ship 3742 m away is 129° 43'. From lighthouse B the bearing of the ship is 39° 43'. Find the distance between the lighthouses.

23. A ship leaves its home port and sails on a bearing of N 28° 10' E. Another ship leaves the same port at the same time and sails on a bearing of S 61° 50' E. If the first ship sails at 24.0 mph and the second sails at 28.0 mph, find the distance between the two ships after 4 hrs.

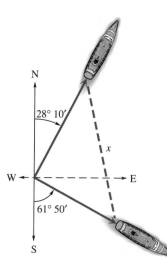

24. Radio direction finders are set up at points A and B, which are 2.50 mi apart on an east-west line. From A it is found that the bearing of the signal from a radio transmitter is N 36° 20' E, while from B the bearing of the same signal is N 53° 40' W. Find the distance of the transmitter from B.

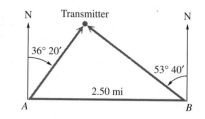

25. The bearing from Winston-Salem, North Carolina, to Danville, Virginia, is N 42° E. The bearing from Danville to Goldsboro, North Carolina, is S 48° E. A small plane piloted by Mark Ferrari, traveling at 60 mph, takes 1 hr to go from Winston-Salem to Danville and 1.8 hr to go from Danville to Goldsboro. Find the distance from Winston-Salem to Goldsboro.

26. The bearing from Atlanta to Macon is S 27° E, and the bearing from Macon to Augusta is N 63° E. A plane traveling at 60 mph needs 1 1/4 hr to go from Atlanta to Macon and 1 3/4 hr to go from Macon to Augusta. Find the distance from Atlanta to Augusta.

CHAPTER 2 SUMMARY ■

SECTION	KEY IDEAS
2.1 Trigonometric Functions of Acute Angles	**Right Triangle-Based Definitions of the Trigonometric Functions** For any acute angle A in standard position, $$\sin A = \frac{y}{r} = \frac{\text{side opposite}}{\text{hypotenuse}} \qquad \csc A = \frac{r}{y} = \frac{\text{hypotenuse}}{\text{side opposite}}$$ $$\cos A = \frac{x}{r} = \frac{\text{side adjacent}}{\text{hypotenuse}} \qquad \sec A = \frac{r}{x} = \frac{\text{hypotenuse}}{\text{side adjacent}}$$ $$\tan A = \frac{y}{x} = \frac{\text{side opposite}}{\text{side adjacent}} \qquad \cot A = \frac{x}{y} = \frac{\text{side adjacent}}{\text{side opposite}}.$$ **Cofunction Identities** For any acute angle A, $$\sin A = \cos(90° - A) \qquad \cot A = \tan(90° - A)$$ $$\cos A = \sin(90° - A) \qquad \csc A = \sec(90° - A)$$ $$\tan A = \cot(90° - A) \qquad \sec A = \csc(90° - A).$$ **Function Values of Special Angles**

Function Values of Special Angles

θ	$\sin\theta$	$\cos\theta$	$\tan\theta$	$\cot\theta$	$\sec\theta$	$\csc\theta$
30°	$\dfrac{1}{2}$	$\dfrac{\sqrt{3}}{2}$	$\dfrac{\sqrt{3}}{3}$	$\sqrt{3}$	$\dfrac{2\sqrt{3}}{3}$	2
45°	$\dfrac{\sqrt{2}}{2}$	$\dfrac{\sqrt{2}}{2}$	1	1	$\sqrt{2}$	$\sqrt{2}$
60°	$\dfrac{\sqrt{3}}{2}$	$\dfrac{1}{2}$	$\sqrt{3}$	$\dfrac{\sqrt{3}}{3}$	2	$\dfrac{2\sqrt{3}}{3}$

SECTION	KEY IDEAS
2.2 Reference Angles; Coterminal Angles	**Reference Angles for θ in $(0°, 360°)$**

θ in Quadrant	θ' Is
I	θ
II	$180° - \theta$
III	$\theta - 180°$
IV	$360° - \theta$

SECTION	KEY IDEAS
	Finding Trigonometric Function Values for Any Angle
	1. Add or subtract 360° as many times as needed to get an angle of at least 0° but less than 360°.
	2. Find the reference angle θ'.
	3. Find the trigonometric function value for θ'.
	4. Find the correct sign.
	Identities for Coterminal Angles
	For any angle θ and any integer n,
	$$\sin \theta = \sin (\theta + n \cdot 360°) \qquad \cot \theta = \cot (\theta + n \cdot 360°)$$ $$\cos \theta = \cos (\theta + n \cdot 360°) \qquad \sec \theta = \sec (\theta + n \cdot 360°)$$ $$\tan \theta = \tan (\theta + n \cdot 360°) \qquad \csc \theta = \csc (\theta + n \cdot 360°).$$
2.5 Further Applications of Right Triangles	**Bearing** 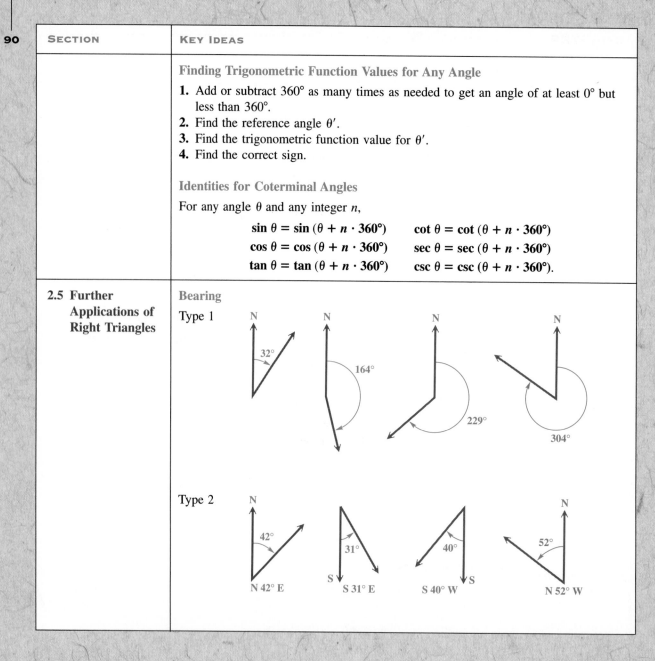

CHAPTER 2 REVIEW EXERCISES ■ ————————————————

Find the values of the trigonometric functions for each angle A.

1.

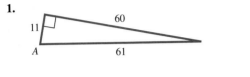

2.

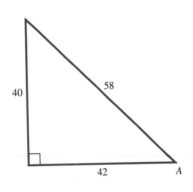

Solve each of the following. Assume that all angles are acute angles.

3. $\sin 4\beta = \cos 5\beta$

4. $\sec (2\gamma + 10°) = \csc (4\gamma + 20°)$

5. $\tan (5x + 11°) = \cot (6x + 2°)$

6. $\cos \left(\dfrac{3\theta}{5} + 11°\right) = \sin \left(\dfrac{7\theta}{10} + 40°\right)$

Tell whether each of the following is true *or* false.

7. $\sin 46° < \sin 58°$ **8.** $\cos 47° < \cos 58°$ **9.** $\sec 48° \geq \cos 42°$ **10.** $\sin 22° \geq \csc 68°$

Find the reference angle θ' for each of the following.

11. $219°$ **12.** $-154°$ **13.** $578.94°$ **14.** $680° \, 30'$

15. Explain in your own words the concept of reference angle.

16. Explain why, in the figure, the cosine of angle A is equal to the sine of angle B.

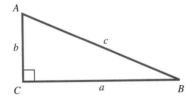

Find the values of the six trigonometric functions for each of the following angles. Give exact values. Do not use a calculator. Rationalize denominators when applicable.

17. $120°$ **18.** $225°$ **19.** $300°$ **20.** $750°$ **21.** $-225°$ **22.** $-390°$

Find all values of θ, if θ is in the interval $[0°, 360°)$ and θ has the given function value.

23. $\sin \theta = -\dfrac{1}{2}$ **24.** $\cos \theta = -\dfrac{1}{2}$ **25.** $\cot \theta = -1$ **26.** $\sec \theta = -\dfrac{2\sqrt{3}}{3}$

Evaluate each of the following. Give exact values.

27. $\cos 60° + 2 \sin^2 30°$

28. $\tan^2 120° - 2 \cot 240°$

29. $\cot^2 300° + \cos^2 120° - 3 \sin^2 240°$

30. $\sec^2 300° - 2 \cos^2 150° + \tan 45°$

Use a calculator to find the value of each of the following.

31. $\sin 72° 30'$

32. $\sec 222° 30'$

33. $\cot 305.6°$

34. $\sin 47° 24'$

35. $\cot 32° 42'$

36. $\csc 78° 21'$

37. $\sec 58.9041°$

38. $\tan 11.7689°$

39. $\sin 89.0043°$

40. $\cot 1.49783°$

41. A test item read "Find the exact value of sin 45°." A student gave the answer .7071067812, obtained with his powerful new calculator, yet did not receive credit. Was the teacher justified in not accepting his answer? Explain.

42. Which one of the following cannot be *exactly* determined using the methods of this chapter?
(a) $\cos 135°$ **(b)** $\cot -45°$ **(c)** $\sin 300°$ **(d)** $\tan 140°$

Use a calculator to find the value of θ, where θ is in the interval [0°, 90°). Give the answers in decimal degrees.

43. $\sin \theta = .82584121$

44. $\cot \theta = 1.1249386$

45. $\cos \theta = .97540415$

46. $\sec \theta = 1.2637891$

47. $\tan \theta = 1.9633124$

48. $\csc \theta = 9.5670466$

Find two angles in the interval [0°, 360°) that satisfy each of the following. Leave your answer in decimal degrees.

49. $\sin \theta = .73254290$

50. $\tan \theta = 1.3865342$

Tell whether each of the following is true *or* false. *If false, tell why. Use a calculator for Exercises 51 and 54.*

51. $\sin 50° + \sin 40° = \sin 90°$

52. $\cos 210° = \cos 180° \cdot \cos 30° - \sin 180° \cdot \sin 30°$

53. $\sin 240° = 2 \sin 120° \cdot \cos 120°$

54. $\sin 42° + \sin 42° = \sin 84°$

Solve each of the following right triangles. In Exercise 56, give angles to the nearest minute. In Exercises 57 and 58, label the triangles as shown in Figure 2.18 in Section 2.4. Use a calculator as necessary.

55.

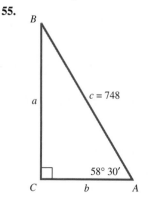

56.

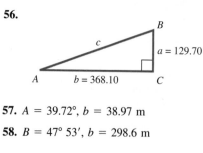

57. $A = 39.72°, b = 38.97$ m

58. $B = 47° 53', b = 298.6$ m

Solve each of the following problems.

59. The angle of elevation from a point 93.2 ft from the base of a tower to the top of the tower is 38° 20′. Find the height of the tower.

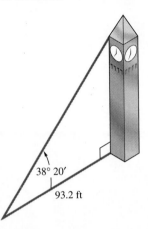

60. The angle of depression of a television tower to a point on the ground 36.0 m from the bottom of the tower is 29.5°. Find the height of the tower.

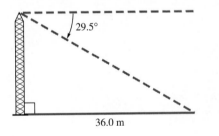

61. A rectangle has adjacent sides measuring 10.93 cm and 15.24 cm. The angle between the diagonal and the longer side is 35.65°. Find the length of the diagonal.

62. An isosceles triangle has a base of length 49.28 m. The angle opposite the base is 58.746°. Find the length of each of the two equal sides.

63. The bearing of *B* from *C* is 254°. The bearing of *A* from *C* is 344°. The bearing of *A* from *B* is 32°. The distance from *A* to *C* is 780 m. Find the distance from *A* to *B*.

64. A ship leaves a pier on a bearing of S 55° E and travels for 80 km. It then turns and continues on a bearing of N 35° E for 74 km. How far is the ship from the pier?

65. Two cars leave an intersection at the same time. One heads due south at 55 mph. The other travels due west. After two hours, the bearing of the car headed west from the car headed south is 324°. How far apart are they at that time?

66. Find a formula for *h* in terms of *k*, *A*, and *B*. Assume *A* < *B*.

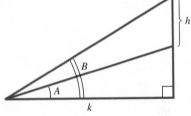

RADIAN MEASURE AND THE CIRCULAR FUNCTIONS

In most work involving applications of trigonometry, angles are measured in degrees, minutes, and seconds. This method of angle measure dates back to the Babylonians, who were the first to subdivide the circumference of a circle into 360 parts. There are various theories as to why the number 360 was chosen. One is that it is approximately the number of days in a year, and it has many divisors which makes it convenient to work with. Another involves a roundabout theory dealing with the length of a Babylonian mile.

However, in more advanced work in mathematics, the use of *radian measure* of angles is preferred. Radian measure also allows us to treat our familiar trigonometric functions as functions with domains of *real numbers,* rather than angles. These ideas are introduced in this chapter.

3.1 ——— RADIAN MEASURE

Figure 3.1 shows an angle θ in standard position along with a circle of radius r. The vertex of θ is at the center of the circle. Angle θ intercepts an arc on the circle equal in length to the radius of the circle. Because of this, angle θ is said to have a measure of one radian.

RADIAN   An angle that has its vertex at the center of a circle and that intercepts an arc on the circle equal in length to the radius of the circle has a measure of **one radian.**

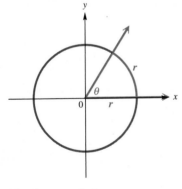

■ FIGURE 3.1

It follows that an angle of measure 2 radians intercepts an arc equal in length to twice the radius of the circle, an angle of measure 1/2 radian intercepts an arc equal in length to half the radius of the circle, and so on.

The circumference of a circle, the distance around the circle, is given by $C = 2\pi r$, where r is the radius of the circle. The formula $C = 2\pi r$ shows that the radius can be laid off 2π times around a circle. Therefore, an angle of 360°, which corresponds to a complete circle, intercepts an arc equal in length to 2π times the radius of the circle. Because of this, an angle of 360° has a measure of 2π radians:

$$360° = 2\pi \text{ radians.}$$

An angle of 180° is half the size of an angle of 360°, so an angle of 180° has half the radian measure of an angle of 360°.

$$180° = \frac{1}{2}(2\pi) \text{ radians}$$

DEGREE/RADIAN RELATIONSHIP $$180° = \pi \text{ radians}$$

Dividing both sides of $180° = \pi$ radians by π leads to

$$1 \text{ radian} = \frac{180°}{\pi},$$

or approximately,

$$1 \text{ radian} \approx \frac{180°}{3.1415927} \approx 57.295779° \approx 57° \ 17' \ 45''.$$

Since $180° = \pi$ radians, dividing both sides by 180 gives

$$1° = \frac{\pi}{180} \text{ radians},$$

or, approximately,

$$1° \approx \frac{3.1415927}{180} \text{ radians} \approx .01745329 \text{ radians}.$$

Angle measures can be converted back and forth between degrees and radians by using one of several methods described below.

CONVERTING BETWEEN DEGREES AND RADIANS		
1. Proportion: $\dfrac{\text{Radian measure}}{\pi} = \dfrac{\text{Degree measure}}{180}$		

2. Formulas:

From	To	Multiply by
Radians	Degrees	$\dfrac{180°}{\pi}$
Degrees	Radians	$\dfrac{\pi}{180°}$

3. If a radian measure involves a multiple of π, replace π with $180°$, and simplify in order to convert to degrees.

■ *Example 1*
CONVERTING DEGREES TO RADIANS

Convert each degree measure to radians.

(a) 45°
By the proportion method,

$$\frac{\text{Radian measure}}{\pi} = \frac{45}{180}.$$

Multiply both sides by π.

$$\text{Radian measure} = \frac{45\pi}{180} = \frac{\pi}{4}$$

To use the formula method, multiply by $\pi/180$.

$$45° = 45°\left(\frac{\pi}{180°}\right) = \frac{45\pi}{180} = \frac{\pi}{4} \text{ radians}$$

(b) 240°

Using the formula, $240° = 240°\left(\dfrac{\pi}{180°}\right) = \dfrac{4\pi}{3}$ radians. ■

■ *Example 2*

CONVERTING
RADIANS TO
DEGREES

Convert each of the following radian measures to degrees.

(a) $\dfrac{9\pi}{4}$

Using the proportion method.

$$\frac{\dfrac{9\pi}{4}}{\pi} = \frac{x}{180°}$$

$$\frac{9}{4} = \frac{x}{180°}$$

$$x = \frac{9}{4}(180°) = 405°.$$

(b) $\dfrac{11\pi}{3}$

Using the formula, we multiply this radian measure by $180°/\pi$ to find the corresponding degree measure.

$$\frac{11\pi}{3} \cdot \frac{180°}{\pi} = \left(\frac{1980\pi}{3\pi}\right)^{\circ} = 660°$$

(c) $\dfrac{-5\pi}{6}$

Fractional multiples of π often appear as radian measures. To convert to degrees, we may replace π with $180°$ and simplify.

$$\frac{-5\pi}{6}\text{ radians} = \frac{-5(180°)}{6} = -150° \quad ■$$

■ *Example 3*

CONVERTING
FRACTIONAL
DEGREES TO
RADIANS

Convert to radians.

(a) $29°\ 40'$

Since $40' = 40/60 = 2/3$ of a degree,

$$29°\ 40' = 29\frac{2}{3}^{\circ}$$

$$= \frac{89°}{3}$$

$$= \frac{89°}{3}\left(\frac{\pi}{180°}\right)\text{ radians} \qquad \text{Multiply by }\frac{\pi}{180°}.$$

$$= \frac{89\pi}{540}\text{ radians}. \qquad \text{Exact radian value}$$

The formula method was used here. This answer is exact. If π is replaced with the approximation 3.1415927, we get

$$29°\ 40' \approx \frac{89(3.1415927)}{540} \approx .518\text{ radians}.$$

$$74.9162° \left(\frac{\pi}{180} \right) =$$

(b) 74.9162°

Using the formula method gives

$$74.9162° = 74.9162° \left(\frac{\pi}{180°} \right) \qquad \text{Multiply by } \frac{\pi}{180°}.$$

$$\approx 74.9162° \left(\frac{3.1415927}{180°} \right)$$

$$\approx 1.30753 \text{ radians.} \quad ■$$

■ *Example 4*
CONVERTING
RADIANS TO
DEGREES

Convert 4.2 radians to degrees. Write the result to the nearest minute.
By the formula,

$$4.2 \text{ radians} = 4.2 \left(\frac{180°}{\pi} \right) \qquad \text{Multiply by } \frac{180°}{\pi}.$$

$$\approx \frac{4.2(180°)}{3.1415927} \qquad \pi \approx 3.1415927$$

$$\approx 240.64°$$

$$\approx 240° + .64°$$

$$= 240° + .64(60') \qquad 1° = 60'$$

$$\approx 240° + 38'$$

$$= 240° \, 38'. \quad ■$$

Scientific calculators have keys to enter the value of π. Depending on the make and model, the number of digits displayed and the number of digits stored internally vary. In Example 4, we could have used the $\boxed{\pi}$ key on a calculator. If this were done, there might be slight discrepancies in the final result from model to model. Such discrepancies are typical in using scientific calculators, and students should not be overly concerned if their answers vary in the last decimal place (or last several decimal places) from those given in this text and in the answers in the back of the book.

Some calculators feature a key that automatically converts between degrees and radians. With such a key, 84.9076° could be converted to radians as follows.

Enter: 84.9076   Use appropriate key   **Display:** $\boxed{1.4819172}$

Converting from radians to degrees can also be done on these calculators.

Enter: 4.2   Use appropriate key   **Display:** $\boxed{240.64227}$

30   $\boxed{1718.8734}$

This last result shows that 30 radians is equivalent to about 1719°.

CAUTION | Figure 3.2 shows angles measuring 30 radians and 30°. These angle measures are not at all close, so be careful not to confuse them.

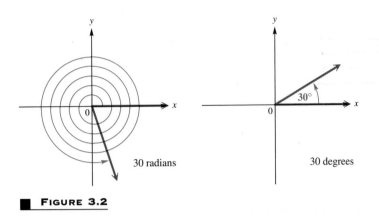

FIGURE 3.2

Trigonometric function values for angles measured in radians can be found by first converting the radian measure to degrees. (You should try to skip this intermediate step as soon as possible, and find the function values directly from the radian measure.)*

■ *Example 5*

FINDING A FUNCTION VALUE OF AN ANGLE IN RADIAN MEASURE

Find $\tan \dfrac{2\pi}{3}$.

First convert $2\pi/3$ radians to degrees.

$$\tan \frac{2\pi}{3} = \tan \left(\frac{2\pi}{3} \cdot \frac{180°}{\pi} \right) \qquad \text{Multiply by } \frac{180°}{\pi}.$$

$$= \tan 120°$$

$$= -\sqrt{3} \qquad \text{Use the methods of Chapter 2.} \quad ■$$

The following table and Figure 3.3 give some equivalent angles measured in degrees and radians. It will be useful to memorize these equivalent values. Keep in mind that $180° = \pi$ radians. Then it will be easy to reproduce the rest of the table.

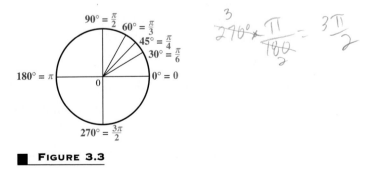

$$270° \cdot \frac{\pi}{180} = \frac{3\pi}{2}$$

FIGURE 3.3

---

*A table giving the values of the trigonometric functions for common radian and degree measures is given inside the front cover of this book.

EQUIVALENT ANGLE MEASURES IN DEGREES AND RADIANS	Degrees	Radians		Degrees	Radians	
		Exact	Approximate		Exact	Approximate
	0°	0	0	90°	$\pi/2$	1.57
	30°	$\pi/6$	.52	180°	π	3.14
	45°	$\pi/4$	.79	270°	$3\pi/2$	4.71
	60°	$\pi/3$	1.05	360°	2π	6.28

■ **Example 6**

FINDING A
FUNCTION VALUE
OF AN ANGLE IN
RADIAN MEASURE

Find $\sin 3\pi/2$.

From the table, $3\pi/2 = 270°$, so

$$\sin \frac{3\pi}{2} = \sin 270° = -1. \quad ■$$

3.1 EXERCISES ■

Convert each of the following degree measures to radians. Leave answers as multiples of π. See Example 1.

1. 60°   **2.** 30°   **3.** 90°   **4.** 120°   **5.** 150°   **6.** 135°   **7.** 210°

8. 270°   **9.** 300°   **10.** 315°   **11.** 450°   **12.** 480°   **13.** 20°   **14.** 80°

15. In your own words, explain how to convert degree measure to radian measure.

16. In your own words, explain how to convert radian measure to degree measure.

Convert each of the following radian measures to degrees. See Example 2.

17. $\frac{\pi}{3}$ **18.** $\frac{8\pi}{3}$ **19.** $\frac{7\pi}{4}$ **20.** $\frac{2\pi}{3}$ **21.** $\frac{11\pi}{6}$ **22.** $\frac{15\pi}{4}$ **23.** $-\frac{\pi}{6}$ **24.** $-\frac{\pi}{4}$

25. $\frac{8\pi}{5}$ **26.** $\frac{7\pi}{10}$ **27.** $\frac{11\pi}{15}$ **28.** $\frac{4\pi}{15}$ **29.** $\frac{7\pi}{20}$ **30.** $\frac{17\pi}{20}$ **31.** $\frac{11\pi}{30}$ **32.** $\frac{15\pi}{32}$

Convert each of the following degree measures to radians. See Example 3.

33. 39°   **34.** 74°   **35.** 42° 30′   **36.** 53° 40′

37. 139° 10′   **38.** 174° 50′   **39.** 64.29°   **40.** 85.04°

41. 56° 25′   **42.** 122° 37′   **43.** 47.6925°   **44.** 23.0143°

45. −29° 42′ 36″   **46.** −157° 11′ 9″   **47.** −209° 46′ 15″   **48.** −387° 05′ 09″

Convert each of the following radian measures to degrees. Write answers to the nearest minute. See Example 4.

49. 2   **50.** 5   **51.** 1.74   **52.** 3.06   **53.** .0912

54. .3417   **55.** 9.84763   **56.** 5.01095   **57.** −3.47189   **58.** −1.28306

59. The value of sin 30 is not 1/2. Explain why.

60. Explain in your own words what is meant by an angle of one radian.

Find the exact value of each of the following without using a calculator. See Examples 5 and 6.

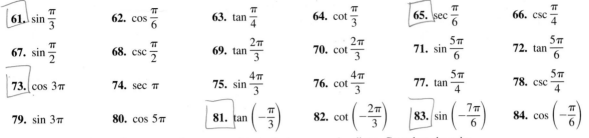

61. $\sin \frac{\pi}{3}$ **62.** $\cos \frac{\pi}{6}$ **63.** $\tan \frac{\pi}{4}$ **64.** $\cot \frac{\pi}{3}$ **65.** $\sec \frac{\pi}{6}$ **66.** $\csc \frac{\pi}{4}$

67. $\sin \frac{\pi}{2}$ **68.** $\csc \frac{\pi}{2}$ **69.** $\tan \frac{2\pi}{3}$ **70.** $\cot \frac{2\pi}{3}$ **71.** $\sin \frac{5\pi}{6}$ **72.** $\tan \frac{5\pi}{6}$

73. $\cos 3\pi$ **74.** $\sec \pi$ **75.** $\sin \frac{4\pi}{3}$ **76.** $\cot \frac{4\pi}{3}$ **77.** $\tan \frac{5\pi}{4}$ **78.** $\csc \frac{5\pi}{4}$

79. $\sin 3\pi$ **80.** $\cos 5\pi$ **81.** $\tan \left(-\frac{\pi}{3} \right)$ **82.** $\cot \left(-\frac{2\pi}{3} \right)$ **83.** $\sin \left(-\frac{7\pi}{6} \right)$ **84.** $\cos \left(-\frac{\pi}{6} \right)$

85. The figure shows the same angles measured in both degrees and radians. Complete the missing measures.

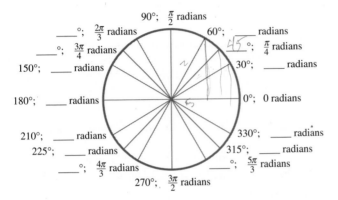

86. A circular pulley is rotating about its center. Through how many radians would it turn in **(a)** 8 rotations, and **(b)** 30 rotations?

87. Through how many radians will the hour hand on a clock rotate in **(a)** 24 hours, and **(b)** 4 hours?

88. A space vehicle is orbiting the earth in a circular orbit. What radian measure corresponds to **(a)** 2.5 orbits, and **(b)** 4/3 of an orbit?

3.2 ——— APPLICATIONS OF RADIAN MEASURE

As mentioned in the previous section, radian measure is used to simplify certain formulas. Two of these formulas are discussed in this section. Both would be more complicated if expressed in degrees.

ARC LENGTH OF A CIRCLE   The first of these formulas is used to find the length of an arc of a circle. The formula comes from the fact (proven in plane geometry) that the length of an arc is proportional to the measure of its central angle.

In Figure 3.4, angle QOP has a measure of 1 radian and intercepts an arc of length r on the circle. Angle ROT has a measure of θ radians and intercepts an arc of length s on the circle. Since the lengths of the arcs are proportional to the measure of their central angles,

$$\frac{s}{r} = \frac{\theta}{1}.$$

Multiplying both sides by r gives the following result.

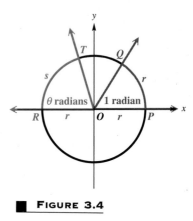

■ **FIGURE 3.4**

LENGTH OF ARC	The length s of the arc intercepted on a circle of radius r by a central angle of measure θ radians is given by the product of the radius and the radian measure of the angle, or $$s = r\theta, \quad \theta \text{ in radians.}$$

This formula is a good example of the usefulness of radian measure. To see why, try to write the equivalent formula for an angle measured in degrees.

CAUTION	When applying the formula $s = r\theta$, the value of θ *must be* expressed in *radians*.

■ *Example 1*
FINDING ARC LENGTH USING $s = r\theta$

A circle has a radius of 18.2 centimeters. Find the length of the arc intercepted by a central angle having each of the following measures.

(a) $\dfrac{3\pi}{8}$ radians

Here $r = 18.2$ cm and $\theta = 3\pi/8$. Since $s = r\theta$,

$$18.2\left(\frac{3\pi}{8}\right) \text{ centimeters}$$

$$s = \frac{54.6\pi}{8} \text{ centimeters} \qquad \text{The exact answer}$$

or $\qquad s \approx 21.4$ centimeters. $\qquad$ Calculator approximation

(b) $144°$

The formula $s = r\theta$ requires that θ be measured in radians. First, convert θ to radians by the methods of the previous section.

$$144° = 144°\left(\frac{\pi}{180°}\right) \text{ radians} \qquad \text{Multiply by } \frac{\pi}{180°}.$$

$$144° = \frac{4\pi}{5} \text{ radians}$$

Now

$$s = 18.2\left(\frac{4\pi}{5}\right) \text{ centimeters} \qquad \text{Use } s = r\theta.$$

$$s = \frac{72.8\pi}{5} \text{ centimeters,}$$

or

$$s \approx 45.7 \text{ centimeters.} \quad \blacksquare$$

■ **Example 2**
**FINDING THE
DISTANCE BETWEEN
TWO CITIES USING
LATITUDES**

Reno, Nevada, is approximately due north of Los Angeles. The latitude of Reno is 40° N, while that of Los Angeles is 34° N. (The N in 34° N means *north* of the equator.) If the radius of the earth is 6400 kilometers, find the north-south distance between the two cities.

Latitude gives the measure of a central angle with vertex at the earth's center whose initial side goes through the earth's equator and whose terminal side goes through the given location. As shown in Figure 3.5, the central angle for Reno and Los Angeles is 6°. The distance between the two cities can thus be found by the formula $s = r\theta$, after 6° is first converted to radians.

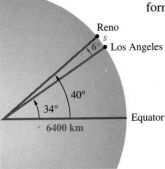

$$6° = 6°\left(\frac{\pi}{180°}\right) = \frac{\pi}{30} \text{ radians}$$

The distance between the two cities is

$$s = r\theta$$

$$s = 6400\left(\frac{\pi}{30}\right) \text{ kilometers} \qquad r = 6400, s = \frac{\pi}{30}$$

$$\approx 670 \text{ kilometers.} \quad \blacksquare$$

■ **FIGURE 3.5**

■ **Example 3**
**FINDING A LENGTH
USING $s = r\theta$**

A rope is being wound around a drum with radius .8725 feet. (See Figure 3.6.) How much rope will be wound around the drum if the drum is rotated through an angle of 39.72°?

The length of rope wound around the drum is just the arc length for a circle of radius .8725 feet and a central angle of 39.72°. Use the formula $s = r\theta$, with the angle converted to radian measure.

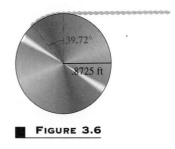

■ **FIGURE 3.6**

$$s = r\theta$$

$$s = (.8725)\left[(39.72°)\left(\frac{\pi}{180°}\right)\right] \quad \text{Multiply by } \frac{\pi}{180°}.$$

$$\approx .6049$$

The length of the rope wound around the drum is approximately .6049 feet. ■

■ *Example 4*

FINDING AN ANGLE MEASURE USING $s = r\theta$

Two gears are adjusted so that the smaller gear drives the larger one as shown in Figure 3.7. If the smaller gear rotates through 225°, through how many degrees will the larger gear rotate?

First find the radian measure of the angle, which will give the arc length on the smaller gear that determines the motion of the larger gear. Since 225° = 5π/4 radians, for the smaller gear,

■ **FIGURE 3.7**

$$s = r\theta = 2.5\left(\frac{5\pi}{4}\right) \approx 9.8 \text{ centimeters.}$$

An arc length of 9.8 centimeters on the larger gear corresponds to an angle measure θ, in radians, of

$$s = r\theta$$

$$9.8 = 4.8\theta$$

$$2.0 \approx \theta.$$

Changing back to degrees shows that the larger gear rotates through

$$2.0\left(\frac{180°}{\pi}\right) \approx 110°,$$

to two significant figures. ■

SECTOR OF A CIRCLE The other useful formula given in this section is used to find the area of a "piece of pie," or sector. A **sector of a circle** is the portion of the interior of a circle intercepted by a central angle. See Figure 3.8.

To find the area of a sector, assume that the radius of the circle is r. A complete circle can be thought of as an angle with a measure of 2π radians. If a central angle for the sector has measure θ radians, then the sector makes up a fraction $\theta/(2\pi)$ of a complete circle. The area of a complete circle is $A = \pi r^2$. Therefore, the area of the sector is given by the

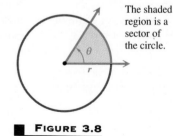

The shaded region is a sector of the circle.

■ **FIGURE 3.8**

product of the fraction $\theta/(2\pi)$ and the total area, πr^2, or

$$\text{area of sector} = \frac{\theta}{2\pi}(\pi r^2) = \frac{1}{2}r^2\theta, \qquad \theta \text{ in radians.}$$

This discussion is summarized as follows.

AREA OF SECTOR The area of a sector of a circle of radius r and central angle θ in radians is given by

$$A = \frac{1}{2}r^2\theta, \quad \theta \text{ in radians.}$$

CAUTION	As in the formula for arc length, the measure of θ must be in radians when using this formula for the area of a sector.

■ **Example 5**
FINDING AREA
USING $A = \frac{1}{2}r^2\theta$

Figure 3.9 shows a field in the shape of a sector of a circle. The central angle is 15° and the radius of the circle is 321 meters. Find the area of the field.

First, convert 15° to radians.

$$15° = 15°\left(\frac{\pi}{180°}\right) = \frac{\pi}{12} \text{ radians}$$

Now use the formula for the area of a sector.

$$A = \frac{1}{2}r^2\theta$$

$$= \frac{1}{2}(321)^2\left(\frac{\pi}{12}\right)$$

$$\approx 13,500 \text{ m}^2. \quad ■$$

■ **FIGURE 3.9**

3.2 EXERCISES ■

Find the length of the arc intercepted by a central angle θ in a circle of radius r in each of the following exercises. Give calculator approximations in your answers. See Example 1. $S = r\theta$

1. $r = 8.00$ in, $\theta = \pi$ radians

2. $r = 72.0$ ft, $\theta = \pi/8$ radians

3. $r = 12.3$ cm, $\theta = 2\pi/3$ radians

4. $r = .892$ cm, $\theta = 11\pi/10$ radians

5. $r = 253$ m, $\theta = 2\pi/5$ radians

6. $r = 120$ mm, $\theta = \pi/9$ radians

7. $r = 4.82$ m, $\theta = 60°$

8. $r = 71.9$ cm, $\theta = 135°$

9. $r = 58.402$ m, $\theta = 52.417°$

10. $r = 39.4$ cm, $\theta = 68.059°$

11. If γ is a central angle whose measure is in degrees, give the formula for the arc length s intercepted in a circle of radius r by the angle.

12. Work Exercise 7 above using the formula found in Exercise 11, replacing θ with γ.

106

Find the distance in kilometers between each of the following pairs of cities whose latitudes are given. Assume that the cities are on a north-south line and that the radius of the earth is 6400 km. See Example 2.

13. Madison, South Dakota, 44° N, and Dallas, Texas, 33° N

14. Charleston, South Carolina, 33° N, and Toronto, Ontario, 43° N

15. Panama City, Panama, 9° N, and Pittsburgh, Pennsylvania, 40° N

16. Farmersville, California, 36° N, and Penticton, British Columbia, 49° N

17. New York City, New York, 41° N, and Lima, Peru, 12° S

18. Halifax, Nova Scotia, 45° N, and Buenos Aires, Argentina, 34° S

Longitude gives the measure of a central angle with vertex at the center of the earth measured east or west of the prime meridian in Greenwich, England. See the figure. Find the distance between the following pairs of cities located along the equator.

19. Nairobi, Kenya, longitude 40° E, and Singapore, Republic of Singapore, 105° E

20. Quito, Ecuador, longitude 80° W, and Libreville, Gabon, 10° E

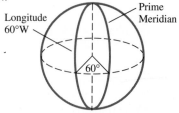

Work the following exercises. See Examples 3 and 4.

21. (a) How many inches will the weight in the figure rise if the pulley is rotated through an angle of 71° 50′?
(b) Through what angle, to the nearest minute, must the pulley be rotated to raise the weight 6 in?

9.27 in

22. Find the radius of the pulley in the figure if a rotation of 51.6° raises the weight 11.4 cm.

r

23. The rotation of the smaller wheel in the figure causes the larger wheel to rotate. Through how many degrees will the larger wheel rotate if the smaller one rotates through 60.0°?

5.23 cm   8.16 cm

24. Find the radius of the larger wheel in the figure if the smaller wheel rotates 80° when the larger wheel rotates 50°.

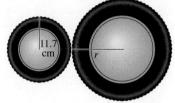

11.7 cm   r

25. The figure shows the chain drive of a bicycle. How far will the bicycle move if the pedals are rotated through 180°? Assume that the radius of the bicycle wheel is 13.6 in.

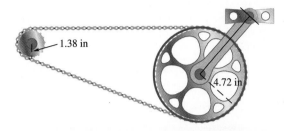

1.38 in

4.72 in

26. The speedometer of a small pickup truck is designed to be accurate with tires of radius 14 in.

(a) Find the number of rotations of a tire in 1 hr if the truck is driven at 55 mph.

(b) Suppose that oversize tires of radius 16 in are placed on the truck. If the truck is now driven for 1 hr with the speedometer reading 55 mph, how far has the truck gone? If the speed limit is 55 mph, does the driver deserve a speeding ticket?

If a central angle is very small, there is little difference in length between an arc and the inscribed chord. See the figure. Approximate each of the following lengths by finding the necessary arc length. (Note: When a central angle intercepts an arc, the arc is said to subtend *the angle.)*

Arc length ≈ length of inscribed chord

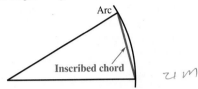

Arc

Inscribed chord

27. A tree 600 m away subtends an angle of 2°. Find the height of the tree.

28. A building 500 m away subtends an angle of 3°. Find the height of the building.

29. A railroad track in the desert is 3.5 km away. A train on the track subtends (horizontally) an angle of 3° 20′. Find the length of the train.

30. An oil tanker 2.3 km at sea subtends a 1° 30′ angle horizontally. Find the length of the ship.

31. The full moon subtends an angle of 1/2°. The moon is 240,000 mi away. Find the diameter of the moon.

32. A building with a height of 58 m subtends an angle of 1° 20′. How far away is the building?

33. The mast of Brent Simon's boat is 32 ft high. If it subtends an angle of 2° 10′, how far away is it?

34. A television tower 530 m high subtends an angle of 2° 40′. How far away is the tower?

Find the area of a sector of a circle having radius r and central angle θ in each of the following. See Example 5.

35. *r* = 29.2 m, θ = 5π/6 radians

36. *r* = 59.8 km, θ = 2π/3 radians

37. *r* = 52 cm, θ = 3π/10 radians

38. *r* = 25 mm, θ = π/15 radians

39. *r* = 12.7 cm, θ = 81°

40. *r* = 18.3 m, θ = 125°

41. *r* = 32.6 m, θ = 38° 40′

42. *r* = 59.8 ft, θ = 74° 30′

43. *r* = 86.243 m, θ = 11.7142°

44. *r* = 111.976 cm, θ = 29.8477°

45. If γ is a central angle whose measure is in degrees, give the formula for the area of a sector with this central angle in a circle with radius *r*.

46. Work Exercise 39 above using the formula found in Exercise 45, replacing θ with γ.

47. The figure shows Medicine Wheel, an Indian structure in northern Wyoming. This circular structure is perhaps 200 years old. There are 32 spokes in the wheel, all equally spaced.

(a) Find the measure of each central angle in degrees and in radians.

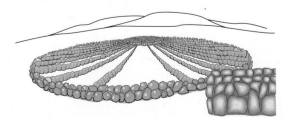

(b) If the radius of the wheel is 76 ft, find the circumference.

(c) Find the length of each arc intercepted by consecutive pairs of spokes.

(d) Find the area of each sector formed by consecutive spokes.

48. The unusual corral in the photograph is separated into 26 areas, many of which approximate sectors of a circle. Assume that the corral has a diameter of 50 m.

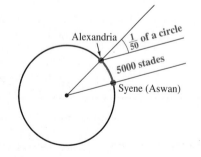

(a) Find the central angle for each region, assuming that the 26 regions are all equal sectors, with the fences meeting at the center.

(b) What is the area of each sector?

49. Eratosthenes (*ca.* 230 B.C.) made a famous measurement of the earth. He observed at Syene (the modern Aswan) at noon and at the summer solstice that a vertical stick had no shadow, while at Alexandria (on the same meridian as Syene) the sun's rays were inclined 1/50 of a complete circle to the vertical. See the figure. He then calculated the circumference of the earth

from the known distance of 5000 stades between Alexandria and Syene. Obtain Eratosthenes' result of 250,000 stades for the circumference of the earth. There is reason to suppose that a stade is about equal to 516.7 ft. Assuming this, use Eratosthenes' result to calculate the polar diameter of the earth in miles. (The actual polar diameter of the earth, to the nearest mile, is 7900 mi.)*

Multiply the area of the base times the height to find the volume of each solid.

50.

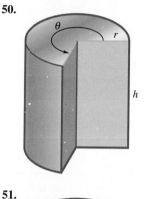

51.

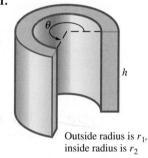

Outside radius is r_1, inside radius is r_2

52. The area of a sector is given approximately by

$$A \approx .0087266463 \; \gamma \, r^2,$$

where r is the radius of the circle and γ is measured in degrees. Show how this formula was obtained.

---

*One mathematical exercise and figure from *A Survey of Geometry*, Vol. 1 by Howard Eves. Reprinted by permission of the author.

So far we have defined the six trigonometric functions for *angles*. The angles can be measured either in degrees or in radians. While the domain of the trigonometric functions is a set of angles, the range is a set of real numbers. In advanced work, such as calculus, it is necessary to modify the trigonometric functions so that the domain contains not angles, but real numbers. Trigonometric functions having a domain of real numbers are called **circular functions.**

To find values of the circular functions for any real number *s*, we use a **unit circle,** as shown in Figure 3.10. This unit circle has its center at the origin and a radius of one unit (hence the name *unit circle*). Recall from algebra that the equation of this circle is

$$x^2 + y^2 = 1.$$

Start at the point (1, 0) and lay off an arc of length *s* along the circle. Go counterclockwise if *s* is positive, and clockwise if *s* is negative. Let the endpoint of the arc be at the point (*x, y*). The six **circular functions** of *s* are defined as follows. (Assume that no denominators are zero.)

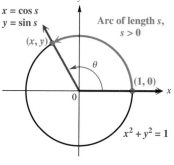

■ **FIGURE 3.10**

CIRCULAR FUNCTIONS			
	$\sin s = y$	$\tan s = \dfrac{y}{x}$	$\sec s = \dfrac{1}{x}$
	$\cos s = x$	$\cot s = \dfrac{x}{y}$	$\csc s = \dfrac{1}{y}$

NOTE   Since $\sin s = y$ and $\cos s = x$, we can replace *x* and *y* in the equation $x^2 + y^2 = 1$ and obtain the familiar Pythagorean identity

$$\cos^2 s + \sin^2 s = 1.$$

Since the ordered pair (*x, y*) represents a point on the unit circle,

$$-1 \le x \le 1 \quad \text{and} \quad -1 \le y \le 1,$$

making

$$-1 \le \cos s \le 1 \quad \text{and} \quad -1 \le \sin s \le 1.$$

For any value of *s*, both sin *s* and cos *s* exist, so the domain of these functions is the set of all real numbers. For tan *s*, defined as *y/x*, *x* ≠ 0. The only way *x* can equal 0 is when the arc length *s* is π/2, 3π/2, −π/2, −3π/2, and so on. To avoid a zero denominator, the domain of tangent must be restricted to those values of *s* satisfying

$$s \ne \frac{\pi}{2} + n\pi, \quad n \text{ any integer.}$$

The definition of secant also has x in the denominator, making the domain of secant the same as the domain of tangent. Both cotangent and cosecant are defined with a denominator of y. To guarantee that $y \neq 0$, the domain of these functions must be the set of all values of s satisfying

$$s \neq 0 + n\pi, \quad n \text{ any integer.}$$

The domains of the circular functions are summarized in the following box.

DOMAINS OF THE CIRCULAR FUNCTIONS

The domains of the circular functions are as follows. Assume that n is any integer.

Sine and Cosine Functions: $(-\infty, \infty)$

Tangent and Secant Functions: $\{s \mid s \neq \frac{\pi}{2} + n\pi\}$

Cotangent and Cosecant Functions: $\{s \mid s \neq n\pi\}$

These circular functions are closely related to the trigonometric functions discussed earlier. To develop this connection, place an angle θ in standard position on a unit circle, as in Figure 3.10. Assume that θ is measured in radians. The arc length intercepted by angle θ is s. From the previous section, the arc length is given by $s = r\theta$. Since $r = 1$, the arc length is $s = 1(\theta) = \theta$, making s the radian measure of angle θ. As shown in the figure, (x, y) is a point on the terminal side of θ, with $r = 1$. Using the definitions of the trigonometric functions,

$$\sin \theta = \frac{y}{r} = \frac{y}{1} = y = \sin s$$

$$\cos \theta = \frac{x}{r} = \frac{x}{1} = x = \cos s.$$

Similar results hold for the other four functions.

As shown above, the trigonometric functions and the circular functions lead to the same function values. Because of this, a value such as $\sin \pi/2$ can be found without worrying about whether $\pi/2$ is a real number or the radian measure of an angle. In either case, $\sin \pi/2 = 1$.

All the formulas developed in this book are valid for either angles or real numbers. For example, $\sin \theta = 1/\csc \theta$ is equally valid for θ as the measure of an angle in degrees or radians or for θ as a real number.

In order to use a calculator to find an approximation of a circular function of a real number, we must first set the calculator to *radian* mode. The next example shows how to find such approximations.

■ *Example 1*

FINDING CIRCULAR FUNCTION VALUES WITH A CALCULATOR

Use a calculator to find an approximation for each of the following circular function values.

(a) cos .5149

First, be sure that the calculator is in radian mode. Enter .5149 and press the cosine key.

Enter: .5149 [cos]   **Display:** .87034197

(b) sin 6.6759

With the calculator in radian mode, enter 6.6759 and press the sine key.

Enter: 6.6759 [sin]   **Display:** .38269786

(c) cot 1.3206

Because scientific calculators do not have keys for cotangent, secant, and cosecant, in order to find these values we must use a combination of the appropriate reciprocal function and the [1/x] key. To find cot 1.3206, first find tan 1.3206 and then find the reciprocal.

Enter: 1.3206 [tan][1/x]   **Display:** .25555106

(d) sec (−2.9234)

To find the secant of a negative number, we must use the change of sign key [+/−], the cosine key, and the reciprocal key as follows.

Enter: 2.9234 [+/−][cos][1/x]   **Display:** −1.0242855   ■

CAUTION One of the most common errors in trigonometry involves using calculators in degree mode when radian mode should be used. Remember that if you are finding a circular function value of a real number, the calculator *must* be in radian mode.

Reference angles were used to find the values of trigonometric functions for angles larger than $\pi/2$ radians or smaller than 0 radians. To find function values of real numbers larger than $\pi/2$ or smaller than 0, use **reference numbers.**

Reference numbers are found in much the same way that reference angles were found. Use the following rules to find reference numbers or angles.

REFERENCE NUMBERS OR ANGLES

θ or s in Quadrant	θ' Is	s' Is
I (0 to $\pi/2$) (0 to 1.5708)	θ	s
II ($\pi/2$ to π) (1.5708 to 3.1416)	$180° - \theta$	$\pi - s$
III (π to $3\pi/2$) (3.1416 to 4.7124)	$\theta - 180°$	$s - \pi$
IV ($3\pi/2$ to 2π) (4.7124 to 6.2832)	$360° - \theta$	$2\pi - s$

112

■ *Example 2*

FINDING
REFERENCE
NUMBERS

Find the reference number for each value of *s*.

(a) $s = \dfrac{2\pi}{3}$

Figure 3.11 shows that $s = 2\pi/3$ is in quadrant II with reference number

$$s' = \pi - \frac{2}{3}\pi = \frac{1}{3}\pi \qquad \text{or} \qquad \frac{\pi}{3}.$$

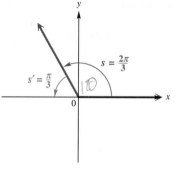

■ FIGURE 3.11

(b) $s = -\dfrac{\pi}{4}$

The angle of $-\pi/4$ radian is coterminal with an angle of $7\pi/4$ radians. Thus the reference number for $-\pi/4$ is the same as that of $7\pi/4$. Since an angle of $7\pi/4$ radians is in quadrant IV, its reference angle is $2\pi - 7\pi/4 = \pi/4$. The reference number for $-\pi/4$ is $\pi/4$. See Figure 3.12.

■ FIGURE 3.12

(c) $s = 1.8638844$

This value of *s* is between $\pi/2$ and π, and thus is a quadrant II number. Its reference number is found by subtracting *s* from π. Using $\pi \approx 3.1415927$, we find the reference number

$$s' = \pi - s$$
$$= 3.1415927 - 1.8638844$$
$$= 1.2777083.$$

See Figure 3.13. ■

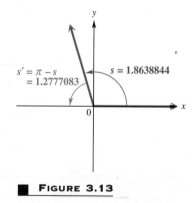

■ FIGURE 3.13

We can use the ideas of the preceding discussion to find exact circular function values of numbers whose reference numbers are $\pi/6$, $\pi/4$, and $\pi/3$. The next example illustrates this.

Find the exact circular fu

(a) $\cos \dfrac{2\pi}{3}$

As seen in Example 2(a), the refe
cosine is negative. Using the methods fi

$$\cos \frac{2\pi}{3} = -\cos \frac{\pi}{3}$$

(b) $\tan\left(-\dfrac{\pi}{4}\right)$

The number $-\pi/4$ is a quadrant IV number so its ta
found in Example 2(b), its reference number is $\pi/4$. Therefore,

$$\tan\left(-\frac{\pi}{4}\right) = -\tan \frac{\pi}{4} = -1.$$

(c) $\csc \dfrac{10\pi}{3}$

Because $10\pi/3$ is not between 0 and 2π, we must first find a number between 0 and 2π with which it is coterminal. To do this, subtract 2π as many times as needed. Here, 2π must be subtracted only once.

$$\frac{10\pi}{3} - 2\pi = \frac{10\pi}{3} - \frac{6\pi}{3} = \frac{4\pi}{3}$$

The number $4\pi/3$ is in quadrant III with reference number $\pi/3$. Because the cosecant is negative in quadrant III, $\csc 10\pi/3 = \csc 4\pi/3$ will be negative. Therefore,

$$\csc \frac{10\pi}{3} = \csc \frac{4\pi}{3} = -\csc \frac{\pi}{3} = \frac{-2\sqrt{3}}{3}. \quad ■$$

Examples 2 and 3 involved finding circular function values of real numbers. The inverse procedure can be performed as well; that is, if we are given a circular function value, we can determine a number in a specified interval having this value.

(a) Find the value of s in the interval $[0, \pi/2]$ that has $\cos s = .96854556$.

The value of s can be found with a calculator set for radian mode.

Enter: .96854556 [INV][cos]   **Display:** [.25147856]

(Recall that other keys such as $\boxed{\cos^{-1}}$ or $\boxed{\text{arccos}}$ may be needed instead of $\boxed{\text{INV}}$ $\boxed{\cos}$.)

n the interval $[3\pi/2, 2\pi]$ for which

reference number for s must be $\pi/4$,
be in the interval $[3\pi/2, 2\pi]$, we must
from 2π. See Figure 3.14. Therefore,

calculator approximations and the [INV]
et the approximation for $-\pi/4$ for s, and
he reason for this is that calculators give
$\pi/2, \pi/2]$. More about this will be said in
ic functions are studied.

aph of the unit circle $x^2 + y^2 = 1$ with a
)egree and radian measures are given for the
l the coordinates of the points on the circle
are also given. Ke... rst coordinate is the cosine of the angle (or
number) and the second coordinate is the sine. This figure should help you in the
rest of your study of trigonometry.

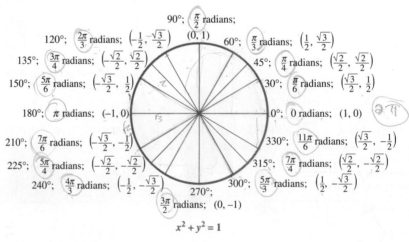

FIGURE 3.15

3.3 EXERCISES ■

Use a calculator to find an approximation for each of the following circular function values. Be sure that your calculator is set in radian mode. See Example 1.

1. tan .4538

2. sin .6109

3. cot 1.0821

4. cos 1.1519

5. sin .8203

6. cot .6632

7. cos .6429

8. tan .9047

9. sin 1.5097

10. cot .0465

11. csc 1.3875

12. tan 1.3032

13. sin 7.5835

14. tan 6.4752

15. cot 7.4526

16. cos 6.6701

17. tan 4.0230

18. cot 3.8426

19. cos 4.2528

20. sin 3.4645

21. sin (−2.2864)

22. cot (−2.4871)

23. cos (−3.0602)

24. tan (−1.7861)

25. cot 6.0301

26. cos 5.2825

27. sin 5.9690

28. tan 5.4513

29. cos 13.8143

30. sin 13.6572

31. cot 12.9795

32. tan 11.0392

33. csc (−9.4946)

34. cos (−13.7881)

35. tan (−23.7744)

36. sin (−17.5784)

Find the reference number for each of the following. In Exercises 51 and 52, use $\pi \approx 3.1415927$. See Example 2.

37. $\dfrac{5\pi}{3}$

38. $\dfrac{7\pi}{6}$

39. $\dfrac{5\pi}{6}$

40. $\dfrac{5\pi}{4}$

41. $\dfrac{13\pi}{3}$

42. $\dfrac{11\pi}{4}$

43. $-\dfrac{3\pi}{4}$

44. $-\dfrac{7\pi}{4}$

45. $-\dfrac{5\pi}{6}$

46. $-\dfrac{5\pi}{3}$

47. $\dfrac{13\pi}{12}$

48. $\dfrac{11\pi}{12}$

49. .28376614

50. .70872930

51. 3.5551212

52. −4.6387146

Find the exact circular function value in each of the following. See Example 3.

53. $\sin \dfrac{7\pi}{6}$

54. $\cos \dfrac{5\pi}{3}$

55. $\tan \dfrac{3\pi}{4}$

56. $\sin \dfrac{5\pi}{3}$

57. $\cos \dfrac{7\pi}{6}$

58. $\tan \dfrac{4\pi}{3}$

59. $\sec \dfrac{2\pi}{3}$

60. $\csc \dfrac{11\pi}{6}$

61. $\cot \dfrac{5\pi}{6}$

62. $\cos \left(-\dfrac{4\pi}{3}\right)$

63. $\sin \left(-\dfrac{5\pi}{6}\right)$

64. $\tan \dfrac{17\pi}{3}$

65. $\sec \dfrac{23\pi}{6}$

66. $\csc \dfrac{13\pi}{3}$

67. $\cos \dfrac{13\pi}{4}$

68. Suppose that a student attempts to work Exercises 53–67 above using a calculator. Can the student expect to get the correct results, according to the directions given for those exercises? Explain.

Find the value of s in the interval [0, π/2] that makes each of the following true. Use a calculator. See Example 4(a).

69. tan s = .21264138

70. cos s = .78269876

71. sin s = .99184065

72. cot s = .29949853

73. cot s = .62084613

74. tan s = 2.6058440

75. cos s = .57834328

76. sin s = .98771924

77. cot s = .09637041

78. csc s = 1.0219553

79. tan s = 1.6213129

80. cos s = .92728460

In Exercises 81–90, find the exact value of s in the given interval that has the given circular function value. Do not use a calculator. See Example 4(b).

81. $\left[0, \dfrac{\pi}{2}\right];$ $\sin s = \dfrac{\sqrt{3}}{2}$

82. $\left[0, \dfrac{\pi}{2}\right];$ $\cos s = \dfrac{\sqrt{2}}{2}$

83. $\left[\dfrac{\pi}{2}, \pi\right];$ $\sin s = \dfrac{1}{2}$

84. $\left[\dfrac{\pi}{2}, \pi\right];$ $\cos s = -\dfrac{1}{2}$

85. $\left[\pi, \dfrac{3\pi}{2}\right]$; $\tan s = \sqrt{3}$

86. $\left[\pi, \dfrac{3\pi}{2}\right]$; $\sin s = -\dfrac{1}{2}$

87. $\left[\dfrac{3\pi}{2}, 2\pi\right]$; $\tan s = -1$

88. $\left[\dfrac{3\pi}{2}, 2\pi\right]$; $\cos s = \dfrac{\sqrt{3}}{2}$

89. $\left[\dfrac{\pi}{2}, \pi\right]$; $\cos s = -\dfrac{\sqrt{2}}{2}$

90. $\left[\dfrac{3\pi}{2}, 2\pi\right]$; $\sin s = -\dfrac{\sqrt{3}}{2}$

Suppose that an arc of length s lies on the unit circle $x^2 + y^2 = 1$, starting at the point $(1, 0)$ and terminating at the point (x, y). (See Figure 3.10.) Use a calculator to find the approximate coordinates for (x, y). (Hint: $x = \cos s$ and $y = \sin s$.)

91. $s = 2.5$ **92.** $s = 3.4$ **93.** $s = -7.4$ **94.** $s = -3.9$

Use a calculator to determine the quadrant in which an arc of length s would terminate on the unit circle $x^2 + y^2 = 1$. (Hint: Find $\cos s$ and $\sin s$, and observe the signs.)

95. $s = 45$ **96.** $s = 23$ **97.** $s = -32.8$ **98.** $s = 15.1$

99. The values of the circular functions repeat every 2π. For this reason, circular functions are used to describe things that repeat periodically. For example, the maximum afternoon temperature in a given city might be approximated by

$$t = 60 - 30 \cos \dfrac{x\pi}{6},$$

where t represents the maximum afternoon temperature in month x, with $x = 0$ representing January, $x = 1$ representing February, and so on. Find the maximum afternoon temperature for each of the following months.

(a) January   (b) April
(c) May   (d) June
(e) August   (f) October

100. The temperature in Fairbanks is approximated by

$$T(x) = 37 \sin \left[\dfrac{2\pi}{365} (x - 101)\right] + 25,$$

where $T(x)$ is the temperature in degrees Fahrenheit on day x, with $x = 1$ corresponding to January 1 and $x = 365$ corresponding to December 31.* Use a calculator to estimate the temperature on the following days.

(a) March 1 (day 60)   (b) April 1 (day 91)
(c) Day 150   (d) June 15
(e) September 1   (f) October 31

---

*Barbara Lando and Clifton Lando, "Is the Graph of Temperature Variation a Sine Curve?" *The Mathematics Teacher,* 70 (September, 1977): 534–37.

3.4 — LINEAR AND ANGULAR VELOCITY

Suppose that point P moves at a constant speed along a circle of radius r and center O. See Figure 3.16. The measure of how fast the position of P is changing is called **linear velocity.** If v represents linear velocity, then

$$v = \dfrac{s}{t},$$

where s is the length of the arc traced by point P at time t. (This formula is just a restatement of the familiar result $d = rt$ with s as distance, v as the rate, and t as time.)

Look at Figure 3.16 again. As point P moves along the circle, ray OP rotates around the origin. Since the ray OP is the

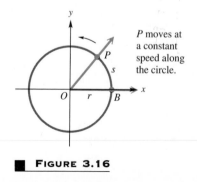

P moves at a constant speed along the circle.

■ **FIGURE 3.16**

terminal side of angle *POB,* the measure of the angle changes as *P* moves along the circle. The measure of how fast angle *POB* is changing is called **angular velocity.** Angular velocity, written ω, can be given as

$$\omega = \frac{\theta}{t}, \quad \theta \text{ in radians,}$$

where θ is the measure of angle *POB* at time *t.* As with the earlier formulas in this chapter, θ must be measured in radians, with ω expressed as radians per unit of time. Angular velocity is used in physics and engineering, among other applications.

In Section 3.2 the length *s* of the arc intercepted on a circle of radius *r* by a central angle of measure θ radians was found to be *s* = *r*θ. Using this formula, the formula for linear velocity, *v* = *s*/*t*, becomes

$$v = \frac{r\theta}{t}$$

or

$$v = r\omega$$

This last formula relates linear and angular velocity.

As mentioned in Section 3.1, a radian is a "pure number," with no units associated with it. This is why the product of the length *r*, measured in units such as centimeters, and ω, measured in units such as radians per second, is velocity, *v*, measured in units such as centimeters per second.

All the formulas given in this section are summarized below.

ANGULAR AND LINEAR VELOCITY	Angular Velocity	Linear Velocity
	$\omega = \dfrac{\theta}{t}$ (ω in radians per unit time, θ in radians)	$v = \dfrac{s}{t}$ $v = \dfrac{r\theta}{t}$ $v = r\omega$

■ *Example 1*
USING THE LINEAR
AND ANGULAR
VELOCITY
FORMULAS

Suppose that point *P* is on a circle with a radius of 10 centimeters, and ray *OP* is rotating with angular velocity of π/18 radians per second.

(a) Find the angle generated by *P* in 6 seconds.

The velocity of ray *OP* is ω = π/18 radians per second. Since ω = θ/*t*, then in 6 seconds

$$\frac{\pi}{18} = \frac{\theta}{6},$$

or θ = 6(π/18) = π/3 radians.

(b) Find the distance traveled by P along the circle in 6 seconds.

In 6 seconds P generates an angle of $\pi/3$ radians. Since $s = r\theta$,

$$s = 10\left(\frac{\pi}{3}\right) = \frac{10\pi}{3} \text{ centimeters.}$$

(c) Find the linear velocity of P.

Since $v = s/t$, in 6 seconds

$$v = \frac{\frac{10\pi}{3}}{6} = \frac{5\pi}{9} \text{ centimeters per second.} \quad ■$$

PROBLEM SOLVING

In practical applications, angular velocity is often given as revolutions per unit of time, which must be converted to radians per unit of time before using the formulas given in this section. ■

■ *Example 2*

USING THE LINEAR
AND ANGULAR
VELOCITY
FORMULAS

A belt runs a pulley of radius 6 centimeters at 80 revolutions per minute.

(a) Find the angular velocity of the pulley in radians per second.

In one minute, the pulley makes 80 revolutions. Each revolution is 2π radians, for a total of

$$80(2\pi) = 160\pi \text{ radians per minute.}$$

Since there are 60 seconds in a minute, ω, the angular velocity in radians per second, is found by dividing 160π by 60.

$$\omega = \frac{160\pi}{60} = \frac{8\pi}{3} \text{ radians per second}$$

(b) Find the linear velocity of the belt in centimeters per second.

The linear velocity of the belt will be the same as that of a point on the circumference of the pulley. Thus,

$$v = r\omega$$

$$v = 6\left(\frac{8\pi}{3}\right)$$

$$v = 16\pi \text{ centimeters per second}$$

$$v \approx 50.3 \text{ centimeters per second.} \quad ■$$

■ *Example 3*

FINDING THE
LINEAR VELOCITY
AND DISTANCE
TRAVELED BY A
SATELLITE

A satellite traveling in a circular orbit 1600 kilometers above earth takes two hours to make an orbit. Assume that the radius of earth is 6400 kilometers.

(a) Find the linear velocity of the satellite.

The distance of the satellite from the center of the earth is

$$r = 1600 + 6400 = 8000 \text{ kilometers.}$$

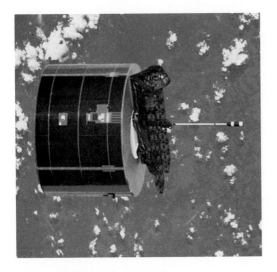

Since it takes 2 hours to complete an orbit, for one orbit $\theta = 2\pi$, and

$$s = r\theta = 8000(2\pi) \text{ km.}$$

Then the linear velocity is

$$v = \frac{s}{t} = \frac{8000(2\pi)}{2}$$

$$= 8000\pi$$

$$\approx 25{,}000 \text{ kilometers per hour.}$$

(b) Find the distance traveled in 4.5 hours.

$$s = vt = (8000\pi)(4.5)$$

$$= 36{,}000\pi$$

$$\approx 110{,}000 \text{ kilometers.} \quad ■$$

3.4 EXERCISES ■

Use the formula $\omega = \theta/t$ to find the value of the missing variable in each of the following. Use a calculator in Exercises 7–8.

1. $\omega = \pi/4$ radians per min, $t = 5$ min

2. $\omega = 2\pi/3$ radians per sec, $t = 3$ sec

3. $\theta = 2\pi/5$ radians, $t = 10$ sec

4. $\theta = 3\pi/4$ radians, $t = 8$ sec

5. $\theta = 3\pi/8$ radians, $\omega = \pi/24$ radians per min

6. $\theta = 2\pi/9$ radians, $\omega = 5\pi/27$ radians per min

7. $\omega = .90674$ radians per min, $t = 11.876$ min

8. $\theta = 3.871142$ radians, $t = 21.4693$ sec

Use the formula $v = r\omega$ to find the value of the missing variable in each of the following. Use a calculator in Exercises 13–14.

9. $r = 8$ cm, $\omega = 9\pi/5$ radians per sec

10. $r = 12$ m, $\omega = 2\pi/3$ radians per sec

11. $v = 18$ ft per sec, $r = 3$ ft

12. $v = 9$ m per sec, $r = 5$ m

13. $r = 24.93215$ cm, $\omega = .372914$ radians per sec

14. $v = 107.692$ m per sec, $r = 58.7413$ m

The formula $\omega = \theta/t$ can be rewritten as $\theta = \omega t$. Using ωt for θ changes $s = r\theta$ to $s = r\omega t$. Use this formula to find the values of the missing variables in each of the following.

15. $r = 6$ cm, $\omega = \pi/3$ radians per sec, $t = 9$ sec

16. $r = 9$ yd, $\omega = 2\pi/5$ radians per sec, $t = 12$ sec

17. $s = 6\pi$ cm, $r = 2$ cm, $\omega = \pi/4$ radians per sec

18. $s = 12\pi/5$ m, $r = 3/2$ m, $\omega = 2\pi/5$ radians per sec

19. $s = 3\pi/4$ km, $r = 2$ km, $t = 4$ sec

20. $s = 8\pi/9$ m, $r = 4/3$ m, $t = 12$ sec

21. Explain the similarities between the familiar $d = rt$ formula and the formula $s = vt$.

22. Suppose that you must convert k radians per second to degrees per minute. Explain how you would do this.

Find ω for each of the following.

23. The hour hand of a clock

24. The minute hand of a clock

25. The second hand of a clock

26. A line from the center to the edge of a phonograph record revolving 33 1/3 times per minute

Find v for each of the following.

27. The tip of the minute hand of a clock, if the hand is 7 cm long

28. The tip of the second hand of a clock, if the hand is 28 mm long

29. A point on the edge of a flywheel of radius 2 m, rotating 42 times per min

30. A point on the tread of a tire of radius 18 cm, rotating 35 times per min

31. The tip of an airplane propeller 3 m long, rotating 500 times per min (*Hint: r* = 1.5 m)

32. A point on the edge of a gyroscope of radius 83 cm, rotating 680 times per minute

Solve the following problems, which review the ideas of this chapter. See Examples 1–3.

33. A railroad track is laid along the arc of a circle of radius 1800 ft. The circular part of the track subtends a central angle of 40°. How long (in seconds) will it take a point on the front of a train traveling 30 mph to go around this portion of the track?

34. Two pulleys of diameter 4 m and 2 m, respectively, are connected by a belt. The larger pulley rotates 80 times per min. Find the speed of the belt in meters per second and the angular velocity of the smaller pulley.

35. The earth revolves on its axis once every 24 hr. Assuming that the earth's radius is 6400 km, find the following.
(a) Angular velocity of the earth in radians per day and radians per hr
(b) Linear velocity at the North Pole or South Pole
(c) Linear velocity at Quito, Equador, a city on the equator
(d) Linear velocity at Salem, Oregon (halfway from the equator to the North Pole)

36. The earth travels about the sun in an orbit that is almost circular. Assume that the orbit is a circle, with a radius of 93,000,000 mi. See the figure.

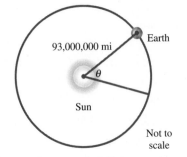

(a) Assume that a year is 365 days, and find θ, the angle formed by the earth's movement in one day.

(b) Give the angular velocity in radians per hour.
(c) Find the linear velocity of the earth in miles per hour.

37. The pulley shown has a radius of 12.96 cm. Suppose that it takes 18 sec for 56 cm of belt to go around the pulley. Find the angular velocity of the pulley in radians per second.

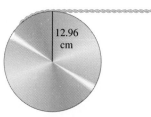

38. The two pulleys in the figure have radii of 15 cm and 8 cm, respectively. The larger pulley rotates 25 times in 36 sec. Find the angular velocity of each pulley in radians per sec.

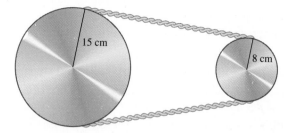

39. A gear is driven by a chain that travels 1.46 m per sec. Find the radius of the gear if it makes 46 revolutions per min.

40. A thread is being pulled off a spool at the rate of 59.4 cm per sec. Find the radius of the spool if it makes 152 revolutions per min.

SECTION	KEY IDEAS
3.1 Radian Measure	**Radian** An angle that has its vertex at the center of a circle and that intercepts an arc on the circle equal in length to the radius of the circle has a measure of **one radian**. **Degree/Radian Relationship** $$180° = \pi \text{ radians}$$ **Converting Between Degrees and Radians** **1.** Proportion: $\dfrac{\text{Radian measure}}{\pi} = \dfrac{\text{Degree measure}}{180}$ **2.** Formulas: <table><tr><th>From</th><th>To</th><th>Multiply by</th></tr><tr><td>Radians</td><td>Degrees</td><td>$\dfrac{180°}{\pi}$</td></tr><tr><td>Degrees</td><td>Radians</td><td>$\dfrac{\pi}{180°}$</td></tr></table> **3.** If a radian measure involves a multiple of π, replace π with 180°, and simplify in order to convert to degrees.
3.2 Applications of Radian Measure	**Length of Arc** The length s of the arc intercepted on a circle of radius r by a central angle of measure θ radians is given by the product of the radius and the radian measure of the angle, or $$s = r\theta, \quad \theta \text{ in radians}.$$ **Area of Sector** The area of a sector of a circle of radius r and central angle θ in radians is given by $$A = \frac{1}{2}r^2\theta, \quad \theta \text{ in radians}.$$

SECTION	KEY IDEAS
3.3 Circular Functions of Real Numbers	**Circular Functions** Start at the point (1, 0) on the unit circle $x^2 + y^2 = 1$ and lay off an arc of length s along the circle, going counterclockwise if s is positive, and clockwise if s is negative. Let the endpoint of the arc be at the point (x, y). The six circular functions of s are defined as follows. (Assume that no denominators are zero.) $\sin s = y \qquad \tan s = \dfrac{y}{x} \qquad \sec s = \dfrac{1}{x}$ $\cos s = x \qquad \cot s = \dfrac{x}{y} \qquad \csc s = \dfrac{1}{y}$ **Domains of the Circular Functions** The domains of the circular functions are as follows. Assume that n is any integer. **Sine and Cosine Functions** $(-\infty, \infty)$ **Tangent and Secant Functions** $\left\{ ss \neq \dfrac{\pi}{2} + n\pi \right\}$ **Cotangent and Cosecant Functions** $\{ s \mid s \neq n\pi \}$ **Reference Numbers or Angles**

θ or s in quadrant	θ' is	s' is
I (0 to $\pi/2$)	θ	s
II ($\pi/2$ to π)	$180° - \theta$	$\pi - s$
III (π to $3\pi/2$)	$\theta - 180°$	$s - \pi$
IV ($3\pi/2$ to 2π)	$360° - \theta$	$2\pi - s$

SECTION	KEY IDEAS
3.4 Linear and Angular Velocity	**Angular and Linear Velocity**

Angular Velocity	Linear Velocity
$\omega = \dfrac{\theta}{t}$	$v = \dfrac{s}{t}$
(ω in radians per unit time, θ in radians)	$v = \dfrac{r\theta}{t}$
	$v = r\omega$

$s = rwt$

Convert each of the following degree measures to radians. Leave answers as multiples of π.

1. 45° **2.** 120° **3.** 80° **4.** 175°

5. 330° **6.** 800° **7.** 1020° **8.** 2000°

Convert each of the following radian measures to degrees.

9. $\dfrac{5\pi}{4}$ **10.** $\dfrac{9\pi}{10}$ **11.** $\dfrac{8\pi}{3}$ **12.** $-\dfrac{6\pi}{5}$

13. $-\dfrac{11\pi}{18}$ **14.** $\dfrac{21\pi}{5}$ **15.** $\dfrac{14\pi}{15}$ **16.** $\dfrac{33\pi}{5}$

Convert each radian measure to degrees in order to find the exact function value. Do not use a calculator.

17. $\tan \dfrac{\pi}{3}$ **18.** $\cos \dfrac{2\pi}{3}$ **19.** $\sin \left(-\dfrac{5\pi}{6}\right)$ **20.** $\cot \dfrac{11\pi}{6}$

21. $\tan \left(-\dfrac{7\pi}{3}\right)$ **22.** $\sec \dfrac{\pi}{3}$ **23.** $\csc \left(-\dfrac{11\pi}{6}\right)$ **24.** $\cot \left(-\dfrac{17\pi}{3}\right)$

Solve each of the following problems. Use a calculator as necessary.

25. The radius of a circle is 15.2 cm. Find the length of an arc of the circle intercepted by a central angle of $3\pi/4$ radians.

26. Find the length of an arc intercepted by a central angle of .769 radians on a circle with a radius of 11.4 cm.

27. A circle has a radius of 8.973 cm. Find the length of an arc on this circle intercepted by a central angle of 49.06°.

28. A central angle of $7\pi/4$ radians forms a sector of a circle. Find the area of the sector if the radius of the circle is 28.69 in.

29. Find the area of a sector of a circle having a central angle of 21° 40′ in a circle of radius 38.0 m.

30. A tree 2000 yd away subtends an angle of 1° 10′. Find the height of the tree to two significant digits.

Assume that the radius of earth is 6400 km in the next two exercises.

31. Find the distance in kilometers between cities on a north-south line that are on latitudes 28° N and 12° S, respectively.

32. Two cities on the equator have longitudes of 72° E and 35° W, respectively. Find the distance between the cities.

Use a calculator to find an approximation for each of the following circular function values. Be sure that your calculator is set in radian mode.

33. $\sin 1.0472$ **34.** $\tan 1.2275$ **35.** $\cos (-.2443)$ **36.** $\cot 3.0543$

37. $\tan 7.3159$ **38.** $\sin 4.8386$ **39.** $\sec .4864$ **40.** $\csc (-.8385)$

Find the reference number for each of the following. In Exercises 45–46, use π ≈ 3.1415927.

41. $\dfrac{7\pi}{6}$ **42.** $\dfrac{5\pi}{3}$ **43.** $-\dfrac{4\pi}{3}$ **44.** $-\dfrac{7\pi}{4}$

45. 1.9169294 **46.** 4.5046264

124 *Find the exact circular function value in each of the following using reference numbers. (Compare your answers to those of Exercises 17–24.)*

47. $\tan \dfrac{\pi}{3}$

48. $\cos \dfrac{2\pi}{3}$

49. $\sin \left(-\dfrac{5\pi}{6}\right)$

50. $\cot \dfrac{11\pi}{6}$

51. $\tan \left(-\dfrac{7\pi}{3}\right)$

52. $\sec \dfrac{\pi}{3}$

53. $\csc \left(-\dfrac{11\pi}{6}\right)$

54. $\cot \left(-\dfrac{17\pi}{3}\right)$

Find the value of s in the interval $[0, \pi/2]$ that makes each of the following true. Use a calculator.

55. $\cos s = .92500448$

56. $\tan s = 4.0112357$

57. $\sin s = .49244294$

58. $\csc s = 1.2361343$

59. $\cot s = .50221761$

60. $\sec s = 4.5600039$

Find the exact value of s in the given interval that has the given circular function value. Do not use a calculator.

61. $\left[0, \dfrac{\pi}{2}\right]$; $\cos s = \dfrac{\sqrt{2}}{2}$ $= \pi/4$ (in book)

62. $\left[\dfrac{\pi}{2}, \pi\right]$; $\tan s = -\sqrt{3}$

63. $\left[\pi, \dfrac{3\pi}{2}\right]$; $\sec s = -\dfrac{2\sqrt{3}}{3}$ $= 7\pi/6$

64. $\left[\dfrac{3\pi}{2}, 2\pi\right]$; $\sin s = -\dfrac{1}{2}$

Solve each of the following problems. Use a calculator in Exercise 69.

65. Find t if $\theta = 5\pi/12$ radians and $\omega = 8\pi/9$ radians per sec.

66. Find θ if $t = 12$ sec and $\omega = 9$ radians per sec.

67. Find ω if $t = 8$ sec and $\theta = 2\pi/5$ radians.

68. Find ω if $s = 12\pi/25$ ft, $r = 3/5$ ft, and $t = 15$ sec.

69. Find s if $r = 11.46$ cm, $\omega = 4.283$ radians per sec, and $t = 5.813$ sec.

70. Find the linear velocity of a point on the edge of a flywheel of radius 7 m if the flywheel is rotating 90 times per sec.

In Section 3.3 we introduced the concept of circular functions, and showed how the unit circle $x^2 + y^2 = 1$ can be used to find the cosine and sine of a real number s, where s is an arc length. We imagined that a point traveled along the circumference of the circle, starting at $(1, 0)$. If it stops after traveling a distance s along the circle, the first coordinate of the terminal point is $\cos s$ and the second coordinate is $\sin s$.

This concept can be illustrated using modern graphing calculators, such as the TI-81 manufactured by Texas Instruments. Begin by using the RANGE function to set the domain and range to be $[-1, 1]$. Then, using the ZOOM function, "square off" the axes. Since the relation $x^2 + y^2 = 1$ does not represent a function, in order to graph the unit circle we must graph the two relations

$$y = \sqrt{1 - x^2}$$

and $$y = -\sqrt{1 - x^2}.$$

Now graph the two relations using the GRAPH function to obtain this unit circle.

The TRACE function of the calculator allows you to find coordinates of points on the graph. Experiment with this tracing feature, and notice that the x and y coordinates are displayed below the graph. The x coordinate represents the cosine of the length of the arc from the point $(1, 0)$ to the point indicated by the cursor. For example, one such point is

$$x = .9 \qquad y = .43588989.$$

While the calculator does not give us the arc length, the length can be found by setting the calculator in radian mode, and finding either $\boxed{\cos^{-1}}$ .9 or $\boxed{\sin^{-1}}$ .43588989. By doing this we find that the arc length is approximately .4510268.

You may want to experiment with your graphing calculator using the exercises below.

---

1. Set both the domain and range to $[0, 1]$ and graph $y = \sqrt{1 - x^2}$ so that only the first quadrant points of the unit circle $x^2 + y^2 = 1$ are displayed. Use the TRACE function to move the cursor along the arc to the point at which x and y are approximately equal. What is the length of the arc from $(1, 0)$ to the cursor? Now compare this to the calculator value of $\pi/4$.

2. Graph the unit circle $x^2 + y^2 = 1$ as described earlier, and move the cursor to any point on the graph. Then, replace the x and y values displayed by the calculator in the equation $x^2 + y^2 = 1$, and verify that $\cos^2 s + \sin^2 s = 1$. (There may be a slight discrepancy due to roundoff by the calculator.)

3. Graph the unit circle $x^2 + y^2 = 1$ as described earlier. Then graph the function $y = x$, and use the ZOOM function to find the point of intersection of the circle and the line in quadrant I. Note that the positive x-axis and the ray of the line in the first quadrant form an angle of $45°$, or $\pi/4$ radians. The decimal approximations for both x and y at the point of intersection should be approximations for what exact number?

4. Graph the unit circle $x^2 + y^2 = 1$ as described earlier. Then graph the function $y = \sqrt{3}x$. What angle does the line make with the positive x-axis? Use the ZOOM function to find the point of intersection of the circle and the line in quadrant I. The decimal approximations for x and y at the point of intersection should be aproximations for what exact numbers?

CHAPTER

GRAPHS OF THE CIRCULAR FUNCTIONS

Many things in daily life repeat with a predictable pattern: the daily newspaper is delivered at the same time each morning, in warm areas electricity use goes up in the summer and down in the winter, the price of fresh fruit goes down in the summer and up in the winter, and attendance at amusement parks increases in the summer and declines in autumn. There are many examples of these *periodic* phenomena (see Figure 4.1). As we shall see in this chapter, the trigonometric functions are periodic and very useful for describing periodic activities.

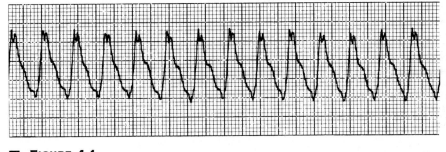

▪ **FIGURE 4.1**

4.1 —————— GRAPHS OF THE SINE AND COSINE I

By the identities for coterminal angles,

$$\sin 0 = \sin 2\pi$$

$$\sin \frac{\pi}{2} = \sin \left(\frac{\pi}{2} + 2\pi \right)$$

and

$$\sin \pi = \sin (\pi + 2\pi),$$

with $\sin x = \sin (x + 2\pi)$ for any real number x. The value of the sine function is the same for x and for $x + 2\pi$, making $y = \sin x$ a *periodic* function with period 2π.

PERIODIC FUNCTION

A **periodic function** is a function f such that

$$f(x) = f(x + p),$$

for every real number x in the domain of f and for some positive real number p. The smallest possible positive value of p is the **period** of the function.

While it is true that $\sin x = \sin (x + 4\pi)$ and $\sin x = \sin (x + 6\pi)$, the *smallest* positive value of p making $\sin x = \sin (x + p)$ is $p = 2\pi$, so 2π is the period for sine.

In Section 3.3 we saw that if an arc of length s is traced along the unit circle $X^2 + Y^2 = 1$, starting at the point $(1, 0)$, the terminal point of the arc has coordinates $(\cos s, \sin s)$. Let us examine the function $y = \sin x$. (Here, we will use the letter x rather than s so that we may draw its graph on the familiar XY-coordinate system.) We saw that this function has domain $(-\infty, \infty)$ and range $[-1, 1]$. Since the sine function has period 2π, one period of $y = \sin x$ can be graphed using values of x from 0 to 2π. To graph this period, look at Figure 4.2, which shows a unit circle with a point (p, q) marked on it. Based on the definitions for circular functions given in Section 3.3, for any angle x, $p = \cos x$ and $q = \sin x$. As x increases from 0 to $\pi/2$ (or 90°), q (or $\sin x$) increases from 0 to 1, while p (or $\cos x$) decreases from 1 to 0.

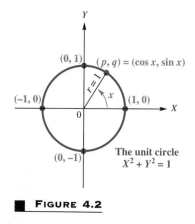

■ **FIGURE 4.2**

As x increases from $\pi/2$ to π (or 180°), q decreases from 1 to 0, while p decreases from 0 to -1. Similar results can be found for the other quadrants, as shown in the following table.

As x Increases From	$\sin x$	$\cos x$
0 to $\pi/2$	Increases from 0 to 1	Decreases from 1 to 0
$\pi/2$ to π	Decreases from 1 to 0	Decreases from 0 to -1
π to $3\pi/2$	Decreases from 0 to -1	Increases from -1 to 0
$3\pi/2$ to 2π	Increases from -1 to 0	Increases from 0 to 1

Selecting key values of x and finding the corresponding values of $\sin x$ give the following results. (Decimals are rounded to the nearest tenth.)

x	0	$\pi/4$	$\pi/2$	$3\pi/4$	π	$5\pi/4$	$3\pi/2$	$7\pi/4$	2π
$\sin x$	0	.7	1	.7	0	$-.7$	-1	$-.7$	0

Plotting the points from the table of values and connecting them with a smooth curve gives the solid portion of the graph in Figure 4.3. Since $y = \sin x$ is

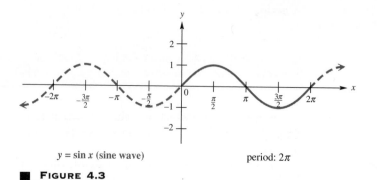

$y = \sin x$ (sine wave)　　　　　period: 2π

■ **FIGURE 4.3**

periodic and has $(-\infty, \infty)$ as its domain, the graph continues in both directions indefinitely, as indicated by the arrows. This graph is sometimes called a **sine wave** or **sinusoid.** You should learn the shape of this graph and be able to sketch it quickly. The key points of the graph are $(0, 0)$, $(\pi/2, 1)$, $(\pi, 0)$, $(3\pi/2, -1)$, and $(2\pi, 0)$. By plotting these five points and connecting them with the characteristic sine wave, you can quickly sketch the graph.

The same scales are used on both the x and y axes of Figure 4.3 so as not to distort the graph. Since the period of $y = \sin x$ is 2π, it is convenient to use subdivisions of 2π on the x-axis. The more familiar x-values, 1, 2, 3, 4, and so on, are still present, but are usually not shown to avoid cluttering the graph. These values are shown in Figure 4.4.

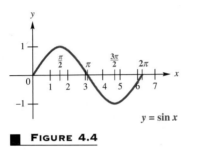

$y = \sin x$

■ **FIGURE 4.4**

Sine graphs occur in many different practical applications. For one application, look back at Figure 4.2 and assume that the line from the origin to the point (p, q) is part of the pedal of a bicycle wheel, with a foot placed at (p, q). As mentioned earlier, q is equal to sin x, showing that the height of the pedal from the horizontal axis in Figure 4.2 is given by sin x. By choosing various angles for the pedal and calculating q for each angle, the height of the pedal leads to the sine curve shown in Figure 4.5.

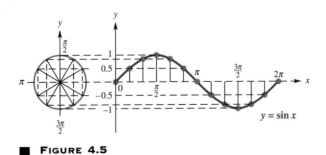

$y = \sin x$

■ **FIGURE 4.5**

The graph of $y = \cos x$ can be found in much the same way as the graph of $y = \sin x$ was found. The domain of cosine is $(-\infty, \infty)$, and the range of $y = \cos x$ is $[-1, 1]$. A table of values is shown below for $y = \cos x$.

x	0	$\pi/4$	$\pi/2$	$3\pi/4$	π	$5\pi/4$	$3\pi/2$	$7\pi/4$	2π
$\cos x$	1	.7	0	−.7	−1	−.7	0	.7	1

Here the key points are $(0, 1)$, $(\pi/2, 0)$, $(\pi, -1)$, $(3\pi/2, 0)$, and $(2\pi, 1)$.

The graph of $y = \cos x$, in Figure 4.6, has the same shape as the graph of $y = \sin x$. In fact, it is the graph of the sine function, shifted $\pi/2$ units to the left.

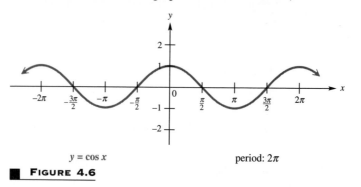

$y = \cos x$ period: 2π

■ **FIGURE 4.6**

The examples in the rest of this section show graphs that are "stretched" either vertically, horizontally, or both when compared with the graphs of $y = \sin x$ or $y = \cos x$.

■ *Example 1*

GRAPHING
$y = a \sin x$

Graph $y = 2 \sin x$.

For a given value of x, the value of y is twice as large as it would be for $y = \sin x$, as shown in the table of values. The only change in the graph is the range, which becomes $[-2, 2]$. See Figure 4.7, which also shows a graph of $y = \sin x$ for comparison.

x	0	$\pi/2$	π	$3\pi/2$	2π
$\sin x$	0	1	0	−1	0
$2 \sin x$	0	2	0	−2	0

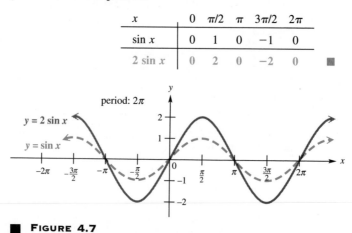

$y = 2 \sin x$
$y = \sin x$
period: 2π

■ **FIGURE 4.7**

| NOTE When graphing periodic functions, it is customary to first graph over one pe-
riod. Then more of the graph can be sketched using the fact that the graph repeats the cycle over and over.

Generalizing from Example 1 and assuming $a \neq 0$ gives the following.

AMPLITUDE OF	The graph of $y = a \sin x$ or $y = a \cos x$ will have the same shape as the graph of
SINE AND COSINE	$y = \sin x$ or $y = \cos x$, respectively, except with range $[-|a|, |a|]$. The number
	$|a|$ is called the **amplitude.**

No matter what the value of the amplitude, the period of $y = a \sin x$ and
$y = a \cos x$ is still 2π. However, the graph of a function of the form $y = \sin bx$ or
$\cos bx$, for $b > 0$, $b \neq 1$, will have a period different from 2π. To see why this is
so, remember that the values of $\sin bx$ or $\cos bx$ will take on all possible values as
bx ranges from 0 to 2π. Therefore, to see what the period of either of these will
be, we must solve the compound inequality

$$0 \leq bx \leq 2\pi.$$

Dividing by the positive number b gives

$$0 \leq x \leq \frac{2\pi}{b}.$$

Therefore, the period is $2\pi/b$. By dividing the interval $[0, 2\pi/b]$ into four
equal parts, we obtain the values for which $\sin bx$ or $\cos bx$ is -1, 0, or 1. These
will give minimum points, x-intercepts, and maximum points on the graph. Once
these points are determined, the graph can be completed by joining the points with
a smooth sinusoidal curve.

| NOTE To divide an interval into four equal parts, find its midpoint by adding the
x-values of the endpoints and dividing by 2. Then use the left endpoint and the
midpoint, repeat this process, and find the first quarter point. Finally, find the
third quarter point by repeating the process with the midpoint and the right endpoint.

▪ *Example 2*

| GRAPHING $y = \sin bx$

Graph $y = \sin 2x$.

For this function, $b = 2$, so the period is $2\pi/2 = \pi$. Therefore, the graph will
complete one cycle over the interval $[0, \pi]$. Dividing this interval into four equal
parts gives the following x-values:

Now make a table of values.

x	0	$\pi/4$	$\pi/2$	$3\pi/4$	π
$2x$	0	$\pi/2$	π	$3\pi/2$	2π
$\sin 2x$	0	1	0	-1	0

These points are joined with a smooth sinusoidal curve. More of the graph can be sketched by repeating this cycle over and over, as shown in Figure 4.8. Notice that the amplitude is not changed. The graph of $y = \sin x$ is included for comparison. ■

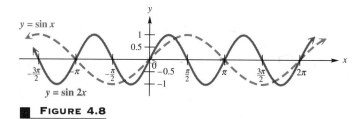

$y = \sin x$

$y = \sin 2x$

■ **FIGURE 4.8**

Generalizing from Example 2 leads to the following result.

PERIOD OF SINE AND COSINE FUNCTIONS	For $b > 0$, the graph of $y = \sin bx$ will look like that of $y = \sin x$, but with a period of $2\pi/b$. Also, the graph of $y = \cos bx$ will look like that of $y = \cos x$, but with a period of $2\pi/b$.

■ *Example 3*
| GRAPHING $y = \cos bx$

Graph $y = \cos \dfrac{2}{3}x$.

For this function the period is

$$\frac{2\pi}{2/3} = 3\pi,$$

and the amplitude is 1. Divide the interval $[0, 3\pi]$ into four equal parts to get the x-values that will yield minimum points, maximum points, and x-intercepts. These x-values are as follows.

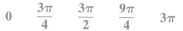

$$0 \qquad \frac{3\pi}{4} \qquad \frac{3\pi}{2} \qquad \frac{9\pi}{4} \qquad 3\pi$$

These values are used to get a table of key points.

x	0	$3\pi/4$	$3\pi/2$	$9\pi/4$	3π
$(2/3)x$	0	$\pi/2$	π	$3\pi/2$	2π
$\cos (2/3)x$	1	0	-1	0	1

$y = \cos \frac{2}{3}x$

■ **FIGURE 4.9**

Now plot these points and join them with a smooth curve. The graph is shown in Figure 4.9. ■

NOTE

Look at the middle row in each of the tables in Examples 2 and 3. The method of dividing the interval $[0, 2\pi/b]$ into four equal parts will always give the values 0, $\pi/2$, π, $3\pi/2$, and 2π for this row, resulting in values of -1, 0, or 1 for the circular function. These lead to key points on the graph, which can then be easily sketched.

Throughout this chapter we assume $b > 0$. If a function has $b < 0$, then the identities given in the next chapter can be used to change the function to one in which $b > 0$. The steps used to graph $y = a \sin bx$ or $y = a \cos bx$, where $b > 0$, are given below.

GRAPHING THE SINE AND COSINE FUNCTIONS

To graph $y = a \sin bx$ or $y = a \cos bx$, with $b > 0$:

1. Find the period, $2\pi/b$. Start at 0 on the x-axis and lay off a distance of $2\pi/b$.
2. Divide the interval into four equal parts.
3. Evaluate the function for each of the five x-values resulting from Step 2. The points will be maximum points, minimum points, and x-intercepts.
4. Plot the points found in Step 3, and join them with a sinusoidal curve.
5. Draw additional cycles of the graph, to the right and to the left, as needed.

The amplitude is $|a|$.

The functions in Examples 4 and 5 have both amplitude and period affected by constants.

■ **Example 4**

GRAPHING
$y = a \sin bx$

Graph $y = -2 \sin 3x$.

Step 1 For this function, $b = 3$, so the period is $2\pi/3$. We will first graph the function over the interval $[0, 2\pi/3]$.

Step 2 Dividing the interval $[0, 2\pi/3]$ into four equal parts gives the x-values 0, $\pi/6$, $\pi/3$, $\pi/2$, and $2\pi/3$.

Step 3 Make a table of points determined by the x-values resulting from Step 2.

x	0	$\pi/6$	$\pi/3$	$\pi/2$	$2\pi/3$
$3x$	0	$\pi/2$	π	$3\pi/2$	2π
$\sin 3x$	0	1	0	-1	0
$-2 \sin 3x$	0	-2	0	2	0

Step 4 Plot the points $(0, 0)$, $(\pi/6, -2)$, $(\pi/3, 0)$, $(\pi/2, 2)$, and $(2\pi/3, 0)$, and join them with a sinusoidal curve. See Figure 4.10.

Step 5 If necessary, the graph in Figure 4.10 can be extended by repeating the cycle over and over.

The amplitude is $|-2| = 2$. ■

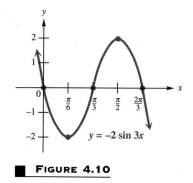

■ **FIGURE 4.10**

■ *Example 5*

GRAPHING
$y = a \cos bx$

Graph $y = 3 \cos \dfrac{1}{2} x$.

The period is $2\pi/(1/2) = 4\pi$. The key points have x-values of

$$0, \qquad \frac{1}{4}(4\pi) = \pi, \qquad \frac{1}{2}(4\pi) = 2\pi,$$

$$\frac{3}{4}(4\pi) = 3\pi, \qquad \text{and} \qquad 4\pi.$$

Evaluating the function for these x-values gives the following points.

$$(0, 3) \qquad (\pi, 0) \qquad (2\pi, -3) \qquad (3\pi, 0) \qquad (4\pi, 3)$$

Figure 4.11 shows these points joined with a smooth curve. The amplitude is 3. ■

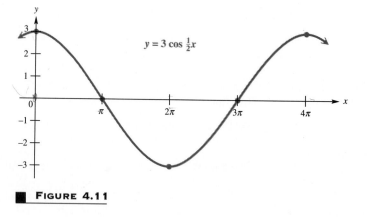

■ **FIGURE 4.11**

4.1 EXERCISES ■

Graph the following functions over the interval $[-2\pi, 2\pi]$. *Give the amplitude. See Example 1.*

1. $y = 2 \cos x$ **2.** $y = 3 \sin x$ **3.** $y = \dfrac{2}{3} \sin x$ **4.** $y = \dfrac{3}{4} \cos x$

5. $y = -\cos x$ **6.** $y = -\sin x$ **7.** $y = -2 \sin x$ **8.** $y = -3 \cos x$

Graph each of the following functions over a two-period interval. Give the period and the amplitude. See Examples 2–5.

9. $y = \sin \dfrac{1}{2} x$ **10.** $y = \sin \dfrac{2}{3} x$ **11.** $y = \cos \dfrac{1}{3} x$ **12.** $y = \cos \dfrac{3}{4} x$

13. $y = \sin 3x$ **14.** $y = \sin 2x$ **15.** $y = \cos 2x$ **16.** $y = \cos 3x$

17. $y = -\sin 4x$ **18.** $y = -\cos 6x$ **19.** $y = 2 \sin \dfrac{1}{4} x$ **20.** $y = 3 \sin 2x$

21. $y = -2 \cos 3x$ **22.** $y = -5 \cos 2x$ **23.** $y = \cos \pi x$ **24.** $y = -\sin \pi x$

Match each function with its graph in Exercises 25–32.

25. $y = \sin x$

26. $y = \cos x$

27. $y = -\sin x$

28. $y = -\cos x$

29. $y = \sin 2x$

30. $y = \cos 2x$

31. $y = 2 \sin x$

32. $y = 2 \cos x$

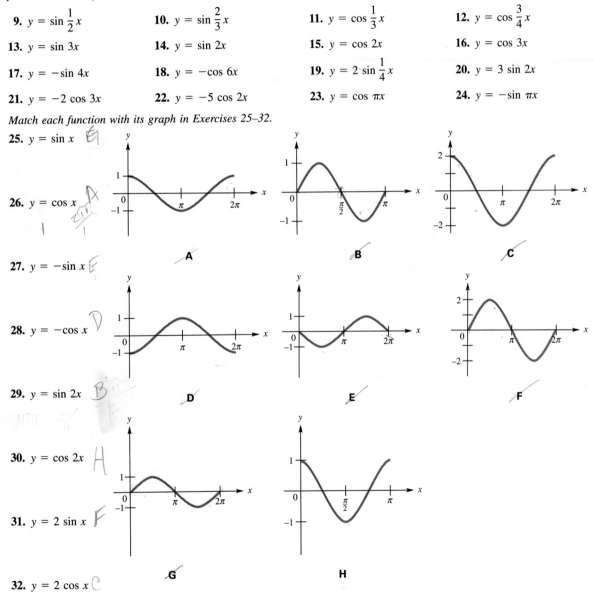

33. Compare the graphs of $y = \sin 2x$ and $y = 2 \sin x$ over the interval $[0, 2\pi]$. Can we say that, in general, $\sin bx = b \sin x$? Explain.

34. Compare the graphs of $y = \cos 3x$ and $y = 3 \cos x$ over the interval $[0, 2\pi]$. Can we say that, in general, $\cos bx = b \cos x$? Explain.

Graph each of the following functions over two periods.

35. $y = (\sin x)^2$ [*Hint:* $(\sin x)^2 = \sin x \cdot \sin x$]

36. $y = (\cos x)^2$

37. $y = (\sin 2x)^2$

38. $y = (\cos 2x)^2$

39. The graph shown gives the variation in blood pressure for a typical person. Systolic and diastolic pressures are the upper and lower limits of the periodic changes in pressure that produce the pulse. The length of time between peaks is called the period of the pulse.

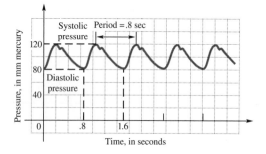

Time, in seconds

(a) Find the amplitude of the graph.
(b) Find the pulse rate (the number of pulse beats in one minute) for this person.

40. Scientists believe that the average annual temperature in a given location is periodic. The overall temperature at a given place during a given season fluctuates as time goes on, from colder to warmer, and back to colder. The graph shows an idealized description of

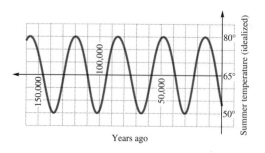

Years ago

the temperature for the last few thousand years of a location at the same latitude as Anchorage.
(a) Find the highest and lowest temperatures recorded.
(b) Use these two numbers to find the amplitude. (*Hint:* An alternative definition of the amplitude is half the difference between the y-values of the highest and lowest points on the graph.)
(c) Find the period of the graph.
(d) What is the trend of the temperature now?

41. Many of the activities of living organisms are periodic. For example, the graph below shows the time that a certain nocturnal animal begins its evening activity.
(a) Find the amplitude of this graph.
(b) Find the period.

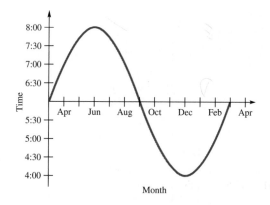

Month

42. The figure shows schematic diagrams of a rhythmically moving arm. The upper arm RO rotates back and forth about the point R; the position of the arm is measured by the angle y between the actual position and the downward vertical position.*

---

*From *Calculus for the Life Sciences* by Rodolfo De Sapio. Copyright © 1978 by W. H. Freeman and Company. Reprinted by permission.

(a) Find an equation of the form $y = a \sin kt$ for the graph shown.
(b) How long does it take for a complete movement of the arm?

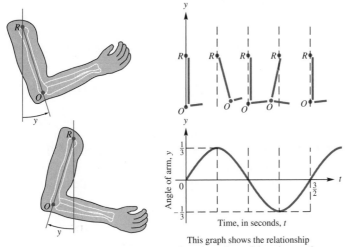

This graph shows the relationship
between angle y and time t in seconds.

*Pure sounds produce single sine waves on an oscilloscope. Find the amplitude and period of
each sine wave in the following photographs. On the vertical scale, each square represents .5,
and on the horizontal scale each square represents 30° or π/6.*

43.

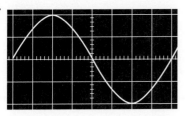

44.

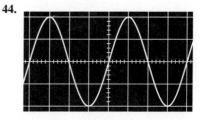

45. The voltage E in an electrical circuit is given by

$$E = 5 \cos 120\pi t,$$

where t is time measured in seconds.
(a) Find the amplitude and the period.
(b) How many cycles are completed in one second?
(The number of cycles (periods) completed in one second is the **frequency** of the function.)
(c) Find E when $t = 0, .03, .06, .09, .12$.
(d) Graph E, for $0 \le t \le 1/30$.

46. For another electrical circuit, the voltage E is given by

$$E = 3.8 \cos 40\pi t,$$

where t is time measured in seconds.
(a) Find the amplitude and the period.
(b) Find the frequency. See Exercise 45(b).
(c) Find E when $t = .02, .04, .08, .12, .14$.
(d) Graph one period of E.

4.2 ─────── GRAPHS OF THE SINE AND COSINE II

In Section 4.1 we studied the basic graphs of $y = \sin x$ and $y = \cos x$, and observed that the constants a and b affect the graphs of

$$y = a \sin bx \quad \text{and} \quad y = a \cos bx$$

by changing the amplitude and period, respectively. In this section we will see how the constants c and d affect the graphs of

$$y = c + a \sin b(x - d) \quad \text{and} \quad y = c + a \cos b(x - d).$$

HORIZONTAL TRANSLATIONS In general, the graph of the function $y = f(x - d)$ is translated *horizontally* when compared to the graph of $y = f(x)$. The translation is d units to the right if $d > 0$ and $|d|$ units to the left if $d < 0$. See Figure 4.12. With circular functions, a horizontal translation is called a **phase shift.** In the function $y = f(x - d)$, the expression $x - d$ is called the **argument.**

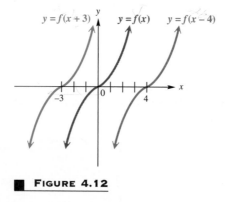

$y = f(x + 3)$ $y = f(x)$ $y = f(x - 4)$

■ **FIGURE 4.12**

In the first example, we show two methods that can be used to graph a circular function involving a phase shift.

■ *Example 1*

GRAPHING
$y = \sin (x - d)$

Graph $y = \sin\left(x - \dfrac{\pi}{3}\right)$.

Method 1   The argument $x - \pi/3$ indicates that the graph will be translated $\pi/3$ units to the *right* (the phase shift) as compared to the graph of $y = \sin x$. Notice that in Figure 4.13 the graph of $y = \sin x$ is shown as a dashed curve, and the graph of $y = \sin (x - \pi/3)$ is shown as a solid curve. Therefore, to graph a function using this method, first graph the basic circular function, and then graph the desired function by using the appropriate translation.

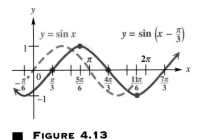

■ FIGURE 4.13

Method 2   For the argument $x - \pi/3$ to result in all possible values throughout one period, it must take on all values between 0 and 2π, inclusive. Therefore, to find an interval of one period, we solve the compound inequality

$$0 \le x - \frac{\pi}{3} \le 2\pi.$$

Add $\pi/3$ to each expression to find the interval

$$\frac{\pi}{3} \le x \le \frac{7\pi}{3} \qquad \text{or} \qquad \left[\frac{\pi}{3}, \frac{7\pi}{3}\right].$$

As first shown in Section 4.1, divide this interval into four equal parts, getting the following values.

$$\frac{\pi}{3} \qquad \frac{5\pi}{6} \qquad \frac{4\pi}{3} \qquad \frac{11\pi}{6} \qquad \frac{7\pi}{3}$$

Make a table of points using the x-values above.

x	$\pi/3$	$5\pi/6$	$4\pi/3$	$11\pi/6$	$7\pi/3$
$x - \pi/3$	0	$\pi/2$	π	$3\pi/2$	2π
$\sin(x - \pi/3)$	0	1	0	-1	0

Join these points to get the graph shown in Figure 4.13. The period is 2π and the amplitude is 1.   ■

■ *Example 2*

GRAPHING
$y = a \cos (x - d)$

Graph $y = 3 \cos \left(x + \dfrac{\pi}{4}\right)$.

Start by writing $3 \cos (x + \pi/4)$ in the form $a \cos (x - d)$.

$$3 \cos \left(x + \frac{\pi}{4}\right) = 3 \cos \left[x - \left(-\frac{\pi}{4}\right)\right]$$

This result shows that $d = -\pi/4$. Since $-\pi/4$ is negative, the phase shift is $|-\pi/4| = \pi/4$ to the left. The period is 2π and the amplitude is 3. The graph is the same as that of $y = 3 \cos x$, except that it is shifted $\pi/4$ units to the left. See Figure 4.14.

Alternatively, the graph can be sketched by first solving the inequality

$$0 \le x + \frac{\pi}{4} \le 2\pi.$$

Adding $-\pi/4$ to each expression gives

$$-\frac{\pi}{4} \le x \le \frac{7\pi}{4},$$

an interval over which one period of the function can be graphed. Dividing this interval into four equal parts gives x-values of $-\pi/4$, $\pi/4$, $3\pi/4$, $5\pi/4$, and $7\pi/4$. A table of points for these x-values once again leads to maximum points, minimum points, and x-intercepts.

x	$-\pi/4$	$\pi/4$	$3\pi/4$	$5\pi/4$	$7\pi/4$
$x + \pi/4$	0	$\pi/2$	π	$3\pi/2$	2π
$3 \cos (x + \pi/4)$	3	0	-3	0	3

This alternative method produces the same graph as the one shown in Figure 4.14. ■

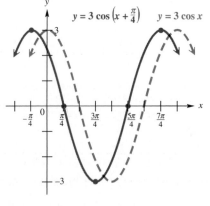

■ **FIGURE 4.14**

The next example shows a function of the form $y = a \cos b(x - d)$. Such functions have both a phase shift (if $d \ne 0$) and a period different from 2π (if $b \ne 1$).

■ *Example 3*

GRAPHING

$y = a \cos b(x - d)$

Graph $y = -2 \cos (3x + \pi)$.

First write the expression in the form $a \cos b(x - d)$ by factoring 3 out of the argument as follows.

$$y = -2 \cos (3x + \pi) = -2 \cos 3\left(x + \frac{\pi}{3}\right)$$

Then $a = -2$, $b = 3$, and $d = -\pi/3$. The amplitude is $|-2| = 2$, and the period is $2\pi/3$ (since the value of b is 3). The phase shift is $|-\pi/3| = \pi/3$ to the left as compared to the graph of $y = -2 \cos 3x$.

The function can be sketched over one period by solving the compound inequality

$$0 \le 3\left(x + \frac{\pi}{3}\right) \le 2\pi$$

to get the interval $[-\pi/3, \pi/3]$. Divide this interval into four equal parts to get the following points.

$$\left(-\frac{\pi}{3}, -2\right) \qquad \left(-\frac{\pi}{6}, 0\right) \qquad (0, 2) \qquad \left(\frac{\pi}{6}, 0\right) \qquad \left(\frac{\pi}{3}, -2\right)$$

Plot these points and then join them with a smooth curve. By graphing an additional half period to the left and to the right, we obtain the sketch shown in Figure 4.15. ∎

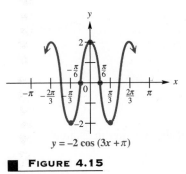

$y = -2 \cos (3x + \pi)$

∎ **FIGURE 4.15**

VERTICAL TRANSLATIONS   The graph of a function of the form $y = c + f(x)$ is shifted *vertically* as compared with the graph of $y = f(x)$. See Figure 4.16. The function $y = c + f(x)$ is called a **vertical translation** of $y = f(x)$. The next example illustrates a vertical translation of a circular function.

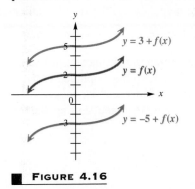

∎ **FIGURE 4.16**

■ *Example 4*

GRAPHING
$y = c + a \cos bx$

Graph $y = 3 - 2 \cos 3x$.

The values of y will be 3 greater than the corresponding values of y in $y = -2 \cos 3x$. This means that the graph of $y = 3 - 2 \cos 3x$ is the same as the graph of $y = -2 \cos 3x$, except with a vertical translation of 3 units upward. Since the period of $y = -2 \cos 3x$ is $2\pi/3$, the key points have the following x-values.

$$0 \qquad \frac{\pi}{6} \qquad \frac{\pi}{3} \qquad \frac{\pi}{2} \qquad \frac{2\pi}{3}$$

The key points are shown on the graph in Figure 4.17, along with more of the graph, which can be sketched by keeping in mind the fact that it is a periodic function. ■

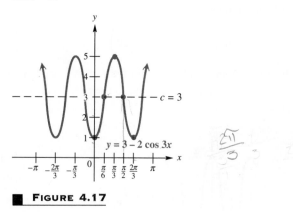

■ **FIGURE 4.17**

The amplitude of the function in Example 4 is 2. A formal definition of amplitude follows.

AMPLITUDE OF A PERIODIC FUNCTION	The amplitude of a periodic function is defined as $$\frac{M - m}{2},$$ where M is the maximum value of the function and m is the minimum value.

Applying this definition to the function in Example 4 ($y = 3 - 2 \cos 3x$), we see that $M = 5$ and $m = 1$, so the amplitude is $(5 - 1)/2 = 2$, as stated earlier.

In the final example of this section, we graph a function that involves all the types of stretching, compressing, and shifting studied in the previous section and this one.

■ *Example 5*

GRAPHING
$y = c + a \sin b(x - d)$

Graph $y = -1 + 2 \sin 4\left(x + \dfrac{\pi}{4}\right)$.

Start by finding an interval over which the graph will complete one cycle. To do this, use the argument $4(x + \pi/4)$ in a compound inequality, with 0 as one endpoint and 2π as the other. Then solve the inequality for x.

$$0 \le 4\left(x + \frac{\pi}{4}\right) \le 2\pi \qquad 0 \le 4 + \frac{4\pi}{6} \le 2\pi$$

$$0 \le x + \frac{\pi}{4} < \frac{\pi}{2} \qquad \text{Divide by 4.}$$

$$-\frac{\pi}{4} \le x \le \frac{\pi}{4} \qquad \text{Subtract } \frac{\pi}{4}.$$

Divide the interval $[-\pi/4, \pi/4]$ into four equal parts to find the key points on the graph, as shown in the following table.

x	$-\pi/4$	$-\pi/8$	0	$\pi/8$	$\pi/4$
$x + \pi/4$	0	$\pi/8$	$\pi/4$	$3\pi/8$	$\pi/2$
$4(x + \pi/4)$	0	$\pi/2$	π	$3\pi/2$	2π
$-1 + 2 \sin 4(x + \pi/4)$	-1	1	-1	-3	-1

Join the key points with a smooth curve as shown in Figure 4.18. This function has period $\pi/2$ and amplitude $[1 - (-3)]/2 = 2$. The phase shift is $\pi/4$ units to the left, and there is a vertical translation of 1 unit down. ■

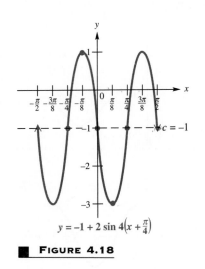

$$y = -1 + 2 \sin 4\left(x + \frac{\pi}{4}\right)$$

■ FIGURE 4.18

> **GRAPHING GENERAL SINE AND COSINE FUNCTIONS**
>
> To graph the general function $y = c + a \sin b(x - d)$ or $y = c + a \cos b(x - d)$, where $b > 0$, follow these steps.
>
> 1. Find an interval whose length is one period $(2\pi/b)$ by solving the compound inequality
> $$0 \le b(x - d) \le 2\pi.$$
>
> 2. Divide the interval into four equal parts.
> 3. Evaluate the function for each of the five x-values resulting from Step 2. The points will be maximum points, minimum points, and points that intersect the line $y = c$ ("middle" points of the wave).
> 4. Plot the points found in Step 3, and join them with a sinusoidal curve.
> 5. Draw additional cycles of the graph, to the right and to the left, as needed.
>
> The amplitude of the function is $|a|$. The vertical translation is c units up if $c > 0$, $|c|$ units down if $c < 0$. The horizontal translation (phase shift) is d units to the right if $d > 0$, and $|d|$ units to the left if $d < 0$.

4.2 EXERCISES ■

Match each function with its graph in Exercises 1–8.

1. $y = \sin\left(x - \dfrac{\pi}{4}\right)$ D

2. $y = \sin\left(x + \dfrac{\pi}{4}\right)$ G

3. $y = \cos\left(x - \dfrac{\pi}{4}\right)$ H

4. $y = \cos\left(x + \dfrac{\pi}{4}\right)$ A

5. $y = 1 + \sin x$ B

6. $y = -1 + \sin x$ E

7. $y = 1 + \cos x$ F

8. $y = -1 + \cos x$ C

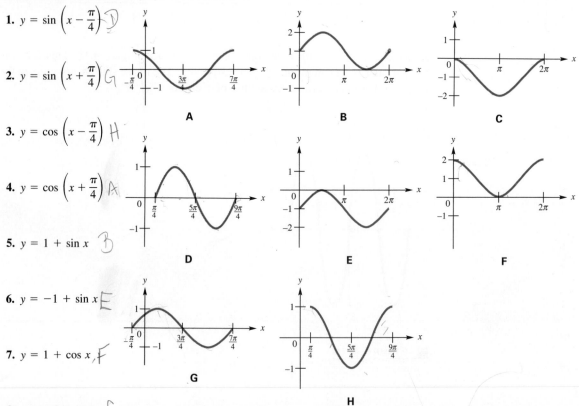

In Exercises 9 and 10, fill in the blanks with the appropriate responses.

9. The function $y = -4 + \sin 4(x + \pi/2)$ has amplitude _____, period _____, phase shift _____ units to the _____, and has vertical translation _____ units _____.

10. If the graph of $y = \cos x$ is shifted $\pi/2$ units horizontally to the _____, it will coincide with the graph of $y = \sin x$.

For each of the following, find the amplitude, the period, any vertical translation, and any phase shift. See Examples 1–5.

11. $y = 2 \sin (x - \pi)$

12. $y = \frac{2}{3} \sin \left(x + \frac{\pi}{2} \right)$

13. $y = 4 \cos \left(\frac{x}{2} + \frac{\pi}{2} \right)$

14. $y = -\cos \frac{2}{3} \left(x - \frac{\pi}{3} \right)$

15. $y = 3 \cos 2 \left(x - \frac{\pi}{4} \right)$

16. $y = \frac{1}{2} \sin \left(\frac{x}{2} + \pi \right)$

17. $y = 2 - \sin \left(3x - \frac{\pi}{5} \right)$

18. $y = -1 + \frac{1}{2} \cos (2x - 3\pi)$

Graph each of the following functions over a two-period interval. See Examples 1 and 2.

19. $y = \cos \left(x - \frac{\pi}{2} \right)$

20. $y = \sin \left(x - \frac{\pi}{4} \right)$

21. $y = \sin \left(x + \frac{\pi}{4} \right)$

22. $y = \cos \left(x - \frac{\pi}{3} \right)$

23. $y = 2 \cos \left(x - \frac{\pi}{3} \right)$

24. $y = 3 \sin \left(x - \frac{3\pi}{2} \right)$

Graph each of the following functions over a one-period interval. See Example 3.

25. $y = \frac{3}{2} \sin 2 \left(x + \frac{\pi}{4} \right)$

26. $y = -\frac{1}{2} \cos 4 \left(x + \frac{\pi}{2} \right)$

27. $y = -4 \sin (2x - \pi)$

28. $y = 3 \cos (4x + \pi)$

29. $y = \frac{1}{2} \cos \left(\frac{1}{2} x - \frac{\pi}{4} \right)$

30. $y = -\frac{1}{4} \sin \left(\frac{3}{4} x + \frac{\pi}{8} \right)$

Graph each of the following functions over a two-period interval. See Example 4.

31. $y = -3 + 2 \sin x$

32. $y = 2 - 3 \cos x$

33. $y = 1 - \frac{2}{3} \sin \frac{3}{4} x$

34. $y = -1 - 2 \cos 5x$

35. $y = 2 - \cos x$

36. $y = 1 + \sin x$

37. $y = 1 - 2 \cos \frac{1}{2} x$

38. $y = -3 + 3 \sin \frac{1}{2} x$

39. $y = -2 + \frac{1}{2} \sin 3x$

40. $y = 1 + \frac{2}{3} \cos \frac{1}{2} x$

Graph each of the following functions over a one-period interval. See Example 5.

41. $y = -3 + 2 \sin \left(x + \frac{\pi}{2} \right)$

42. $y = 4 - 3 \cos (x - \pi)$

43. $y = \frac{1}{2} + \sin 2 \left(x + \frac{\pi}{4} \right)$

44. $y = -\frac{5}{2} + \cos 3 \left(x - \frac{\pi}{6} \right)$

45. Consider the function $y = -4 - 3 \sin 2(x - \pi/6)$. Without actually graphing the function, write an explanation of how the constants -4, -3, 2, and $\pi/6$ affect the graph, using the graph of $y = \sin x$ as a basis for comparison.

46. Explain how, by an appropriate translation, the graph of $y = \sin x$ can be made to coincide with the graph of $y = \cos x$.

4.3 ———— # GRAPHS OF THE OTHER CIRCULAR FUNCTIONS

In this section we discuss the graphs of the four remaining circular functions: cosecant, secant, tangent, and cotangent.

GRAPHS OF COSECANT AND SECANT   Since cosecant values are reciprocals of the corresponding sine values, the period of the function $y = \csc x$ is 2π, the same as for $y = \sin x$. When $\sin x = 1$, the value of $\csc x$ is also 1, and when $0 < \sin x < 1$, then $\csc x > 1$. Also, if $-1 < \sin x < 0$, then $\csc x < -1$. (Verify these statements with a calculator set in radian mode.) As $|x|$ approaches 0, $|\sin x|$ approaches 0, and $|\csc x|$ gets larger and larger. The graph of $\csc x$ approaches the vertical line $x = 0$ but never touches it. The line $x = 0$ is called a **vertical asymptote**. In fact, the lines $x = n\pi$, where n is any integer, are all vertical asymptotes. Using this information and plotting a few points shows that the graph takes the shape of the solid curve shown in Figure 4.19. To show how the two graphs are related, the graph of $y = \sin x$ is also shown, as a dashed curve.

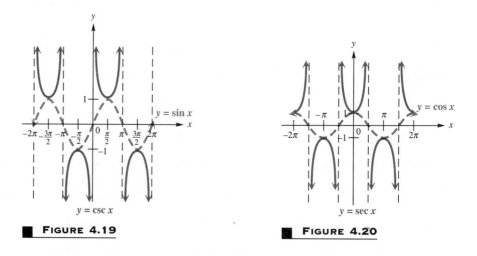

■ **FIGURE 4.19**

■ **FIGURE 4.20**

The domain of the function $y = \csc x$ is $\{x | x \neq n\pi$, where n is any integer$\}$, and the range is $(-\infty, -1] \cup [1, \infty)$.

The graph of $y = \sec x$, shown in Figure 4.20, is related to the cosine graph in the same way that the graph of $y = \csc x$ is related to the sine graph, because $\sec x = 1/\cos x$.

The domain of the function $y = \sec x$ is $\{x | x \neq \pi/2 + n\pi$, where n is any integer$\}$, and the range is $(-\infty, -1] \cup [1, \infty)$.

In order to graph functions based on the cosecant and secant, see the summary that follows.

| GRAPHING THE COSECANT AND SECANT FUNCTIONS | To graph $y = a \csc bx$ or $y = a \sec bx$, with $b > 0$, follow these steps.

1. Graph the corresponding reciprocal function as a guide, using a dashed curve. That is, |

To Graph	Use as a Guide
$y = a \csc bx$	$y = a \sin bx$
$y = a \sec bx$	$y = a \cos bx.$

2. Sketch the vertical asymptotes. They will have equations of the form $x = k$, where k is an x-intercept of the graph of the guide function.

3. Sketch the graph of the desired function by drawing the typical U-shaped branches between the adjacent asymptotes. The branches will be above the graph of the guide function when the guide function values are positive, and below the graph of the guide function when the guide function values are negative. The graph will resemble the graphs in Figures 4.19 and 4.20.

Like the sine and cosine functions, the secant and cosecant function graphs may be translated vertically as well as horizontally. The period of both functions is 2π.

∎ *Example 1*

GRAPHING
$y = a \sec bx$

Graph $y = 2 \sec \dfrac{1}{2} x$.

Use the guidelines above.

Step 1 This function involves the secant, so the corresponding reciprocal function will involve the cosine. The function that we will graph as a guide is

$$y = 2 \cos \frac{1}{2} x.$$

Using the guidelines of Section 4.1, we find that one period of the graph lies along the interval that satisfies the inequality $0 \le (1/2)x \le 2\pi$, or $[0, 4\pi]$. Dividing this interval into four equal parts gives the following key points.

$$(0, 2) \quad (\pi, 0) \quad (2\pi, -2) \quad (3\pi, 0) \quad (4\pi, 2)$$

These are joined with a smooth curve; it is dashed to indicate that this graph is only a guide. An additional period is graphed as seen in Figure 4.21(a).

Step 2 Sketch the vertical asymptotes. These occur at x-values for which the guide function equals 0. A few of them have these equations:

$$x = -3\pi \quad x = -\pi \quad x = \pi \quad x = 3\pi.$$

See Figure 4.21(a).

Step 3   Sketch the graph of $y = 2 \sec (1/2)x$ by drawing in the typical U-shaped branches, approaching the asymptotes. See Figure 4.21(b).   ■

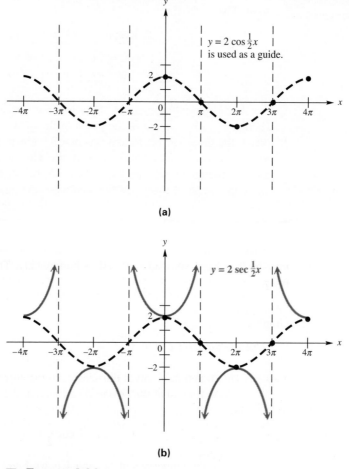

(a)

(b)

■ **FIGURE 4.21**

■ *Example 2*
GRAPHING
$y = a \csc (x - d)$

Graph $y = \dfrac{3}{2} \csc \left(x - \dfrac{\pi}{2} \right).$

This function can be graphed using the method of Example 1, by first graphing the corresponding reciprocal function $y = (3/2) \sin (x - \pi/2)$. We can alternatively analyze the function as follows. Compared with the graph of $y = \csc x$, this graph has a phase shift of $\pi/2$ units to the right. Thus, the asymptotes are the lines $x = \pi/2$, $3\pi/2$, and so on. Also, there are no values of y between $-3/2$ and $3/2$. As shown in Figure 4.22, this is related to the increased amplitude of $y = (3/2) \sin x$ compared with $y = \sin x$. (Amplitude does not apply to the secant or cosecant functions; it enters only indirectly from the corresponding cosine or sine graphs.)

This means that the graph goes through the points $(\pi, 3/2)$, $(2\pi, -3/2)$, and so on. **149**
Two periods are shown in Figure 4.22. (The graph of the "guide" function, $y = (3/2) \sin(x - \pi/2)$, is shown in black.) ■

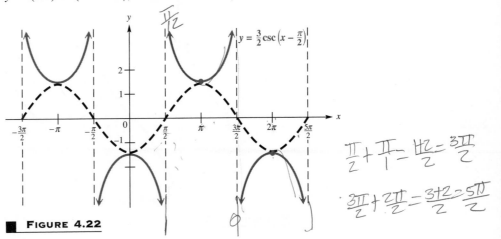

FIGURE 4.22

GRAPHS OF TANGENT AND COTANGENT We will now study the graphs of the two remaining circular functions, $y = \tan x$ and $y = \cot x$.

Since the values of $y = \tan x$ are positive in quadrants I and III, and negative in quadrants II and IV,

$$\tan (x + \pi) = \tan x,$$

so the period of $y = \tan x$ is π. Thus, the tangent function need be investigated only within an interval of π units. A convenient interval for this purpose is $(-\pi/2, \pi/2)$ because, although the endpoints $-\pi/2$ and $\pi/2$ are not in the domain of $y = \tan x$ (why?), $\tan x$ exists for all other values in the interval. In the interval $(0, \pi/2)$, $\tan x$ is positive. As x goes from 0 to $\pi/2$, a calculator shows that $\tan x$ gets larger and larger without bound. As x goes from $-\pi/2$ to 0, the values of $\tan x$ approach 0 through negative values. These results are summarized in the following table.

As x Increases From	$\tan x$
0 to $\pi/2$	Increases from 0, without bound
$-\pi/2$ to 0	Increases to 0

Based on these results, the graph of $y = \tan x$ will approach the vertical line $x = \pi/2$ but never touch it, so the line $x = \pi/2$ is a vertical asymptote. The lines $x = \pi/2 + n\pi$, where n is any integer, are all vertical asymptotes. These asymptotes are indicated with light dashed lines on the graph in Figure 4.23. In the interval $(-\pi/2, 0)$, which corresponds to quadrant IV on the unit circle, $\tan x$ is negative, and as x goes from 0 to $-\pi/2$, $\tan x$ gets smaller and smaller. A table of values for $\tan x$, where $-\pi/2 < x < \pi/2$, follows.

x	−π/3	−π/4	−π/6	0	π/6	π/4	π/3
tan x	−1.7	−1	−.6	0	.6	1	1.7

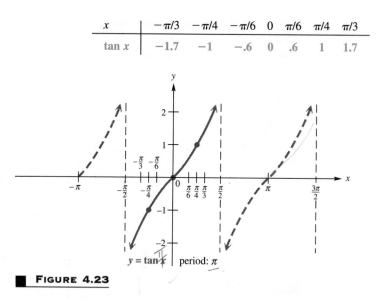

$y = \tan x$ period: π

■ **FIGURE 4.23**

Plotting the points from the table and letting the graph approach the asymptotes at $x = \pi/2$ and $x = -\pi/2$ gives the portion of the graph shown with a solid curve in Figure 4.23. More of the graph can be sketched by repeating the same curve, also as shown in the figure. This graph, like the graphs for the sine and cosine functions, should be learned well enough so that a quick sketch can easily be made. Convenient key points are $(-\pi/4, -1)$, $(0, 0)$, and $(\pi/4, 1)$. These points are shown in Figure 4.23. The lines $x = \pi/2$ and $x = -\pi/2$ are vertical asymptotes. (The idea of *amplitude,* discussed earlier, applies only to the sine and cosine functions, and so is not used here.)

The domain of the tangent function is $\{x | x \neq \pi/2 + n\pi$, where n is any integer$\}$. The range is $(-\infty, \infty)$.

The definition $\cot x = 1/(\tan x)$ can be used to find the graph of $y = \cot x$. The period of the cotangent, like that of the tangent, is π. The domain of $y = \cot x$ excludes $0 + n\pi$, where n is any integer, since $1/\tan x$ is undefined for these values of x. Thus, the vertical lines $x = n\pi$ are asymptotes. Values of x that lead to asymptotes for tan x will make $\cot x = 0$, so $\cot(-\pi/2) = 0$, $\cot \pi/2 = 0$, $\cot 3\pi/2 = 0$, and so on. The values of tan x increase as x goes from $-\pi/2$ to $\pi/2$, so the values of cot x will *decrease* as x goes from $-\pi/2$ to $\pi/2$. A table of values for cot x, where $0 < x < \pi$, is shown below

x	π/6	π/4	π/3	π/2	2π/3	3π/4	5π/6
cot x	1.7	1	.6	0	−.6	−1	−1.7

Plotting these points and using the information discussed above gives the graph of $y = \cot x$ shown in Figure 4.24. (The graph shows two periods.)

The domain of the cotangent function is $\{x | x \neq n\pi$, where n is any integer$\}$. The range is $(-\infty, \infty)$.

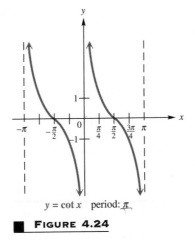

$y = \cot x$ period: π

■ **FIGURE 4.24**

GRAPHING THE TANGENT AND COTANGENT FUNCTIONS

To graph $y = a \tan bx$ or $y = a \cot bx$, with $b > 0$:

1. The period is π/b. To locate two adjacent vertical asymptotes, solve the following equations for x:

$$\text{For } y = a \tan bx: \quad bx = -\frac{\pi}{2} \quad \text{and} \quad bx = \frac{\pi}{2}$$

$$\text{For } y = a \cot bx: \quad bx = 0 \quad \text{and} \quad bx = \pi.$$

2. Sketch the two vertical asymptotes found in Step 1.
3. Divide the interval formed by the vertical asymptotes into four equal parts.
4. Evaluate the function for the first-quarter point, mid-point, and third-quarter point, using the x-values found in Step 3.
5. Join the points with a smooth curve, approaching the vertical asymptotes. Draw additional asymptotes and periods of the graph as necessary.

Like the other circular functions, the graphs of the tangent and cotangent functions may be shifted horizontally as well as vertically.

■ *Example 3*
| **GRAPHING y = tan bx**

Graph $y = \tan 2x$.

Step 1   The period of this function is $\pi/2$. To locate two adjacent vertical asymptotes, solve $2x = -\pi/2$ and $2x = \pi/2$ (since this is a tangent function). We find that two asymptotes have equations

$$x = -\frac{\pi}{4} \quad \text{and} \quad x = \frac{\pi}{4}.$$

Step 2   Sketch the two vertical asymptotes $x = \pm\pi/4$, as shown in Figure 4.25.

Step 3 Divide the interval $(-\pi/4, \pi/4)$ into four equal parts. This gives the following key *x*-values:

$$\text{first-quarter value: } -\frac{\pi}{8}$$

$$\text{middle value: } 0$$

$$\text{third-quarter value: } \frac{\pi}{8}$$

Step 4 Evaluate the function for the *x*-values found in Step 3, as shown in the following table.

x	$-\pi/8$	0	$\pi/8$
$2x$	$-\pi/4$	0	$\pi/4$
$\tan 2x$	-1	0	1

Step 5 Join these points with a smooth curve, approaching the vertical asymptotes. See Figure 4.25. Another period has been graphed as well, one-half period to the left and one-half period to the right. ■

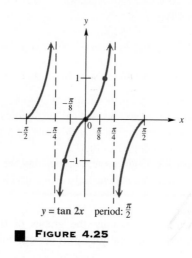

$y = \tan 2x$ period: $\frac{\pi}{2}$

■ **FIGURE 4.25**

■ *Example 4*

GRAPHING

$y = a \tan bx$

Graph $y = -3 \tan \frac{1}{2}x$.

The period is $\pi/(1/2) = 2\pi$. Adjacent asymptotes are at $x = -\pi$ and $x = \pi$. Dividing the interval $-\pi < x < \pi$ into four equal parts gives key *x*-values of $-\pi/2$, 0, and $\pi/2$. Evaluating the function at these *x* values gives these key points.

$$\left(-\frac{\pi}{2}, 3\right) \qquad (0, 0) \qquad \left(\frac{\pi}{2}, -3\right)$$

Plotting these points and joining them with a smooth curve gives the graph shown in Figure 4.26. ■

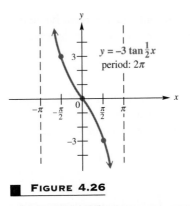

■ **FIGURE 4.26**

NOTE

The function $y = -3 \tan (1/2)x$ of Example 4, graphed in Figure 4.26, has a graph that compares to the graph of $y = \tan x$ as follows:

1. The period is larger, because $b = 1/2$, and $1/2 < 1$.
2. The graph is "stretched," because $a = -3$, and $|-3| > 1$.
3. Each branch of the graph goes down from left to right (that is, the function decreases) between each pair of adjacent asymptotes, because $a = -3 < 0$. When $a < 0$, the graph is reflected about the x-axis.

Graphing a cotangent function is done in a manner similar to graphing a tangent function, as shown in the next example.

■ *Example 5*

GRAPHING
$y = a \cot bx$

Graph $y = \dfrac{1}{2} \cot 2x$.

Because this function involves the cotangent, we can locate two adjacent asymptotes by solving the equations $2x = 0$ and $2x = \pi$. We find that the lines $x = 0$ (the y-axis) and $x = \pi/2$ are two such asymptotes. Divide the interval $0 < x < \pi/2$ into four equal parts, getting key x-values of $\pi/8$, $\pi/4$, and $3\pi/8$. Evaluating the function at these x-values gives the key following points.

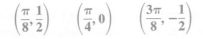

Joining these points with a smooth curve approaching the asymptotes gives the graph shown in Figure 4.27. ■

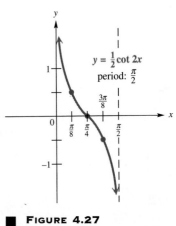

■ **FIGURE 4.27**

As stated earlier, tangent and cotangent function graphs may be translated vertically, horizontally, or both, as shown in the next two examples.

■ *Example 6*

GRAPHING A
TANGENT FUNCTION
WITH A VERTICAL
TRANSLATION

Graph $y = 2 + \tan x$.

Every value of y for this function will be 2 units more than the corresponding value of y in $y = \tan x$, causing the graph of $y = 2 + \tan x$ to be translated 2 units upward as compared with the graph of $y = \tan x$. See Figure 4.28. ■

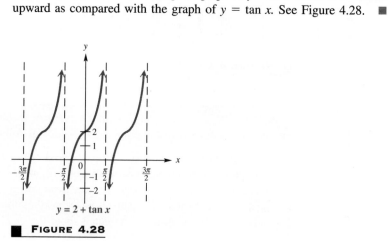

$y = 2 + \tan x$

■ **FIGURE 4.28**

■ *Example 7*

GRAPHING A
COTANGENT
FUNCTION WITH
VERTICAL AND
HORIZONTAL
TRANSLATIONS

Graph $y = -2 - \cot\left(x - \dfrac{\pi}{4}\right)$.

Here $b = 1$, so the period is π. The graph will be translated down 2 units (because $c = -2$), reflected about the x-axis (because of the minus sign in front of the cotangent) and will have a phase shift (horizontal translation) of $\pi/4$ units to the right (because of the argument $(x - \pi/4)$). To locate adjacent asymptotes, since this function involves the cotangent, we solve the following equations:

$$x - \frac{\pi}{4} = 0 \qquad x - \frac{\pi}{4} = \pi$$

$$x = \frac{\pi}{4} \qquad x = \frac{5\pi}{4}.$$

Dividing the interval $\pi/4 < x < 5\pi/4$ into four equal parts and evaluating the function at the three key x-values within the interval gives these key points.

$$\left(\frac{\pi}{2}, -3\right) \qquad \left(\frac{3\pi}{4}, -2\right) \qquad (\pi, -1)$$

These points are joined by a smooth curve. This period of the graph, along with one in the interval $-3\pi/4 < x < -\pi/4$, is shown in Figure 4.29. ■

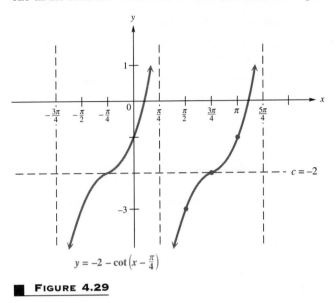

$$y = -2 - \cot\left(x - \frac{\pi}{4}\right)$$

■ **FIGURE 4.29**

4.3 EXERCISES ■

Match each function with its graph in Exercises 1–6.

1. $y = -\csc x$ B

2. $y = -\sec x$ C

3. $y = -\tan x$ E

4. $y = -\cot x$ A

5. $y = \tan\left(x - \frac{\pi}{4}\right)$ D

6. $y = \cot\left(x - \frac{\pi}{4}\right)$ F

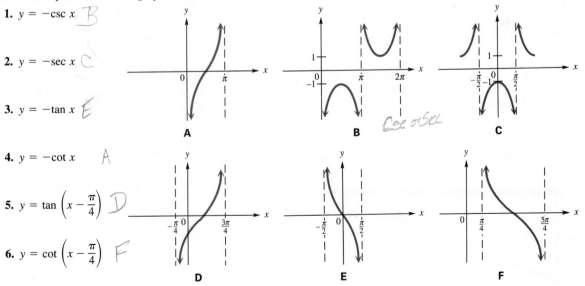

7. Between each pair of successive asymptotes, a portion of the graph of $y = \sec x$ or $y = \csc x$ resembles a parabola. Can each of these portions actually be a parabola? Explain.

8. The quotient $y = \dfrac{\sin x}{\cos x}$ does not exist if $x = -\pi/2$ or if $x = \pi/2$. Start at $x = -1.4$ and evaluate the quotient (using a calculator) with x increasing by .2 until $x = 1.4$ is reached. Plot the values obtained. What graph is suggested?

Graph each of the following functions over a one-period interval. See Examples 1 and 2.

9. $y = \csc\left(x - \dfrac{\pi}{4}\right)$

10. $y = \sec\left(x + \dfrac{3\pi}{4}\right)$

11. $y = \sec\left(x + \dfrac{\pi}{4}\right)$

12. $y = \csc\left(x + \dfrac{\pi}{3}\right)$

13. $y = \sec\left(\dfrac{1}{2}x + \dfrac{\pi}{3}\right)$

14. $y = \csc\left(\dfrac{1}{2}x - \dfrac{\pi}{4}\right)$

15. $y = 2 + 3\sec(2x - \pi)$

16. $y = 1 - 2\csc\left(x + \dfrac{\pi}{2}\right)$

17. $y = 1 - \dfrac{1}{2}\csc\left(x - \dfrac{3\pi}{4}\right)$

18. $y = 2 + \dfrac{1}{4}\sec\left(\dfrac{1}{2}x - \pi\right)$

Graph each of the following functions over a one-period interval. See Examples 3–5.

19. $y = 2\tan x$

20. $y = 2\cot x$

21. $y = \dfrac{1}{2}\cot x$

22. $y = 2\tan\dfrac{1}{4}x$

23. $y = \cot 3x$

24. $y = -\cot\dfrac{1}{2}x$

Graph each of the following functions over a two-period interval. See Examples 6 and 7.

25. $y = \tan(2x - \pi)$

26. $y = \tan\left(\dfrac{x}{2} + \pi\right)$

27. $y = \cot\left(3x + \dfrac{\pi}{4}\right)$

28. $y = \cot\left(2x - \dfrac{3\pi}{2}\right)$

29. $y = 1 + \tan x$

30. $y = -2 + \tan x$ ·

31. $y = 1 - \cot x$

32. $y = -2 - \cot x$

33. $y = -1 + 2\tan x$

34. $y = 3 + \dfrac{1}{2}\tan x$

35. $y = -1 + \dfrac{1}{2}\cot(2x - 3\pi)$

36. $y = -2 + 3\tan(4x + \pi)$

37. $y = \dfrac{2}{3}\tan\left(\dfrac{3}{4}x - \pi\right) - 2$

38. $y = 1 - 2\cot 2\left(x + \dfrac{\pi}{2}\right)$

39. A rotating beacon is located at point A next to a long wall. (See the figure.) The beacon is 4 m from the wall. The distance d is given by

$$d = 4\tan 2\pi t,$$

where t is time measured in seconds since the beacon started rotating. (When $t = 0$, the beacon is aimed at point R. When the beacon is aimed to the right of R, the value of d is positive; d is negative if the beacon is aimed to the left of R.) Find d for the following times.
(a) $t = 0$ (b) $t = .4$
(c) $t = .8$ (d) $t = 1.2$
(e) Why is .25 a meaningless value for t?
(f) What is a meaningful domain for t?

40. In the figure for Exercise 39, the distance a is given by

$$a = 4|\sec 2\pi t|.$$

Find a for the following times.
(a) $t = 0$ (b) $t = .86$ (c) $t = 1.24$

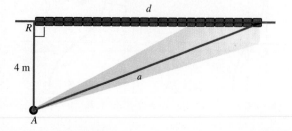

SECTION	KEY IDEAS
4.1, 4.2 Graphs of the Sine and Cosine I, Graphs of the Sine and Cosine II	**Cosine and Sine Functions** 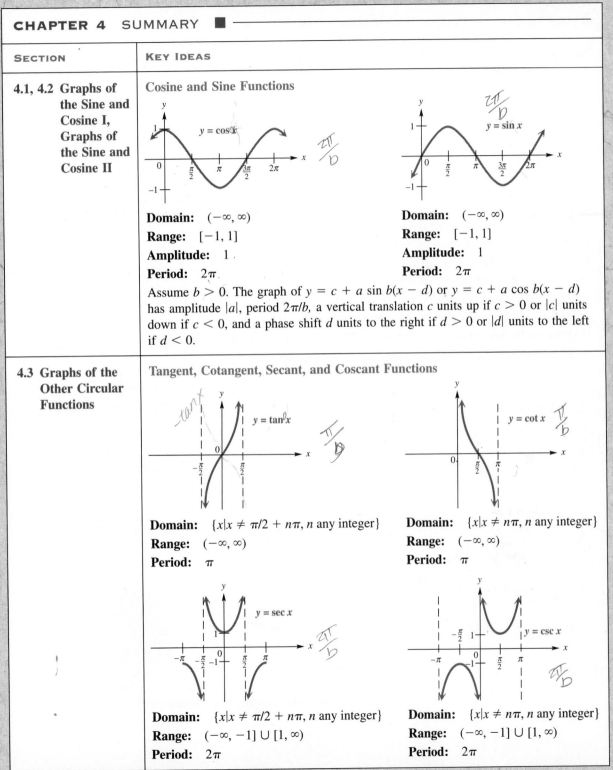 **Domain:** $(-\infty, \infty)$ **Domain:** $(-\infty, \infty)$ **Range:** $[-1, 1]$ **Range:** $[-1, 1]$ **Amplitude:** 1 **Amplitude:** 1 **Period:** 2π **Period:** 2π Assume $b > 0$. The graph of $y = c + a \sin b(x - d)$ or $y = c + a \cos b(x - d)$ has amplitude $|a|$, period $2\pi/b$, a vertical translation c units up if $c > 0$ or $|c|$ units down if $c < 0$, and a phase shift d units to the right if $d > 0$ or $|d|$ units to the left if $d < 0$.
4.3 Graphs of the Other Circular Functions	**Tangent, Cotangent, Secant, and Coscant Functions** **Domain:** $\{x \mid x \neq \pi/2 + n\pi, n \text{ any integer}\}$ **Domain:** $\{x \mid x \neq n\pi, n \text{ any integer}\}$ **Range:** $(-\infty, \infty)$ **Range:** $(-\infty, \infty)$ **Period:** π **Period:** π **Domain:** $\{x \mid x \neq \pi/2 + n\pi, n \text{ any integer}\}$ **Domain:** $\{x \mid x \neq n\pi, n \text{ any integer}\}$ **Range:** $(-\infty, -1] \cup [1, \infty)$ **Range:** $(-\infty, -1] \cup [1, \infty)$ **Period:** 2π **Period:** 2π

For each of the following circular functions, give the amplitude, period, vertical translation, and phase shift, as applicable.

1. $y = 2 \sin x$

2. $y = \tan 3x$

3. $y = -\dfrac{1}{2} \cos 3x$

4. $y = 2 \sin 5x$

5. $y = 1 + 2 \sin \dfrac{1}{4} x$

6. $y = 3 - \dfrac{1}{4} \cos \dfrac{2}{3} x$

7. $y = 3 \cos \left(x + \dfrac{\pi}{2} \right)$

8. $y = -\sin \left(x - \dfrac{3\pi}{4} \right)$

9. $y = \dfrac{1}{2} \csc \left(2x - \dfrac{\pi}{4} \right)$

10. $y = 2 \sec (\pi x - 2\pi)$

11. $y = \dfrac{1}{3} \tan \left(3x - \dfrac{\pi}{3} \right)$

12. $y = \cot \left(\dfrac{x}{2} + \dfrac{3\pi}{4} \right)$

Identify the one of the six circular functions that satisfies the description.

13. Period is π, x intercepts are of the form $n\pi$, where n is an integer

14. Period is 2π, passes through the origin

15. Period is 2π, passes through the point $(\pi/2, 0)$

16. Period is 2π, domain is $\{x \mid x \neq n\pi$, where n is an integer$\}$

17. Period is π, function is decreasing on the interval $0 < x < \pi$

18. Period is 2π, has vertical asymptotes of the form $x = \pi/2 + n\pi$, where n is an integer

Graph each of the following functions over a one-period interval.

19. $y = 3 \sin x$

20. $y = \dfrac{1}{2} \sec x$

21. $y = -\tan x$

22. $y = -2 \cos x$

23. $y = 2 + \cot x$

24. $y = -1 + \csc x$

25. $y = \sin 2x$

26. $y = \tan 3x$

27. $y = 3 \cos 2x$

28. $y = \dfrac{1}{2} \cot 3x$

29. $y = \cos \left(x - \dfrac{\pi}{4} \right)$

30. $y = \tan \left(x - \dfrac{\pi}{2} \right)$

31. $y = \sec \left(2x + \dfrac{\pi}{3} \right)$
$2x + \pi$

32. $y = \sin \left(3x + \dfrac{\pi}{2} \right)$

33. $y = 1 + 2 \cos 3x$

34. $y = -1 - 3 \sin 2x$

35. $y = 2 \sin \pi x$

36. $y = -\dfrac{1}{2} \cos (\pi x - \pi)$

37. $y = 1 - 2 \sec \left(x - \dfrac{\pi}{4} \right)$

38. $y = -\csc (2x - \pi) + 1$

39. Let a person h_1 ft tall stand d ft from an object h_2 ft tall, where $h_2 > h_1$. Let θ be the angle of elevation to the top of the object. (See the figure.)
   (a) Show that $d = (h_2 - h_1) \cot \theta$.
   (b) Let $h_2 = 55$ and $h_1 = 5$. Graph d for the interval $0 < \theta < \pi/2$.

40. The amount of pollution in the air fluctuates with the seasons. It is lower after heavy spring rains and higher after periods of little rain. In addition to this seasonal fluctuation, the long-term trend is upward. An idealized graph of this situation is shown in the figure. Circular functions can be used to decribe the fluctuating part of the pollution levels. Powers of the number e (e is the base of natural logarithms; to six decimal places, $e = 2.718282$) can be used to show the long-term growth. In fact, the pollution level in a certain area might be given by

$$P(t) = 7(1 - \cos 2\pi t)(t + 10) + 100e^{.2t},$$

where t is time in years, with $t = 0$ representing January 1 of the base year. Thus, July 1 of the same year would be represented by $t = .5$, and October 1 of the following year would be represented by $t = 1.75$. Find the pollution levels on the following dates.

(a) January 1, base year

(b) July 1, base year

(c) January 1, following year

(d) July 1, following year

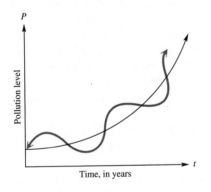

41. The figure shows the population of lynx and hares in Canada for the years 1847–1903. The hares are food for the lynx. An increase in hare population causes an increase in lynx population some time later. The increasing lynx population then causes a decline in hare population.

(a) Estimate the length of one period.

(b) Estimate maximum and minimum hare population.

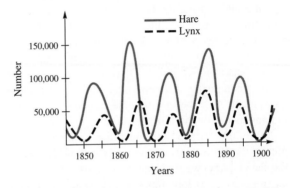

42. Explain how the graph of $y = 2 \cos (3x + 1)$ differs from the graph of $y = 2 \cos 3x + 1$.

■ THE GRAPHING CALCULATOR ■

The graphing calculator can provide an enlightening look at how the constants *a, b, c,* and *d* affect the graph of a function of the form $y = c + a \cdot f[b(x - d)]$, where *f* is a circular function.

One popular model, the TI-81 by Texas Instruments, allows us to set maximum and minimum values of *x* and *y* using the RANGE function. For trigonometric (or circular) functions, the ZOOM feature allows us to set a "Trig" domain and range where *x* takes on values between -2π and 2π, and *y* takes on values between -3 and 3. Of course, there may be times when using different intervals will be more convenient. Also, be sure that the calculator is set for radians, using the MODE function.

With the "Trig" range and radian mode, let us begin our observation using the sine function. By pressing the [Y=] key, we find that the calculator will allow us to enter up to four functions, denoted Y_1, Y_2, Y_3, and Y_4. Enter $Y_1 = \sin x$, and press the [GRAPH] key. The calculator will graph the sine function. Using the TRACE feature, we can obtain *x* and *y* coordinates of points on the graph. The keys marked [◁] and [▷] allow us to move the tracing cursor left and right, and the *x* and *y* coordinates are displayed on the lower part of the screen. One such point is

$$X = 1.1243595, \quad Y = .90199123.$$

This means $\sin 1.1243595 \approx .90199123$.

Now enter $Y_2 = 2 \sin x$. Based on the discussion in Section 4.1, the graph of this function will have the same basic shape as that of Y_1, but the *y*-values will be twice those of Y_1. Now, use the GRAPH function to graph both Y_1 and Y_2 on the same screen. If the graphing mode is set to "Sequence," the graph of $Y_1 = \sin x$ will be plotted first. Then watch carefully as the calculator plots $Y_2 = 2 \sin x$, and notice the effect of the constant 2. You will get a final screen that resembles Figure A. Compare this to Figure 4.7 in the text.

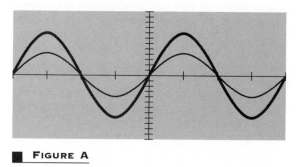

■ FIGURE A

Using the TRACE feature, move the cursor to the graph of Y_2 (using the [△] key) and locate the point whose *x*-coordinate is 1.1243595. The corresponding *y*-value is shown to be 1.8039825, which is twice the *y*-value found earlier. This makes sense, because $Y_2 = 2 \cdot Y_1$.

Keeping Y_1 as sin *x*, re-enter Y_2 as $-2 \sin x$. (You will need to clear the original Y_2 using the [CLEAR] key.) Before graphing, review the discussion in Section 4.1 to predict what the screen will look like after Y_1 and Y_2 are plotted. Were you correct? Now, predict the *y*-value you will find when the tracing cursor is moved to the point on Y_2 that has *x*-coordinate 1.1243595.

To determine the effect of the positive constant *b* on the graph of $y = \cos bx$ or $y = \sin bx$, let Y_1 remain as sin *x* and enter Y_2 as sin 2*x*. Plot these on the calculator and notice how the graph of Y_2 has the same basic shape, but the period is half that of Y_1, confirming our discussion in Section 4.1. See Figure B, and compare it to Figure 4.8 in the text.

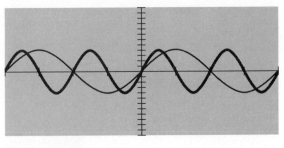

■ FIGURE B

The graphing calculator can also help to reinforce the concepts of horizontal and vertical translations as described in Section 4.2. Again, let Y_1 remain as sin x, and enter $Y_2 = \sin(x - \pi/3)$, taking care to insert parentheses as necessary. Recall that the graph of Y_2 will be the same as that of Y_1, shifted $\pi/3$ units to the right. Press the [GRAPH] key and observe the effect as the graph of Y_2 "leads" the graph of Y_1 by $\pi/3$ units. See Figure C. Now compare the display to Figure 4.13 in Section 4.2.

of Y_1 three units upward. Adjust the range so that the minimum and maximum values of y are -2 and 5, respectively. Now graph these two functions, and observe the results, shown in Figure D. Compare this to Figure 4.17 in the text.

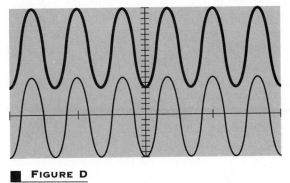

■ **FIGURE D**

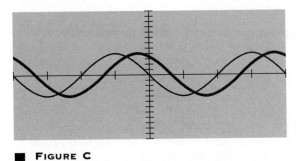

■ **FIGURE C**

To see the effect of a vertical translation, enter $Y_1 = -2 \cos 3x$ and $Y_2 = 3 - 2 \cos 3x$. The constant $c = 3$ in Y_2 will have the effect of shifting the graph

You may wish to experiment with your graphing calculator using the following functions. First, enter Y_1 and Y_2. Before plotting the graphs, describe in your own words how the graph of Y_2 will compare with that of Y_1. Then plot the graphs on the calculator, observing carefully the graphing process. Compare the graph of Y_2 to the text figure referenced.

---

1. $Y_1 = \cos x;$ $Y_2 = \cos \dfrac{2}{3}x$ (Figure 4.9)

2. $Y_1 = \sin x;$ $Y_2 = -2 \sin 3x$ (Figure 4.10)

3. $Y_1 = \cos x;$ $Y_2 = 3 \cos \dfrac{1}{2}x$ (Figure 4.11)

4. $Y_1 = \cos x;$ $Y_2 = 3 \cos \left(x + \dfrac{\pi}{4}\right)$ (Figure 4.14)

5. $Y_1 = \cos x;$ $Y_2 = -2 \cos(3x + \pi)$ (Figure 4.15)

6. $Y_1 = \sin x;$ $Y_2 = -1 + 2 \sin 4\left(x + \dfrac{\pi}{4}\right)$
(Figure 4.18)

7. $Y_1 = \sec x$ $\left(\text{Enter sec } x \text{ as } \dfrac{1}{\cos x}.\right);$ $Y_2 =$
$2 \sec \dfrac{1}{2}x$ (Figure 4.21(b))

8. $Y_1 = \tan x;$ $Y_2 = \tan 2x$ (Figure 4.25)

9. $Y_1 = \cot x$ $\left(\text{Enter cot } x \text{ as } \dfrac{1}{\tan x}.\right);$ $Y_2 = \dfrac{1}{2} \cot 2x$
(Figure 4.27)

10. $Y_1 = \tan x;$ $Y_2 = 2 + \tan x$ (Figure 4.28)

A function formed by combining two other functions, such as

$$y = \cos x + \sin x,$$

has historically been graphed using a method known as *addition of ordinates*. (The ordinate of a point is its *y*-coordinate.) To apply this method to this function, we would graph the functions $y = \cos x$ and $y = \sin x$. Then, for selected values of *x*, we would add $\cos x$ and $\sin x$, and plot the points $(x, \cos x + \sin x)$. Connecting the selected points with a typical circular function-type curve would give the graph of the desired function. While this method illustrates some valuable concepts involving the arithmetic of functions, it is very time-consuming.

With the technology of the graphing calculator, such an exercise can easily be accomplished. Using the "Trig" range, enter $Y_1 = \cos x$, $Y_2 = \sin x$, and $Y_3 = Y_1 + Y_2$, using the *y*-variables menu. Then press the |GRAPH| key and observe carefully as the calculator plots, in order Y_1, Y_2, and $Y_3 = Y_1 + Y_2$. See Figure E.

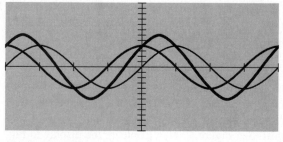

■ **FIGURE E**

Move the tracing cursor to locate the points on each graph as follows:

Y_1:	$X = 1.2566371$	$Y = .30901699$
Y_2:	$X = 1.2566371$	$Y = .95105652$
Y_3:	$X = 1.2566371$	$Y = 1.2600735$

Notice that the sum of the first two *y*-values is (with a slight discrepancy in the final decimal place) equal to the third *y*-value. Do you now see how the term "addition of ordinates" came about?

The following functions may be used to experiment with your graphing calculator. They may either be graphed directly as Y_1, or you may wish to express them as a combination of several functions as described above.

11. $y = \cos x - \tan x$

12. $y = \sin x + \sin 2x$

13. $y = \cos x - \cos \dfrac{1}{2} x$

14. $y = \sin x + \tan x$

15. $y = \sin x + \csc x$

16. $y = 2 \cos x - \sec x$

17. $y = 2 \sec x + \sin x$

18. $y = \cos x + \cot x$

19. $y = \sin x - 2 \cos x$

20. $y = \cos x + \cos 2x$

C ■ FIVE
CHAPTER

TRIGONOMETRIC IDENTITIES

A conditional equation, such as $2x + 1 = 9$ or $m^2 - 2m = 3$, is true for only certain values in the domain of its variable. For example, $2x + 1 = 9$ is true only for $x = 4$, and $m^2 - 2m = 3$ is true only for $m = 3$ and $m = -1$. On the other hand, an **identity** is an equation that is true for *every* value in the domain of its variable. Examples of identities include

$$5(x + 3) = 5x + 15 \qquad \text{and} \qquad (a + b)^2 = a^2 + 2ab + b^2.$$

This chapter discusses identities involving trigonometric and circular functions. The variables in these functions represent either angles or real numbers. The domain of the variable is assumed to be all values for which a given function is defined.

5.1 ———— FUNDAMENTAL IDENTITIES

This section reviews the fundamental trigonometric identities first introduced in Chapter 1 and discusses some of their uses.* We repeat the basic definitions of the trigonometric functions of an angle θ in standard position. See Figure 5.1.

TRIGONOMETRIC FUNCTIONS

Let (x, y) be a point other than the origin on the terminal side of an angle θ in standard position. Then $r = \sqrt{x^2 + y^2}$ is the distance from the point to the origin. The six trigonometric functions of θ are defined as follows.

$$\sin \theta = \frac{y}{r} \qquad \csc \theta = \frac{r}{y} \quad (y \neq 0)$$

$$\cos \theta = \frac{x}{r} \qquad \sec \theta = \frac{r}{x} \quad (x \neq 0)$$

$$\tan \theta = \frac{y}{x} \quad (x \neq 0) \qquad \cot \theta = \frac{x}{y} \quad (y \neq 0)$$

In Chapter 1, these definitions were used to derive the following **reciprocal identities,** which are true for all suitable replacements of the variable.

$$\cot \theta = \frac{1}{\tan \theta} \qquad \csc \theta = \frac{1}{\sin \theta} \qquad \sec \theta = \frac{1}{\cos \theta}$$

Each of these reciprocal identities leads to other forms of the identity. For example, $\csc \theta = 1/\sin \theta$ gives $\sin \theta = 1/\csc \theta$.

From the definitions of the trigonometric functions,

$$\frac{\sin \theta}{\cos \theta} = \frac{y/r}{x/r} = \frac{y}{x} = \tan \theta$$

or

$$\tan \theta = \frac{\sin \theta}{\cos \theta}.$$

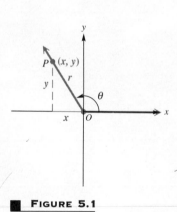

■ **FIGURE 5.1**

*All the identities given in this chapter are summarized at the end of the chapter and inside the back cover.

In a similar manner,

$$\cot \theta = \frac{\cos \theta}{\sin \theta}.$$

These last two identities, also derived in Chapter 1, are called the **quotient identities.**

We also saw in Chapter 1 that the definitions of the trigonometric functions were used to derive the identity

$$\sin^2 \theta + \cos^2 \theta = 1.$$

Dividing both sides by $\cos^2 \theta$ then leads to

$$\tan^2 \theta + 1 = \sec^2 \theta,$$

while dividing through by $\sin^2 \theta$ gives

$$1 + \cot^2 \theta = \csc^2 \theta.$$

These last three identities are the **Pythagorean identities.**

As suggested by the circle shown in Figure 5.2, an angle θ having the point (x, y) on its terminal side has a corresponding angle $-\theta$ with a point $(x, -y)$ on its terminal side. From the definition of sine,

$$\sin (-\theta) = \frac{-y}{r} \quad \text{and} \quad \sin \theta = \frac{y}{r},$$

so that $\sin (-\theta)$ and $\sin \theta$ are negatives of each other, or

$$\sin (-\theta) = -\sin \theta.$$

Figure 5.2 shows an angle θ in quadrant II, but the same result holds for θ in any quadrant. Also, by definition,

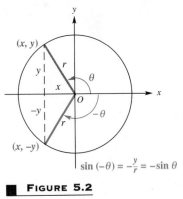

$$\cos (-\theta) = \frac{x}{r} \quad \text{and} \quad \cos \theta = \frac{x}{r},$$

so that $\cos (-\theta) = \cos \theta.$

These formulas for $\sin (-\theta)$ and $\cos (-\theta)$ can be used to find $\tan (-\theta)$ in terms of $\tan \theta$:

$$\tan (-\theta) = \frac{\sin (-\theta)}{\cos (-\theta)} = \frac{-\sin \theta}{\cos \theta} = -\frac{\sin \theta}{\cos \theta}$$

▪ **FIGURE 5.2**

or

$$\tan (-\theta) = -\tan \theta.$$

The preceding three identities are **negative-angle identities.**

The identities given in this section are summarized below. As a group, these are called the **fundamental identities.**

FUNDAMENTAL IDENTITIES

Reciprocal Identities

$$\cot \theta = \frac{1}{\tan \theta} \qquad \sec \theta = \frac{1}{\cos \theta} \qquad \csc \theta = \frac{1}{\sin \theta}$$

Quotient Identities

$$\tan \theta = \frac{\sin \theta}{\cos \theta} \qquad \cot \theta = \frac{\cos \theta}{\sin \theta}$$

Pythagorean Identities

$$\sin^2 \theta + \cos^2 \theta = 1 \qquad \tan^2 \theta + 1 = \sec^2 \theta \qquad 1 + \cot^2 \theta = \csc^2 \theta$$

Negative-Angle Identities

$$\sin (-\theta) = -\sin \theta \qquad \cos (-\theta) = \cos \theta \qquad \tan (-\theta) = -\tan \theta$$

NOTE The forms of the identities given above are the most commonly recognized forms. Throughout this chapter it will be necessary to recognize alternate forms of these identities as well. For example, two other forms of $\sin^2 \theta + \cos^2 \theta = 1$ are

$$\sin^2 \theta = 1 - \cos^2 \theta$$

and

$$\cos^2 \theta = 1 - \sin^2 \theta.$$

You should be able to transform the basic identities using algebraic transformations.

One use for trigonometric identities is to find the values of other trigonometric functions from the value of a given trigonometric function. For example, given a value of $\tan \theta$, the value of $\cot \theta$ can be found from the identity $\cot \theta = 1/\tan \theta$. In fact, given any trigonometric function value and the quadrant in which θ lies, the values of all the other trigonometric functions can be found by using identities, as in the following example.

▪ *Example 1*

FINDING ALL
TRIGONOMETRIC
FUNCTION VALUES,
GIVEN ONE VALUE
AND THE QUADRANT

If tan θ = −5/3 and θ is in quadrant II, find the values of the other trigonometric functions using fundamental identities.

The identity cot θ = 1/tan θ leads to cot θ = −3/5. Next, find sec θ from the identity $\tan^2 \theta + 1 = \sec^2 \theta$.

$$\left(-\frac{5}{3}\right)^2 + 1 = \sec^2 \theta$$

$$\frac{25}{9} + 1 = \sec^2 \theta$$

$$\frac{34}{9} = \sec^2 \theta$$

$$-\sqrt{\frac{34}{9}} = \sec \theta$$

$$-\frac{\sqrt{34}}{3} = \sec \theta$$

Choose the negative square root since sec θ is negative in quadrant II. Now find cos θ:

$$\cos \theta = \frac{1}{\sec \theta} = \frac{-3}{\sqrt{34}} = -\frac{3\sqrt{34}}{34},$$

after rationalizing the denominator. Find sin θ by using the identity $\sin^2 \theta + \cos^2 \theta = 1$, with cos θ = −3/√34.

$$\sin^2 \theta + \left(\frac{-3}{\sqrt{34}}\right)^2 = 1$$

$$\sin^2 \theta = 1 - \frac{9}{34}$$

$$\sin^2 \theta = \frac{25}{34}$$

$$\sin \theta = \frac{5}{\sqrt{34}}$$

$$\sin \theta = \frac{5\sqrt{34}}{34} \qquad \text{Rationalize.}$$

The positive square root is used since sin θ is positive in quadrant II. Finally, since csc θ is the reciprocal of sin θ,

$$\csc \theta = \frac{\sqrt{34}}{5}. \quad ▪$$

CAUTION

Several comments can be made concerning Example 1.

1. We are given $\tan \theta = -5/3$. Although $\tan \theta = (\sin \theta)/(\cos \theta)$, we should *not* assume that $\sin \theta = -5$ and $\cos \theta = 3$. (Why can these values not possibly be correct?)
2. Problems of this type can usually be worked in more than one way. For example, after finding $\cot \theta = -3/5$, we could have then found $\csc \theta$ using the identity $1 + \cot^2 \theta = \csc^2 \theta$. The remaining function values could then be found as well.
3. The most common error made in problems like this is an incorrect sign choice for the functions. When taking the square root, be sure to choose the sign based on the quadrant of θ and the function being found.

Every trigonometric function of an angle θ or a number x can be expressed in terms of every other function. One such case is shown in the next example.

■ *Example 2*

EXPRESSING ONE FUNCTION IN TERMS OF ANOTHER

Express $\cos x$ in terms of $\tan x$.

Since $\sec x$ is related to both $\cos x$ and $\tan x$ by identities, start with $\tan^2 x + 1 = \sec^2 x$. Then take reciprocals to get

$$\frac{1}{\tan^2 x + 1} = \frac{1}{\sec^2 x}$$

or

$$\frac{1}{\tan^2 x + 1} = \cos^2 x$$

$$\pm\sqrt{\frac{1}{\tan^2 x + 1}} = \cos x \qquad \text{Take the square root of both sides.}$$

$$\cos x = \frac{\pm 1}{\sqrt{\tan^2 x + 1}}.$$

Rationalize the denominator to get

$$\cos x = \frac{\pm\sqrt{\tan^2 x + 1}}{\tan^2 x + 1}.$$

Choose the $+$ sign or the $-$ sign, depending on the quadrant of x. ■

Each of $\tan \theta$, $\cot \theta$, $\sec \theta$, and $\csc \theta$ can easily be expressed in terms of $\sin \theta$ and/or $\cos \theta$. For this reason, we often make such substitutions in an expression so that the expression can be simplified. The next example shows such substitutions.

■ *Example 3*

SIMPLIFYING AN
EXPRESSION BY
WRITING IN TERMS
OF SINE AND
COSINE

Use the fundamental identities to write $\tan \theta + \cot \theta$ in terms of $\sin \theta$ and $\cos \theta$, and then simplify the expression.

From the fundamental identities,

$$\tan \theta + \cot \theta = \frac{\sin \theta}{\cos \theta} + \frac{\cos \theta}{\sin \theta}.$$

Simplify this expression by adding the two fractions on the right side, using the common denominator $\cos \theta \sin \theta$.

$$\tan \theta + \cot \theta = \frac{\sin^2 \theta}{\cos \theta \sin \theta} + \frac{\cos^2 \theta}{\cos \theta \sin \theta}$$

$$= \frac{\sin^2 \theta + \cos^2 \theta}{\cos \theta \sin \theta}$$

Now substitute 1 for $\sin^2 \theta + \cos^2 \theta$.

$$\tan \theta + \cot \theta = \frac{1}{\cos \theta \sin \theta} \quad ■$$

CAUTION

> When working with trigonometric expressions and identities, be sure to write the argument of the function. For example, we would *not* write $\sin^2 + \cos^2 = 1$; an argument such as θ is necessary in this identity.

Some problems in calculus are simplified by making an appropriate trigonometric substitution, as in the next example.

■ *Example 4*

MAKING A
TRIGONOMETRIC
SUBSTITUTION

Remove the radical in the expression $\sqrt{9 + x^2}$ by replacing x with $3 \tan \theta$, where θ is in the interval $(0, \pi/2)$.

Letting $x = 3 \tan \theta$ gives

$$\sqrt{9 + x^2} = \sqrt{9 + (3 \tan \theta)^2}$$

$$= \sqrt{9 + 9 \tan^2 \theta}$$

$$= \sqrt{9(1 + \tan^2 \theta)}$$

$$= 3\sqrt{1 + \tan^2 \theta}$$

$$= 3\sqrt{\sec^2 \theta}.$$

In the interval $(0, \pi/2)$, the value of $\sec \theta$ is positive, giving

$$\sqrt{9 + x^2} = 3 \sec \theta. \quad ■$$

The result of Example 4 could be written as

$$\sec \theta = \frac{\sqrt{9 + x^2}}{3}.$$

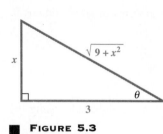

■ **FIGURE 5.3**

In a right triangle, sec θ is the ratio of the length of the hypotenuse to the side adjacent to the angle. This definition was used to label the right triangle in Figure 5.3, and then the Pythagorean theorem was used to find the length of the side opposite angle θ. From the right triangle in Figure 5.3,

$$\sin \theta = \frac{x}{\sqrt{9 + x^2}}, \qquad \cos \theta = \frac{3}{\sqrt{9 + x^2}},$$

and so on.

5.1 EXERCISES ■

Find sin s for each of Exercises 1–6. See Example 1.

1. cos s = 3/4, s in quadrant I

2. cot s = −1/3, s in quadrant IV

3. cos s = √5/5, tan s < 0

4. tan s = −√7/2, sec s > 0

5. sec s = 11/4, tan s < 0

6. csc s = −8/5

7. Why is it not necessary to give the quadrant of s in Exercise 6?

8. What is wrong with this problem? "Find sin s if csc s = −9/5 and s is in quadrant II."

9. Find tan θ if cos θ = −2/5, and sin θ < 0.

10. Find csc α if tan α = 6, and cos α > 0.

Use the fundamental identities to find the remaining five trigonometric functions of θ. See Example 1.

11. $\sin \theta = \frac{2}{3}$, θ in quadrant II

12. $\cos \theta = \frac{1}{5}$, θ in quadrant I

13. $\tan \theta = -\frac{1}{4}$, θ in quadrant IV

14. $\tan \theta = \frac{2}{3}$, θ in quadrant III

15. sec θ = −3, θ in quadrant II

16. $\csc \theta = -\frac{5}{2}$, θ in quadrant III

17. $\cot \theta = \frac{4}{3}$, sin θ > 0

18. $\sin \theta = -\frac{4}{5}$, cos θ < 0

19. $\sec \theta = \frac{4}{3}$, sin θ < 0

20. $\cos \theta = -\frac{1}{4}$, sin θ > 0

For each trigonometric expression in Column I, choose the expression from Column II that completes a fundamental identity.

Column I	Column II
21. $\dfrac{\cos x}{\sin x}$	**(a)** $\sin^2 x + \cos^2 x$
22. tan x	**(b)** cot x
23. cos (−x)	**(c)** $\sec^2 x$
24. $\tan^2 x + 1$	**(d)** $\dfrac{\sin x}{\cos x}$
25. 1	**(e)** cos x

For each expression in Column I, choose the expression from Column II that completes an identity. You will have to rewrite one or both expressions, using a fundamental identity, to recognize the matches.

Column I | Column II

26. $-\tan x \cos x$ **(a)** $\dfrac{\sin^2 x}{\cos^2 x}$

27. $\sec^2 x - 1$ **(b)** $\dfrac{1}{\sec^2 x}$

28. $\dfrac{\sec x}{\csc x}$ **(c)** $\sin(-x)$

29. $1 + \sin^2 x$ **(d)** $\csc^2 x - \cot^2 x + \sin^2 x$

30. $\cos^2 x$ **(e)** $\tan x$

31. A student writes " $1 + \cot^2 = \csc^2$." Comment on this student's work.

32. Another student makes the following claim: "Since $\sin^2 \theta + \cos^2 \theta = 1$, I should be able to also say $\sin \theta + \cos \theta = 1$ if I take the square root of both sides." Comment on this student's statement.

Complete this chart, so that each trigonometric function in the column at the left is expressed in terms of the functions given across the top. See Example 2.

	$\sin \theta$	$\cos \theta$	$\tan \theta$	$\cot \theta$	$\sec \theta$	$\csc \theta$
33. $\sin \theta$	$\sin \theta$	$\pm\sqrt{1 - \cos^2 \theta}$	$\dfrac{\pm\tan \theta \sqrt{1 + \tan^2 \theta}}{1 + \tan^2 \theta}$	$\dfrac{\pm\sqrt{1+\cot^2}}{1+\cot^2}$	$\dfrac{\pm\sqrt{\sec^2\theta+1}}{\sec}$	$\dfrac{1}{\csc \theta}$
34. $\cos \theta$	$\pm\sqrt{1-\sin^2}$	$\cos \theta$	$\dfrac{\pm\sqrt{\tan^2 \theta + 1}}{\tan^2 \theta + 1}$		$\dfrac{1}{\sec \theta}$	
35. $\tan \theta$			$\tan \theta$	$\dfrac{1}{\cot \theta}$		
36. $\cot \theta$			$\dfrac{1}{\tan \theta}$	$\cot \theta$	$\dfrac{\pm\sqrt{\sec^2 \theta - 1}}{\sec^2 \theta - 1}$	
37. $\sec \theta$		$\dfrac{1}{\cos \theta}$			$\sec \theta$	
38. $\csc \theta$	$\dfrac{1}{\sin \theta}$					$\csc \theta$

39. Suppose that $\cos \theta = x/(x + 1)$. Find $\sin \theta$.

40. Find $\tan \alpha$ if $\sec \alpha = (p + 4)/p$.

In each of the following, use the fundamental identities to get an equivalent expression involving only sines and cosines, and then simplify it. See Example 3.

41. $\csc^2 \beta - \cot^2 \beta$

42. $\dfrac{\tan(-\theta)}{\sec \theta}$

43. $\tan(-\alpha) \cos(-\alpha)$

44. $\cot^2 x(1 + \tan^2 x)$

45. $\tan^2 \theta - \dfrac{\sec^2 \theta}{\csc^2 \theta}$

46. $\dfrac{\tan x \csc x}{\sec x}$

47. $\sec \theta + \tan \theta$

48. $\dfrac{\sec \alpha}{\tan \alpha + \cot \alpha}$

49. $\sec^2 t - \tan^2 t$

50. $\csc^2 \gamma + \sec^2 \gamma$

51. $\cot^2 \beta - \csc^2 \beta$

52. $1 + \cot^2 \alpha$

53. $\dfrac{1 + \tan^2 \theta}{\cot^2 \theta}$

54. $\dfrac{1 - \sin^2 t}{\csc^2 t}$

55. $\cot^2 \beta \sin^2 \beta + \tan^2 \beta \cos^2 \beta$

56. $\sec^2 x + \cos^2 x$

57. $\dfrac{\cot^2 \alpha + \csc^2 \alpha}{\cos^2 \alpha}$

58. $1 - \tan^4 \theta$

59. $1 - \cot^4 s$

60. $\tan^4 \gamma - \cot^4 \gamma$

Use the indicated substitution to remove the radical in the given expression in Exercises 61–66. Assume θ is in the interval $(0, \pi/2)$. Then find the indicated functions. See Example 4.

61. $\sqrt{16 + 9x^2}$, let $x = \dfrac{4}{3} \tan \theta$;   find $\sin \theta$ and $\cos \theta$

62. $\sqrt{x^2 - 25}$, let $x = 5 \sec \theta$;   find $\sin \theta$ and $\tan \theta$

63. $\sqrt{(1 - x^2)^3}$, let $x = \cos \theta$;   find $\sin \theta$ and $\tan \theta$

64. $\dfrac{\sqrt{x^2 - 9}}{x}$, let $x = 3 \sec \theta$;   find $\sin \theta$ and $\tan \theta$

65. $x^2 \sqrt{1 + 16x^2}$, let $x = \dfrac{1}{4} \tan \theta$;   find $\sin \theta$ and $\cos \theta$

66. $x^2 \sqrt{9 + x^2}$, let $x = 3 \tan \theta$;   find $\sin \theta$ and $\cos \theta$

67. Let $\cos x = 1/5$. Find all possible values for $\dfrac{\sec x - \tan x}{\sin x}$.

68. Let $\csc x = -3$. Find all possible values for $\dfrac{\sin x + \cos x}{\sec x}$.

69. A function f is defined to be an **even function** if $f(-x) = f(x)$ for all x in the domain of f. A function f is defined to be an **odd function** if $f(-x) = -f(x)$ for all x in the domain of f. Use these definitions to identify each of the following trigonometric functions as even or odd. (*Hint:* Refer to the negative-angle identities.)
    (a) $y = \sin x$ **(b)** $y = \cos x$ **(c)** $y = \tan x$ **(d)** $y = \cot x$ **(e)** $y = \csc x$ **(f)** $y = \sec x$

In Chapter 4 we graphed functions of the form $y = c + a \cdot f[b(x - c)]$ with the assumption that $b > 0$. Using the negative-angle identities, we can graph functions of this form with $b < 0$ by making an appropriate transformation. For example, the graph of $y = \sin(-2x)$ is the same as that of $y = -\sin 2x$, and the graph of $y = \cos(-4x)$ is the same as that of $y = \cos 4x$. Graph one period for each of the following functions by first using a negative-angle identity to rewrite it in the form $y = a \sin bx$ or $y = a \cos bx$, where $b > 0$.

70. $y = \sin(-2x)$

71. $y = \cos(-4x)$

72. $y = 2 \cos(-2x)$

73. $y = 3 \sin(-4x)$

74. $y = -4 \sin(-3x)$

75. How does the graph of $y = \sec(-x)$ compare to that of $y = \sec x$?

76. How does the graph of $y = \csc(-x)$ compare to that of $y = \csc x$?

77. How does the graph of $y = \tan(-x)$ compare to that of $y = \tan x$?

78. How does the graph of $y = \cot(-x)$ compare to that of $y = \cot x$?

Show that each of the following is not an identity by replacing the variables with numbers that show the result to be false.

79. $(\sin s + \cos s)^2 = 1$

80. $(\tan s + 1)^2 = \sec^2 s$

81. $2 \sin s = \sin 2s$

82. $\sin x = \sqrt{1 - \cos^2 x}$

83. $\sin^3 x + \cos^3 x = 1$

84. $\sin x + \sin y = \sin(x + y)$

For students who have studied logarithms: Verify each of the following identities for first-quadrant values of s.

85. $\log \sin s = -\log \csc s$

86. $\log \tan s = \log \sin s - \log \cos s$

87. $\log \sec s = \dfrac{1}{2} \log(\tan^2 s + 1)$

88. $\log \csc s = -\log \sin s$

One of the skills required for more advanced work in mathematics (and especially in calculus) is the ability to use the trigonometric identities to write trigonometric expressions in alternate forms. This skill is developed by using the fundamental identities to verify that a trigonometric equation is an identity (for those values of the variable for which it is defined). Here are some hints that may help you get started.

VERIFYING IDENTITIES

1. Learn the fundamental identities given in the last section. Whenever you see either side of a fundamental identity, the other side should come to mind. Also, be aware of equivalent forms of the fundamental identities. For example $\sin^2 \theta = 1 - \cos^2 \theta$ is an alternate form of $\sin^2 \theta + \cos^2 \theta = 1$.

2. Try to rewrite the more complicated side of the equation so that it is identical to the simpler side.

3. It is often helpful to express all trigonometric functions in the equation in terms of sine and cosine and then simplify the result.

4. Usually any factoring or indicated algebraic operations should be performed. For example, the expression $\sin^2 x + 2 \sin x + 1$ can be factored as follows: $(\sin x + 1)^2$. The sum or difference of two trigonometric expressions, such as

$$\frac{1}{\sin \theta} + \frac{1}{\cos \theta},$$

can be added or subtracted in the same way as any other rational expressions:

$$\frac{1}{\sin \theta} + \frac{1}{\cos \theta} = \frac{\cos \theta}{\sin \theta \cos \theta} + \frac{\sin \theta}{\sin \theta \cos \theta}$$

$$= \frac{\cos \theta + \sin \theta}{\sin \theta \cos \theta}.$$

5. As you select substitutions, keep in mind the side you are not changing, because it represents your goal. For example, to verify the identity

$$\tan^2 x + 1 = \frac{1}{\cos^2 x},$$

try to think of an identity that relates $\tan x$ to $\cos x$. Here, since $\sec x = 1/\cos x$ and $\sec^2 x = \tan^2 x + 1$, the secant function is the best link between the two sides.

6. If an expression contains $1 + \sin x$, multiplying both numerator and denominator by $1 - \sin x$ would give $1 - \sin^2 x$, which could be replaced with $\cos^2 x$. Similar results for $1 - \sin x$, $1 + \cos x$, and $1 - \cos x$ may be useful.

These hints are used in the examples of this section.

CAUTION	Verifying identities is not the same as solving equations. Techniques used in solving equations, such as adding the same terms to both sides, or multiplying both sides by the same term, are not valid when working with identities since you are starting with a statement (to be verified) that may not be true.

■ *Example 1*

VERIFYING AN IDENTITY (WORKING WITH ONE SIDE)

Verify that

$$\cot s + 1 = \csc s(\cos s + \sin s)$$

is an identity.

Use the fundamental identities to rewrite one side of the equation so that it is identical to the other side. Since the right side is more complicated, it is probably a good idea to work with it. Here we use the method of changing all the trigonometric functions to sine or cosine.

Steps

Reasons

$$\csc s(\cos s + \sin s) = \frac{1}{\sin s}(\cos s + \sin s) \qquad \csc s = \frac{1}{\sin s}$$

$$= \frac{\cos s}{\sin s} + \frac{\sin s}{\sin s} \qquad \text{Distributive property}$$

$$= \cot s + 1 \qquad \frac{\cos s}{\sin s} = \cot s; \; \frac{\sin s}{\sin s} = 1$$

The given equation is an identity since the right side equals the left side. ■

■ *Example 2*

VERIFYING AN IDENTITY (WORKING WITH ONE SIDE)

Verify that

$$\tan^2 \alpha(1 + \cot^2 \alpha) = \frac{1}{1 - \sin^2 \alpha}$$

is an identity.

Working with the left side gives the following.

$$\tan^2 \alpha (1 + \cot^2 \alpha) = \tan^2 \alpha + \tan^2 \alpha \cot^2 \alpha \qquad \text{Distributive property}$$

$$= \tan^2 \alpha + \tan^2 \alpha \cdot \frac{1}{\tan^2 \alpha} \qquad \cot^2 \alpha = \frac{1}{\tan^2 \alpha}$$

$$= \tan^2 \alpha + 1 \qquad \tan^2 \alpha \cdot \frac{1}{\tan^2 \alpha} = 1$$

$$= \sec^2 \alpha \qquad \tan^2 \alpha + 1 = \sec^2 \alpha$$

$$= \frac{1}{\cos^2 \alpha} \qquad \sec^2 \alpha = \frac{1}{\cos^2 \alpha}$$

$$= \frac{1}{1 - \sin^2 \alpha} \qquad \cos^2 \alpha = 1 - \sin^2 \alpha$$

Since the left side equals the right side, the given equation is an identity. ■

■ *Example 3*

VERIFYING AN IDENTITY (WORKING WITH ONE SIDE)

Verify that

$$\frac{\tan t - \cot t}{\sin t \cos t} = \sec^2 t - \csc^2 t$$

is an identity.

Since the left side is the more complicated one, transform the left side to equal the right side.

$$\frac{\tan t - \cot t}{\sin t \cos t} = \frac{\tan t}{\sin t \cos t} - \frac{\cot t}{\sin t \cos t} \qquad \frac{a - b}{c} = \frac{a}{c} - \frac{b}{c}$$

$$= \tan t \cdot \frac{1}{\sin t \cos t} - \cot t \cdot \frac{1}{\sin t \cos t} \qquad \frac{a}{b} = a \cdot \frac{1}{b}$$

$$= \frac{\sin t}{\cos t} \cdot \frac{1}{\sin t \cos t} - \frac{\cos t}{\sin t} \cdot \frac{1}{\sin t \cos t} \qquad \tan t = \frac{\sin t}{\cos t}; \ \cot t = \frac{\cos t}{\sin t}$$

$$= \frac{1}{\cos^2 t} - \frac{1}{\sin^2 t}$$

$$= \sec^2 t - \csc^2 t \qquad \frac{1}{\cos^2 t} = \sec^2 t; \ \frac{1}{\sin^2 t} = \csc^2 t$$

Here, writing in terms of sine and cosine only was used in the third line. ■

■ *Example 4*

VERIFYING AN IDENTITY (WORKING WITH ONE SIDE)

Verify that

$$\frac{\cos x}{1 - \sin x} = \frac{1 + \sin x}{\cos x}$$

is an identity.

This time we will work on the right side. Use the suggestion given at the beginning of the section to multiply the numerator and denominator on the right by $1 - \sin x$.

$$\frac{1 + \sin x}{\cos x} = \frac{(1 + \sin x)(1 - \sin x)}{\cos x (1 - \sin x)} \qquad \text{Multiply by 1.}$$

$$= \frac{1 - \sin^2 x}{\cos x (1 - \sin x)}$$

$$= \frac{\cos^2 x}{\cos x (1 - \sin x)} \qquad 1 - \sin^2 x = \cos^2 x$$

$$= \frac{\cos x}{1 - \sin x} \qquad \text{Reduce to lowest terms.} \ ■$$

If both sides of an identity appear to be equally complex, the identity can be verified by working independently on the left side and on the right side, until each side is changed into some common third result. *Each step, on each side, must be reversible.* With all steps reversible, the procedure is as follows.

$$\text{left} \quad = \quad \text{right}$$

common third
expression

The left side leads to the third expression, which leads back to the right side. This procedure is just a shortcut for the procedure used in the first examples of this section: the left side is changed into the right side, but by going through an intermediate step.

■ *Example 5*

VERIFYING AN
IDENTITY (WORKING
WITH BOTH SIDES)

Verify that

$$\frac{\sec \alpha + \tan \alpha}{\sec \alpha - \tan \alpha} = \frac{1 + 2 \sin \alpha + \sin^2 \alpha}{\cos^2 \alpha}$$

is an identity.

Both sides appear equally complex, so verify the identity by changing each side into a common third expression. Work first on the left, multiplying numerator and denominator by $\cos \alpha$.

$$\frac{\sec \alpha + \tan \alpha}{\sec \alpha - \tan \alpha} = \frac{(\sec \alpha + \tan \alpha)\cos \alpha}{(\sec \alpha - \tan \alpha)\cos \alpha}$$

$\dfrac{\cos \alpha}{\cos \alpha} = 1;$ multiplicative identity

$$= \frac{\sec \alpha \cos \alpha + \tan \alpha \cos \alpha}{\sec \alpha \cos \alpha - \tan \alpha \cos \alpha}$$

Distributive property

$$= \frac{1 + \tan \alpha \cos \alpha}{1 - \tan \alpha \cos \alpha}$$

$\sec \alpha \cos \alpha = 1$

$$= \frac{1 + \dfrac{\sin \alpha}{\cos \alpha} \cdot \cos \alpha}{1 - \dfrac{\sin \alpha}{\cos \alpha} \cdot \cos \alpha}$$

$\tan \alpha = \dfrac{\sin \alpha}{\cos \alpha}$

$$= \frac{1 + \sin \alpha}{1 - \sin \alpha}$$

On the right side of the original statement, begin by factoring.

$$\frac{1 + 2\sin\alpha + \sin^2\alpha}{\cos^2\alpha} = \frac{(1 + \sin\alpha)^2}{\cos^2\alpha} \qquad a^2 + 2ab + b^2 = (a + b)^2$$

$$= \frac{(1 + \sin\alpha)^2}{1 - \sin^2\alpha} \qquad \cos^2\alpha = 1 - \sin^2\alpha$$

$$= \frac{(1 + \sin\alpha)^2}{(1 + \sin\alpha)(1 - \sin\alpha)} \qquad 1 - \sin^2\alpha = (1 + \sin\alpha)(1 - \sin\alpha)$$

$$= \frac{1 + \sin\alpha}{1 - \sin\alpha} \qquad \text{Reduce to lowest terms.}$$

We now have shown that

$$\frac{\sec\alpha + \tan\alpha}{\sec\alpha - \tan\alpha} = \frac{1 + \sin\alpha}{1 - \sin\alpha} = \frac{1 + 2\sin\alpha + \sin^2\alpha}{\cos^2\alpha},$$

verifying that the original equation is an identity. ■

NOTE There are usually several ways to verify a given identity. You may wish to go through the examples of this section and verify each using a method different from the one given.

5.2 EXERCISES ■

For each of the following, perform the indicated operations and simplify the result.

1. $\tan\theta + \dfrac{1}{\tan\theta}$

2. $\dfrac{\cos x}{\sin x} + \dfrac{\sin x}{\cos x}$

3. $\cot s(\tan s + \sin s)$

4. $\sec\beta(\cos\beta + \sin\beta)$

5. $\dfrac{1}{\csc^2\theta} + \dfrac{1}{\sec^2\theta}$

6. $\dfrac{1}{\sin\alpha - 1} - \dfrac{1}{\sin\alpha + 1}$

7. $\dfrac{\cos x}{\sec x} + \dfrac{\sin x}{\csc x}$

8. $\dfrac{\cos\gamma}{\sin\gamma} + \dfrac{\sin\gamma}{1 + \cos\gamma}$

9. $(1 + \sin t)^2 + \cos^2 t$

10. $(1 + \tan s)^2 - 2\tan s$

11. $\dfrac{1}{1 + \cos x} - \dfrac{1}{1 - \cos x}$

12. $(\sin\alpha - \cos\alpha)^2$

Factor each of the following trigonometric expressions.

13. $\sin^2\gamma - 1$

14. $\sec^2\theta - 1$

15. $(\sin x + 1)^2 - (\sin x - 1)^2$

16. $(\tan x + \cot x)^2 - (\tan x - \cot x)^2$

17. $2\sin^2 x + 3\sin x + 1$

18. $4\tan^2\beta + \tan\beta - 3$

19. $4\sec^2 x + 3\sec x - 1$

20. $2\csc^2 x + 7\csc x - 30$

21. $\cos^4 x + 2\cos^2 x + 1$

22. $\cot^4 x + 3\cot^2 x + 2$

23. $\sin^3 x - \cos^3 x$

24. $\sin^3\alpha + \cos^3\alpha$

Use the fundamental identities to simplify each of the given expressions.

25. $\tan \theta \cos \theta$ **26.** $\cot \alpha \sin \alpha$ **27.** $\sec r \cos r$ **28.** $\cot t \tan t$

29. $\dfrac{\sin \beta \tan \beta}{\cos \beta}$ **30.** $\dfrac{\csc \theta \sec \theta}{\cot \theta}$ **31.** $\sec^2 x - 1$ **32.** $\csc^2 t - 1$

33. $\dfrac{\sin^2 x}{\cos^2 x} + \sin x \csc x$ **34.** $\dfrac{1}{\tan^2 \alpha} + \cot \alpha \tan \alpha$

Given a trigonometric identity such as those discussed in this section, it is possible to form another identity by replacing each function with its cofunction. Therefore, for example, since the equation

$$\frac{\cot \theta}{\csc \theta} = \cos \theta,$$

in Exercise 35, represents an identity, the equation

$$\frac{\tan \theta}{\sec \theta} = \sin \theta$$

is also an identity. After verifying the identities in Exercises 35–80, you may wish to construct other identities using this method, in order to provide more practice exercises for yourself.

Verify each of the following trigonometric identities. See Examples 1–5.

35. $\dfrac{\cot \theta}{\csc \theta} = \cos \theta$ **36.** $\dfrac{\tan \alpha}{\sec \alpha} = \sin \alpha$ **37.** $\dfrac{1 - \sin^2 \beta}{\cos \beta} = \cos \beta$

38. $\dfrac{\tan^2 \gamma + 1}{\sec \gamma} = \sec \gamma$ **39.** $\cos^2 \theta(\tan^2 \theta + 1) = 1$ **40.** $\sin^2 \beta(1 + \cot^2 \beta) = 1$

41. $\sin^2 \alpha + \tan^2 \alpha + \cos^2 \alpha = \sec^2 \alpha$ **42.** $\cot s + \tan s = \sec s \csc s$

43. $\dfrac{\sin^2 \gamma}{\cos \gamma} = \sec \gamma - \cos \gamma$ **44.** $\dfrac{\cos \alpha}{\sec \alpha} + \dfrac{\sin \alpha}{\csc \alpha} = \sec^2 \alpha - \tan^2 \alpha$

45. $\dfrac{\cos \theta}{\sin \theta \cot \theta} = 1$ **46.** $\sin^4 \theta - \cos^4 \theta = 2 \sin^2 \theta - 1$

47. $\tan^2 \gamma \sin^2 \gamma = \tan^2 \gamma + \cos^2 \gamma - 1$ **48.** $(1 - \cos^2 \alpha)(1 + \cos^2 \alpha) = 2 \sin^2 \alpha - \sin^4 \alpha$

49. $\dfrac{(\sec \theta - \tan \theta)^2 + 1}{\sec \theta \csc \theta - \tan \theta \csc \theta} = 2 \tan \theta$ **50.** $\dfrac{\cos \theta + 1}{\tan^2 \theta} = \dfrac{\cos \theta}{\sec \theta - 1}$

51. $\dfrac{1}{\sec \alpha - \tan \alpha} = \sec \alpha + \tan \alpha$ **52.** $\dfrac{1}{1 - \sin \theta} + \dfrac{1}{1 + \sin \theta} = 2 \sec^2 \theta$

53. $\dfrac{1 - \cos x}{1 + \cos x} = (\cot x - \csc x)^2$ **54.** $\dfrac{\tan s}{1 + \cos s} + \dfrac{\sin s}{1 - \cos s} = \cot s + \sec s \csc s$

55. $\dfrac{1}{\tan \alpha - \sec \alpha} + \dfrac{1}{\tan \alpha + \sec \alpha} = -2 \tan \alpha$ **56.** $\dfrac{\cot \alpha + 1}{\cot \alpha - 1} = \dfrac{1 + \tan \alpha}{1 - \tan \alpha}$

57. $\dfrac{\csc \theta + \cot \theta}{\tan \theta + \sin \theta} = \cot \theta \csc \theta$ **58.** $\sin^2 \alpha \sec^2 \alpha + \sin^2 \alpha \csc^2 \alpha = \sec^2 \alpha$

59. $\sec^4 x - \sec^2 x = \tan^4 x + \tan^2 x$

60. $\dfrac{1 - \sin \theta}{1 + \sin \theta} = \sec^2 \theta - 2 \sec \theta \tan \theta + \tan^2 \theta$

61. $\sin \theta + \cos \theta = \dfrac{\sin \theta}{1 - \dfrac{\cos \theta}{\sin \theta}} + \dfrac{\cos \theta}{1 - \dfrac{\sin \theta}{\cos \theta}}$

62. $\dfrac{\sin \theta}{1 - \cos \theta} - \dfrac{\sin \theta \cos \theta}{1 + \cos \theta} = \csc \theta (1 + \cos^2 \theta)$

63. $\dfrac{\sec^4 s - \tan^4 s}{\sec^2 s + \tan^2 s} = \sec^2 s - \tan^2 s$

64. $\dfrac{\cot^2 t - 1}{1 + \cot^2 t} = 1 - 2 \sin^2 t$

65. $\dfrac{\tan^2 t - 1}{\sec^2 t} = \dfrac{\tan t - \cot t}{\tan t + \cot t}$

66. $(1 + \sin x + \cos x)^2 = 2(1 + \sin x)(1 + \cos x)$

67. $(\sin s + \cos s)^2 \cdot \csc s = 2 \cos s + \dfrac{1}{\sin s}$

68. $\dfrac{\sin^3 t - \cos^3 t}{\sin t - \cos t} = 1 + \sin t \cos t$

69. $\dfrac{1 + \cos x}{1 - \cos x} - \dfrac{1 - \cos x}{1 + \cos x} = 4 \cot x \csc x$

70. $(\sec \alpha - \tan \alpha)^2 = \dfrac{1 - \sin \alpha}{1 + \sin \alpha}$

71. $(\sec \alpha + \csc \alpha)(\cos \alpha - \sin \alpha) = \cot \alpha - \tan \alpha$

72. $\dfrac{\sin^4 \alpha - \cos^4 \alpha}{\sin^2 \alpha - \cos^2 \alpha} = 1$

73. $\dfrac{\cot^2 x + \sec^2 x + 1}{\cot^2 x} = \sec^4 x$

74. $\dfrac{\cos x - (\sin x - 1)}{\cos x + (\sin x - 1)} = \dfrac{\sin x}{1 - \cos x}$ [*Hint:* Multiply numerator and denominator on the left by $\cos x - (\sin x - 1)$.]

75. $\dfrac{\cot \theta}{1 + \csc \theta} + \dfrac{1 + \csc \theta}{\cot \theta} = 2 \sec \theta$

76. $\dfrac{\cot \theta}{1 - \tan \theta} + \dfrac{\tan \theta}{1 - \cot \theta} - 1 = \sec \theta \csc \theta$

77. $\dfrac{\sec x}{1 + \tan x} = \dfrac{\csc x}{1 + \cot x}$

78. $\sin \beta (1 + \tan \beta) + \cos \beta (1 + \cot \beta) = \sec \beta + \csc \beta$

79. $\dfrac{1 + \tan x}{1 - \tan x} = \dfrac{\sec^2 x + 2 \tan x}{2 - \sec^2 x}$

80. $\dfrac{1}{\cos^2 t} + \dfrac{1}{\sin^2 t} = \dfrac{1}{\cos^2 t - \cos^4 t}$

81. A student claims that the equation

$$\cos \theta + \sin \theta = 1$$

is an identity, since by letting $\theta = 90°$ (or $\pi/2$ radians) we get $0 + 1 = 1$, a true statement. Comment on this student's reasoning.

82. Explain why the method described in the text involving working on both sides of an identity to show that each side is equal to the same expression is a valid method of verifying an identity. When using this method, what must be true about each step taken? (*Hint:* See the discussion preceding Example 5.)

Given a complicated equation involving trigonometric functions, it is a good idea to decide whether it really is an identity before trying to prove that it is. Substitute $s = 1$ and $s = 2$ into each of the following (with the calculator set for radian measure). If you get the same results on both sides of the equation, it may be an identity. Then prove that it is.

83. $\dfrac{2 + 5 \cos s}{\sin s} = 2 \csc s + 5 \cot s$

84. $1 + \cot^2 s = \dfrac{\sec^2 s}{\sec^2 s - 1}$

85. $\dfrac{\tan s - \cot s}{\tan s + \cot s} = 2 \sin^2 s$

86. $\dfrac{1}{1 + \sin s} + \dfrac{1}{1 - \sin s} = \sec^2 s$

87. $\dfrac{1 - \tan^2 s}{1 + \tan^2 s} = \cos^2 s - \sin s$

88. $\dfrac{\sin^3 s - \cos^3 s}{\sin s - \cos s} = \sin^2 s + 2 \sin s \cos s + \cos^2 s$

89. $\sin^2 s + \cos^2 s = \dfrac{1}{2}(1 - \cos 4s)$

90. $\cos 3s = 3 \cos s + 4 \cos^3 s$

Show that the following are not identities for all real numbers s and t.

91. $\sin (\csc s) = 1$ **92.** $\sqrt{\cos^2 s} = \cos s$ **93.** $\csc t = \sqrt{1 + \cot^2 t}$ **94.** $\sin t = \sqrt{1 - \cos^2 t}$

95. Let $\tan \theta = t$ and show that

$$\sin \theta \cos \theta = \frac{t}{t^2 + 1}.$$

96. When does $\sin x = \sqrt{1 - \cos^2 x}$?

For students who have studied natural logarithms: Verify each identity.

97. $\ln e^{|\sin x|} = |\sin x|$ **98.** $\ln |\tan x| = -\ln |\cot x|$

99. $\ln |\sec x - \tan x| = -\ln |\sec x + \tan x|$ **100.** $-\ln |\csc t - \cot t| = \ln |\csc t + \cot t|$

5.3 —————— SUM AND DIFFERENCE IDENTITIES FOR COSINE

Several examples presented throughout this book should have convinced you by now that $\cos (A - B)$ does *not* equal $\cos A - \cos B$. For example, if $A = \pi/2$ and $B = 0$,

$$\cos (A - B) = \cos \left(\frac{\pi}{2} - 0 \right) = \cos \frac{\pi}{2} = 0,$$

while

$$\cos A - \cos B = \cos \frac{\pi}{2} - \cos 0 = 0 - 1 = -1.$$

The actual formula for $\cos (A - B)$ is derived in this section. Start by locating angles A and B in standard position on a unit circle, with $B < A$. Let S and Q be the points where angles A and B, respectively, intersect the circle. Locate point R on the unit circle so that angle POR equals the difference $A - B$. See Figure 5.4.

Point Q is on the unit circle, so by the work with circular functions in Chapter 3, the x-coordinate of Q is given by the cosine of angle B, while the y-coordinate of Q is given by the sine of angle B:

Q has coordinates $(\cos B, \sin B)$.

In the same way,

S has coordinates $(\cos A, \sin A)$,

and

R has coordinates $(\cos (A - B), \sin (A - B))$.

Angle SOQ also equals $A - B$. Since the central angles SOQ and POR are equal, chords PR and SQ are equal. By the distance formula, since $PR = SQ$,

$$\sqrt{[\cos (A - B) - 1]^2 + [\sin (A - B) - 0]^2}$$

$$= \sqrt{(\cos A - \cos B)^2 + (\sin A - \sin B)^2}.$$

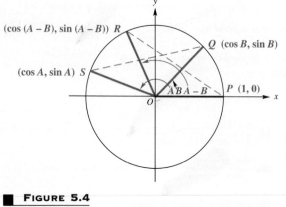

■ **FIGURE 5.4**

Squaring both sides and clearing parentheses gives

$$\cos^2 (A - B) - 2 \cos (A - B) + 1 + \sin^2 (A - B)$$
$$= \cos^2 A - 2 \cos A \cos B + \cos^2 B + \sin^2 A - 2 \sin A \sin B + \sin^2 B.$$

Since $\sin^2 x + \cos^2 x = 1$ for any value of x, rewrite the equation as

$$2 - 2 \cos (A - B) = 2 - 2 \cos A \cos B - 2 \sin A \sin B$$
$$\cos (A - B) = \cos A \cos B + \sin A \sin B.$$

This is the identity for $\cos (A - B)$. Although Figure 5.4 shows angles A and B in the second and first quadrants, respectively, it can be shown that this result is the same for any values of these angles.

To find a similar expression for $\cos (A + B)$, rewrite $A + B$ as $A - (-B)$ and use the identity for $\cos (A - B)$ found above, along with the fact that $\cos (-B) = \cos B$ and $\sin (-B) = -\sin B$.

$$\cos (A + B) = \cos [A - (-B)]$$
$$= \cos A \cos (-B) + \sin A \sin (-B)$$
$$= \cos A \cos B + \sin A (-\sin B)$$
$$\cos (A + B) = \cos A \cos B - \sin A \sin B$$

The two formulas we have just derived are summarized as follows.

COSINE OF SUM OR DIFFERENCE	$\cos (A - B) = \cos A \cos B + \sin A \sin B$ $\cos (A + B) = \cos A \cos B - \sin A \sin B$

These identities are important in calculus and other areas of mathematics and useful in certain applications. Although a calculator can be used to find an approximation for $\cos 15°$, for example, the method shown below can be applied to give practice using the sum and difference identities, as well as to get an exact value.

■ *Example 1*

USING THE COSINE
SUM AND
DIFFERENCE
IDENTITIES TO FIND
EXACT VALUES

Find the *exact* value of the following.

(a) $\cos 15°$

To find $\cos 15°$, write $15°$ as the sum or difference of two angles with known function values. Since we know the trigonometric function values of both $45°$ and $30°$, write $15°$ as $45° - 30°$. (We could also use $60° - 45°$.) Then use the identity for the cosine of the difference of two angles.

$$\cos 15° = \cos (45° - 30°)$$

$$= \cos 45° \cos 30° + \sin 45° \sin 30° \qquad \text{Use the cosine of the difference identity.}$$

$$= \frac{\sqrt{2}}{2} \cdot \frac{\sqrt{3}}{2} + \frac{\sqrt{2}}{2} \cdot \frac{1}{2}$$

$$= \frac{\sqrt{6} + \sqrt{2}}{4}$$

(b) $\cos \dfrac{5}{12}\pi = \cos\left(\dfrac{\pi}{6} + \dfrac{\pi}{4}\right)$

$\quad\quad\quad\quad = \cos\dfrac{\pi}{6}\cos\dfrac{\pi}{4} - \sin\dfrac{\pi}{6}\sin\dfrac{\pi}{4}$ Use the cosine of the sum identity.

$\quad\quad\quad\quad = \dfrac{\sqrt{3}}{2}\cdot\dfrac{\sqrt{2}}{2} - \dfrac{1}{2}\cdot\dfrac{\sqrt{2}}{2}$

$\quad\quad\quad\quad = \dfrac{\sqrt{6} - \sqrt{2}}{4}$

(c) $\cos 87° \cos 93° - \sin 87° \sin 93°$

From the identity for the cosine of the sum of two angles,

$$\cos 87° \cos 93° - \sin 87° \sin 93° = \cos(87° + 93°)$$
$$= \cos(180°)$$
$$= -1. \quad ■$$

NOTE

In Example 1(b) we used the fact that $5\pi/12 = \pi/6 + \pi/4$. At first glance, this sum may not be obvious; but think of the values $\pi/6$ and $\pi/4$ in terms of fractions with denominator 12: $\pi/6 = 2\pi/12$ and $\pi/4 = 3\pi/12$. The list below may help you with problems of this type.

$$\dfrac{\pi}{3} = \dfrac{4\pi}{12}$$

$$\dfrac{\pi}{4} = \dfrac{3\pi}{12}$$

$$\dfrac{\pi}{6} = \dfrac{2\pi}{12}$$

Using this list, for example, we see that $\pi/12 = \pi/3 - \pi/4$ (or $\pi/4 - \pi/6$).

The identities for the cosine of the sum and difference of two angles can be used to derive other identities. Recall the *cofunction identities,* which were presented earlier for values of θ in the interval $[0°, 90°]$.

COFUNCTION IDENTITIES

$\cos(90° - \theta) = \sin\theta$	$\cot(90° - \theta) = \tan\theta$	
$\sin(90° - \theta) = \cos\theta$	$\sec(90° - \theta) = \csc\theta$	
$\tan(90° - \theta) = \cot\theta$	$\csc(90° - \theta) = \sec\theta$	

Similar identities can be obtained for a real number domain by replacing $90°$ by $\pi/2$.

These identities now can be generalized for any angle θ, not just those between $0°$ and $90°$. For example, substituting $90°$ for A and θ for B in the identity given above for $\cos (A - B)$ gives

$$\cos (90° - \theta) = \cos 90° \cos \theta + \sin 90° \sin \theta$$
$$= 0 \cdot \cos \theta + 1 \cdot \sin \theta$$
$$= \sin \theta.$$

This result is true for *any* value of θ since the identity for $\cos (A - B)$ is true for any values of A and B. For the derivations of other cofunction identities, see Exercises 79 and 80.

■ *Example 2*

USING THE
COFUNCTION
IDENTITIES

Find an angle θ that satisfies each of the following.

(a) $\cot \theta = \tan 25°$

Since tangent and cotangent are cofunctions,

$$\cot \theta = \tan (90° - \theta).$$

This means that

$$\tan (90° - \theta) = \tan 25°,$$

or

$$90° - \theta = 25°$$
$$\theta = 65°.$$

(b) $\sin \theta = \cos (-30°)$

In the same way,

$$\sin \theta = \cos (90° - \theta) = \cos (-30°),$$

giving

$$90° - \theta = -30°$$
$$\theta = 120°.$$

(c) $\csc \dfrac{3\pi}{4} = \sec \theta$

Cosecant and secant are cofunctions, so

$$\csc \frac{3\pi}{4} = \sec \left(\frac{\pi}{2} - \frac{3\pi}{4} \right) = \sec \theta$$

$$\sec \left(-\frac{\pi}{4} \right) = \sec \theta$$

$$-\frac{\pi}{4} = \theta. \quad ■$$

NOTE Because trigonometric (and circular) functions are periodic, the solutions in Example 2 are not unique. In each case, we give only one of infinitely many possibilities.

If one of the angles A or B in the identities for cos $(A + B)$ and cos $(A - B)$ is a quadrantal angle, then the identity allows us to write the expression in terms of a single function of A or B. The next example illustrates this.

■ *Example 3*

REDUCING
cos $(A - B)$ TO A
FUNCTION OF A
SINGLE VARIABLE

Write cos $(180° - \theta)$ as a trigonometric function of θ.
Use the difference identity. Replace A with $180°$ and B with θ.

$$\cos (180° - \theta) = \cos 180° \cos \theta + \sin 180° \sin \theta$$
$$= (-1) \cos \theta + (0) \sin \theta$$
$$= -\cos \theta \quad ■$$

■ *Example 4*

FINDING cos $(x + y)$
GIVEN INFORMATION
ABOUT x AND y

Suppose that sin $x = 3/5$, cos $y = -12/13$, and both x and y are in quadrant II. Find cos $(x + y)$.
By the identity above, cos $(x + y) = \cos x \cos y - \sin x \sin y$. The values of sin x and cos y are given, so that cos $(x + y)$ can be found if cos x and sin y are known. To find cos x, use the identity $\sin^2 x + \cos^2 x = 1$, and substitute 3/5 for sin x.

$$\sin^2 x + \cos^2 x = 1 \qquad \text{Pythagorean identity}$$
$$\left(\frac{3}{5}\right)^2 + \cos^2 x = 1 \qquad \sin x = \frac{3}{5}$$
$$\frac{9}{25} + \cos^2 x = 1$$
$$\cos^2 x = \frac{16}{25}$$
$$\cos x = \pm\frac{4}{5}$$

Since x is in quadrant II, cos x is negative, so

$$\cos x = -\frac{4}{5}.$$

Find sin y as follows.

$$\sin^2 y + \cos^2 y = 1 \qquad \text{Pythagorean identity}$$
$$\sin^2 y + \left(-\frac{12}{13}\right)^2 = 1 \qquad \cos y = -\frac{12}{13}$$
$$\sin^2 y + \frac{144}{169} = 1$$
$$\sin y = \pm\frac{5}{13}$$

Since y is in quadrant II, choose the positive value to get

$$\sin y = \frac{5}{13}.$$

Now find cos $(x + y)$.

$$\cos (x + y) = \cos x \cos y - \sin x \sin y$$

$$= -\frac{4}{5} \cdot \left(-\frac{12}{13}\right) - \frac{3}{5} \cdot \frac{5}{13}$$

$$= \frac{48}{65} - \frac{15}{65}$$

$$= \frac{33}{65} \quad \blacksquare$$

| NOTE | In Example 4, the values of cos x and sin y could also be found by sketching right triangles in quadrant II, labeling the known sides, and using the Pythagorean theorem to find the unknown sides. The problem could then be solved using the identity for cos $(x + y)$ in the same way as shown in the example. |

5.3 EXERCISES ■

Use the sum and difference identities for cosine to find the exact value of each of the following.
(Do not use a calculator.) See Example 1.

1. cos 75°

2. cos (−15°)

3. cos (−75°)

4. cos (105°) (*Hint:* 105° = 60° + 45°)

5. cos (−105°) [*Hint:* −105° = −60° + (−45°)]

6. cos (7π/12)

7. cos (−π/12)

8. cos (−5π/12)

9. cos 40° cos 50° − sin 40° sin 50°

10. cos (−10°) cos 35° + sin (−10°) sin 35°

11. cos 2π/5 cos π/10 − sin 2π/5 sin π/10

12. cos 7π/9 cos 2π/9 − sin 7π/9 sin 2π/9

Write each of the following in terms of the cofunction of a complementary angle. See Example 2.

13. tan 87°

14. sin 15°

15. cos π/12

16. sin 2π/5

17. csc (−14° 24′)

18. sin 142° 14′

19. sin 5π/8

20. cot 9π/10

21. sec 146° 42′

22. tan 174° 3′

23. cot 176.9814°

24. sin 98.0142°

Use the cofunction identities to fill in each of the following blanks with the appropriate trigonometric function name. See Example 2.

25. cot $\frac{\pi}{3}$ = _____ $\frac{\pi}{6}$

26. sin $\frac{2\pi}{3}$ = _____ $\left(-\frac{\pi}{6}\right)$

27. _____ 33° = sin 57°

28. _____ 72° = cot 18°

29. cos 70° = $\dfrac{1}{\text{_____ } 20°}$

30. tan 24° = $\dfrac{1}{\text{_____ } 66°}$

Use the cofunction identities to find an angle θ that makes each of the following true. See Example 2.

31. tan θ = cot (45° + 2θ)

32. sin θ = cos (2θ − 10°)

33. sec θ = csc (θ/2 + 20°)

34. cos θ = sin (3θ + 10°)

35. sin (3θ − 15°) = cos (θ + 25°)

36. cot (θ − 10°) = tan (2θ + 20°)

Use the identities for the cosine of a sum or a difference to reduce each expression to a single function of θ. See Example 3.

37. $\cos (0° - \theta)$ **38.** $\cos (90° - \theta)$ **39.** $\cos (180° - \theta)$ **40.** $\cos (270° - \theta)$

41. $\cos (0° + \theta)$ **42.** $\cos (90° + \theta)$ **43.** $\cos (180° + \theta)$ **44.** $\cos (270° + \theta)$

For each of the following, find $\cos (s + t)$ and $\cos (s - t)$. See Example 4.

45. $\cos s = -1/5$ and $\sin t = 3/5$, s and t in quadrant II

46. $\sin s = 2/3$ and $\sin t = -1/3$, s in quadrant II and t in quadrant IV

47. $\sin s = 3/5$ and $\sin t = -12/13$, s in quadrant I and t in quadrant III

48. $\cos s = -8/17$ and $\cos t = -3/5$, s and t in quadrant III

49. $\cos s = -15/17$ and $\sin t = 4/5$, s in quadrant II and t in quadrant I

50. $\sin s = -8/17$ and $\cos t = -8/17$, s and t in quadrant III

51. $\sin s = \sqrt{5}/7$ and $\sin t = \sqrt{6}/8$, s and t in quadrant I

52. $\cos s = \sqrt{2}/4$ and $\sin t = -\sqrt{5}/6$, s and t in quadrant IV

Tell whether each of the following is true or false.

53. $\cos 42° = \cos (30° + 12°)$

54. $\cos (-24°) = \cos 16° - \cos 40°$

55. $\cos 74° = \cos 60° \cos 14° + \sin 60° \sin 14°$

56. $\cos 140° = \cos 60° \cos 80° - \sin 60° \sin 80°$

57. $\cos \pi/3 = \cos \pi/12 \cos \pi/4 - \sin \pi/12 \sin \pi/4$

58. $\cos 2\pi/3 = \cos 11\pi/12 \cos \pi/4 + \sin 11\pi/12 \sin \pi/4$

59. $\cos 70° \cos 20° - \sin 70° \sin 20° = 0$

60. $\cos 85° \cos 40° + \sin 85° \sin 40° = \sqrt{2}/2$

Verify each of the following identities.

61. $\cos (\pi/2 + x) = -\sin x$

62. $\sec (\pi - x) = -\sec x$

63. $\cos 2x = \cos^2 x - \sin^2 x$ [*Hint:* $\cos 2x = \cos (x + x)$.]

64. $\cos (x + y) + \cos (x - y) = 2 \cos x \cos y$

65. $\dfrac{\cos (\alpha - \theta) - \cos (\alpha + \theta)}{\cos (\alpha - \theta) + \cos (\alpha + \theta)} = \tan \theta \tan \alpha$

66. $1 + \cos 2x - \cos^2 x = \cos^2 x$ (*Hint:* Use the result in Exercise 63.)

67. $\cos (\pi + s - t) = -\sin s \sin t - \cos s \cos t$

68. $\cos (\pi/2 + s - t) = \sin (t - s)$

69. $\cos (\alpha + \beta) \cos (\alpha - \beta) = 1 - \sin^2 \alpha - \sin^2 \beta$

70. $\cos 4x \cos 7x - \sin 4x \sin 7x = \cos 11x$

71. Suppose a fellow student tells you that the cosine of the sum of two angles is the sum of their cosines. Write in your own words how you would correct this student's statement.

72. By a cofunction identity, $\cos 20° = \sin 70°$. What are some values other than $70°$ that make $\cos 20° = \sin \theta$ a true statement?

The identities for $\cos (A + B)$ and $\cos (A - B)$ can be used to find exact values of expressions like $\cos 195°$ and $\cos 255°$, where the angle is not in the first quadrant. For example, to find $\cos 195°$, first write $195°$ as $180° + 15°$ (a quadrantal angle plus an acute angle). Then use the identity for $\cos (A + B)$:

$$\cos 195° = \cos (180° + 15°)$$
$$= \cos 180° \cos 15° - \sin 180° \sin 15°$$
$$= (-1) \cos 15° - 0 \sin 15°$$
$$= -\cos 15°.$$

Then, using the identity for cos (A − B) and writing 15° as 45° − 30°, we can complete the work.

$$\cos 195° = -\cos 15° = -\cos (45° - 30°)$$
$$= -[\cos 45° \cos 30° + \sin 45° \sin 30°]$$
$$= -\left[\frac{\sqrt{2}}{2} \cdot \frac{\sqrt{3}}{2} + \frac{\sqrt{2}}{2} \cdot \frac{1}{2}\right]$$
$$= \frac{-\sqrt{6} - \sqrt{2}}{4}$$

Use this method to find the exact value of each expression in Exercises 73–78.

73. cos 255°   **74.** cos 165°   **75.** cos 285°   **76.** cos 345°   **77.** cos $11\pi/12$ **78.** cos $13\pi/12$

79. Use the identity cos $(90° - \theta) = \sin \theta$, and replace θ with $90° - A$, to derive the identity cos $A = \sin (90° - A)$.

80. Use the results of Exercise 79 to derive the identity tan $A = \cot (90° - A)$.

81. Let $f(x) = \cos x$. Prove that $\dfrac{f(x + h) - f(x)}{h} = \cos x\left(\dfrac{\cos h - 1}{h}\right) - \sin x\left(\dfrac{\sin h}{h}\right)$.

82. Without using a calculator, show that cos 140° + cos 100° + cos 20° = 0.

5.4 —— SUM AND DIFFERENCE IDENTITIES FOR SINE AND TANGENT

Formulas for sin $(A + B)$ and sin $(A - B)$ can be developed from the results in Section 5.3. Start with the cofunction relationship

$$\sin \theta = \cos (90° - \theta).$$

Replace θ with $A + B$.

$$\sin (A + B) = \cos [90° - (A + B)]$$
$$= \cos [(90° - A) - B]$$

Using the formula for cos $(A - B)$ from the previous section gives

$$\sin (A + B) = \cos (90° - A) \cos B + \sin (90° - A) \sin B$$

or $$\sin (A + B) = \sin A \cos B + \cos A \sin B.$$

(The cofunction relationships were used in the last step.)

Now write sin $(A - B)$ as sin $[A + (-B)]$ and use the identity for sin $(A + B)$ to get

$$\sin (A - B) = \sin [A + (-B)]$$
$$= \sin A \cos (-B) + \cos A \sin (-B)$$
$$= \sin A \cos B - \cos A \sin B$$

since cos $(-B) = \cos B$ and sin $(-B) = -\sin B$. In summary,

$$\sin (A - B) = \sin A \cos B - \cos A \sin B.$$

Using the identities for sin $(A + B)$, cos $(A + B)$, sin $(A - B)$, and cos $(A - B)$, and the identity tan θ = sin θ/cos θ, gives the following identities.

$$\tan (A + B) = \frac{\tan A + \tan B}{1 - \tan A \tan B}$$

$$\tan (A - B) = \frac{\tan A - \tan B}{1 + \tan A \tan B}$$

We show the proof for the first of these two identities. The proof for the other is very similar. Start with

$$\tan (A + B) = \frac{\sin (A + B)}{\cos (A + B)}$$

$$= \frac{\sin A \cos B + \cos A \sin B}{\cos A \cos B - \sin A \sin B}.$$

To express this result in terms of the tangent function, multiply both numerator and denominator by $1/(\cos A \cos B)$.

$$\tan (A + B) = \frac{\dfrac{\sin A \cos B + \cos A \sin B}{1}}{\dfrac{\cos A \cos B - \sin A \sin B}{1}} \cdot \frac{\dfrac{1}{\cos A \cos B}}{\dfrac{1}{\cos A \cos B}}$$

$$= \frac{\dfrac{\sin A \cos B}{\cos A \cos B} + \dfrac{\cos A \sin B}{\cos A \cos B}}{\dfrac{\cos A \cos B}{\cos A \cos B} - \dfrac{\sin A \sin B}{\cos A \cos B}}$$

$$= \frac{\dfrac{\sin A}{\cos A} + \dfrac{\sin B}{\cos B}}{1 - \dfrac{\sin A}{\cos A} \cdot \dfrac{\sin B}{\cos B}}$$

Using the identity tan θ = sin θ/cos θ,

$$\tan (A + B) = \frac{\tan A + \tan B}{1 - \tan A \tan B}.$$

The identities given in this section are summarized below.

SINE AND TANGENT OF SUM OR DIFFERENCE	
$\sin (A + B) = \sin A \cos B + \cos A \sin B$	
$\sin (A - B) = \sin A \cos B - \cos A \sin B$	
$\tan (A + B) = \dfrac{\tan A + \tan B}{1 - \tan A \tan B}$	
$\tan (A - B) = \dfrac{\tan A - \tan B}{1 + \tan A \tan B}$	

Again, the following examples and the corresponding exercises are given primarily to offer practice in using these new identities.

■ *Example 1*

USING THE SINE AND TANGENT SUM AND DIFFERENCE IDENTITIES TO FIND EXACT VALUES

Find the *exact* value of the following.

(a) $\sin 75° = \sin (45° + 30°)$

$= \sin 45° \cos 30° + \cos 45° \sin 30°$

$= \dfrac{\sqrt{2}}{2} \cdot \dfrac{\sqrt{3}}{2} + \dfrac{\sqrt{2}}{2} \cdot \dfrac{1}{2}$

$= \dfrac{\sqrt{6}}{4} + \dfrac{\sqrt{2}}{4} = \dfrac{\sqrt{6} + \sqrt{2}}{4}$

(b) $\tan \dfrac{7\pi}{12} = \tan \left(\dfrac{\pi}{3} + \dfrac{\pi}{4} \right)$

$= \dfrac{\tan \dfrac{\pi}{3} + \tan \dfrac{\pi}{4}}{1 - \tan \dfrac{\pi}{3} \tan \dfrac{\pi}{4}}$

$= \dfrac{\sqrt{3} + 1}{1 - \sqrt{3} \cdot 1}$

$= \dfrac{\sqrt{3} + 1}{1 - \sqrt{3}} \cdot \dfrac{1 + \sqrt{3}}{1 + \sqrt{3}}$

$= \dfrac{\sqrt{3} + 3 + 1 + \sqrt{3}}{1 - 3}$

$= \dfrac{4 + 2\sqrt{3}}{-2}$

$= -2 - \sqrt{3}$

(c) $\sin 40° \cos 160° - \cos 40° \sin 160° = \sin (40° - 160°)$

$= \sin (-120°)$

$= -\sin 120°$

$= -\dfrac{\sqrt{3}}{2}$ ■

■ *Example 2*

WRITING A FUNCTION AS AN EXPRESSION INVOLVING FUNCTIONS OF θ

Write each of the following as an expression involving functions of θ.

(a) $\sin (30° + \theta)$

Using the identity for $\sin (A + B)$,

$$\sin (30° + \theta) = \sin 30° \cos \theta + \cos 30° \sin \theta$$

$$= \dfrac{1}{2} \cos \theta + \dfrac{\sqrt{3}}{2} \sin \theta.$$

(b) $\tan (45° - \theta) = \dfrac{\tan 45° - \tan \theta}{1 + \tan 45° \tan \theta} = \dfrac{1 - \tan \theta}{1 + \tan \theta}$

(c) $\sin (180° + \theta) = \sin 180° \cos \theta + \cos 180° \sin \theta$

$$= 0 \cdot \cos \theta + (-1) \sin \theta$$

$$= -\sin \theta \quad ■$$

■ *Example 3*

FINDING FUNCTIONS AND THE QUADRANT OF *A* + *B* GIVEN INFORMATION ABOUT *A* AND *B*

If $\sin A = 4/5$ and $\cos B = -5/13$, where A is in quadrant II and B is in quadrant III, find each of the following.

(a) $\sin (A + B)$

The identity for $\sin (A + B)$ requires $\sin A$, $\cos A$, $\sin B$, and $\cos B$. Two of these values are given. The two missing values, $\cos A$ and $\sin B$, must be found first. These values can be found with the identity $\sin^2 x + \cos^2 x = 1$. To find $\cos A$, use

$$\sin^2 A + \cos^2 A = 1$$

$$\frac{16}{25} + \cos^2 A = 1$$

$$\cos^2 A = \frac{9}{25}$$

$$\cos A = -\frac{3}{5}. \qquad \text{Since } A \text{ is in quadrant II, } \cos A < 0.$$

In the same way, $\sin B = -12/13$. Now use the formula for $\sin (A + B)$.

$$\sin (A + B) = \frac{4}{5}\left(-\frac{5}{13}\right) + \left(-\frac{3}{5}\right)\left(-\frac{12}{13}\right)$$

$$= -\frac{20}{65} + \frac{36}{65} = \frac{16}{65}$$

(b) $\tan (A + B)$

Use the values of sine and cosine from part (a) to get $\tan A = -4/3$ and $\tan B = 12/5$. Then

$$\tan (A + B) = \frac{-\dfrac{4}{3} + \dfrac{12}{5}}{1 - \left(-\dfrac{4}{3}\right)\left(\dfrac{12}{5}\right)} = \frac{\dfrac{16}{15}}{1 + \dfrac{48}{15}} = \frac{\dfrac{16}{15}}{\dfrac{63}{15}} = \frac{16}{63}.$$

(c) the quadrant of $A + B$

From the results of parts (a) and (b), we find that $\sin (A + B)$ is positive and $\tan (A + B)$ is also positive. Therefore, $A + B$ must be in quadrant I, since it is the only quadrant in which both sine and tangent are positive. ■

The final example shows how to verify identities using the results of this section and the previous one.

■ *Example 4*

VERIFYING AN
IDENTITY USING
SUM AND
DIFFERENCE
IDENTITIES

Verify the identity

$$\sin\left(\frac{\pi}{6} + s\right) + \cos\left(\frac{\pi}{3} + s\right) = \cos s.$$

Work on the left side, using the identities for $\sin(A + B)$ and $\cos(A + B)$.

$$\sin\left(\frac{\pi}{6} + s\right) + \cos\left(\frac{\pi}{3} + s\right)$$

$$= \left(\sin\frac{\pi}{6}\cos s + \cos\frac{\pi}{6}\sin s\right) \qquad \begin{array}{l}\sin(A + B) \\ \quad = \sin A \cos B + \cos A \sin B\end{array}$$

$$+ \left(\cos\frac{\pi}{3}\cos s - \sin\frac{\pi}{3}\sin s\right) \qquad \begin{array}{l}\cos(A + B) \\ \quad = \cos A \cos B - \sin A \sin B\end{array}$$

$$= \left(\frac{1}{2}\cos s + \frac{\sqrt{3}}{2}\sin s\right) \qquad \sin\frac{\pi}{6} = \frac{1}{2}, \quad \cos\frac{\pi}{6} = \frac{\sqrt{3}}{2}$$

$$+ \left(\frac{1}{2}\cos s - \frac{\sqrt{3}}{2}\sin s\right) \qquad \cos\frac{\pi}{3} = \frac{1}{2}, \quad \sin\frac{\pi}{3} = \frac{\sqrt{3}}{2}$$

$$= \frac{1}{2}\cos s + \frac{1}{2}\cos s$$

$$= \cos s \quad ■$$

5.4 EXERCISES ■

Use the identities of this section to find the exact value of each of the following. See Example 1.

1. $\sin 15°$ **2.** $\sin 105°$ **3.** $\tan 15°$ **4.** $\tan 105°$ **5.** $\sin(-105°)$ **6.** $\tan(-105°)$

7. $\sin 5\pi/12$ **8.** $\tan 5\pi/12$ **9.** $\tan \pi/12$ **10.** $\sin \pi/12$ **11.** $\sin(-7\pi/12)$ **12.** $\tan(-7\pi/12)$

13. $\sin 76° \cos 31° - \cos 76° \sin 31°$ **14.** $\sin 40° \cos 50° + \cos 40° \sin 50°$

15. $\dfrac{\tan 80° + \tan 55°}{1 - \tan 80° \tan 55°}$ **16.** $\dfrac{\tan 80° - \tan(-55°)}{1 + \tan 80° \tan(-55°)}$

17. $\dfrac{\tan 100° + \tan 80°}{1 - \tan 100° \tan 80°}$ **18.** $\sin 100° \cos 10° - \cos 100° \sin 10°$

19. $\sin \pi/5 \cos 3\pi/10 + \cos \pi/5 \sin 3\pi/10$ **20.** $\dfrac{\tan 5\pi/12 + \tan \pi/4}{1 - \tan 5\pi/12 \tan \pi/4}$

Use the identities of this section and the previous one to express each of the following as an expression involving functions of x or θ. See Example 2.

21. $\cos(30° + \theta)$ **22.** $\cos(45° - \theta)$ **23.** $\cos(60° + \theta)$ **24.** $\cos(\theta - 30°)$ **25.** $\cos(3\pi/4 - x)$

26. $\cos(x + \pi/4)$ **27.** $\sin(45° + \theta)$ **28.** $\sin(\theta - 30°)$ **29.** $\tan(\theta + 30°)$ **30.** $\tan(60° - \theta)$

31. $\tan(\pi/4 + x)$ **32.** $\sin(\pi/4 + x)$ **33.** $\sin(180° - \theta)$ **34.** $\sin(270° - \theta)$ **35.** $\tan(180° + \theta)$

36. $\tan(360° - \theta)$ **37.** $\sin(\pi + \theta)$ **38.** $\tan(\pi - \theta)$

39. Why is it not possible to follow Example 2 and find a formula for $\tan(270° - \theta)$?

40. What happens when you try to evaluate $\dfrac{\tan 65.902° + \tan 24.098°}{1 - \tan 65.902° \tan 24.098°}$?

For each of the following, find $\sin(s+t)$, $\sin(s-t)$, $\tan(s+t)$, $\tan(s-t)$, *the quadrant of* $s+t$, *and the quadrant of* $s-t$. *See Example 3.*

41. $\cos s = 3/5$ and $\sin t = 5/13$, s and t in quadrant I

42. $\cos s = -1/5$ and $\sin t = 3/5$, s and t in quadrant II

43. $\sin s = 2/3$ and $\sin t = -1/3$, s in quadrant II and t in quadrant IV

44. $\sin s = 3/5$ and $\sin t = -12/13$, s in quadrant I and t in quadrant III

45. $\cos s = -8/17$ and $\cos t = -3/5$, s and t in quadrant III

46. $\cos s = -15/17$ and $\sin t = 4/5$, s in quadrant II and t in quadrant I

47. $\sin s = -4/5$ and $\cos t = 12/13$, s in quadrant III and t in quadrant IV

48. $\sin s = -5/13$ and $\sin t = 3/5$, s in quadrant III and t in quadrant II

49. $\sin s = -8/17$ and $\cos t = -8/17$, s and t in quadrant III

50. $\sin s = 2/3$ and $\sin t = 2/5$, s and t in quadrant I

51. $\cos s = -\sqrt{7}/4$ and $\sin t = \sqrt{3}/5$, s and t in quadrant II

52. $\cos s = \sqrt{11}/5$ and $\cos t = \sqrt{2}/6$, s and t in quadrant IV

Verify that each of the following is an identity. See Example 4.

53. $\sin\left(\dfrac{\pi}{2}+x\right) = \cos x$

54. $\sin\left(\dfrac{3\pi}{2}+x\right) = -\cos x$

55. $\tan\left(\dfrac{\pi}{2}+x\right) = -\cot x$ $\left(Hint\text{: } \tan\theta = \dfrac{\sin\theta}{\cos\theta}\right)$

56. $\tan\left(\dfrac{\pi}{4}+x\right) = \dfrac{1+\tan x}{1-\tan x}$

57. $\sin 2x = 2\sin x\cos x$ [*Hint:* $\sin 2x = \sin(x+x)$]

58. $\sin(x+y)+\sin(x-y) = 2\sin x\cos y$

59. $\tan(x-y)-\tan(y-x) = \dfrac{2(\tan x-\tan y)}{1+\tan x\tan y}$

60. $\sin(210°+x)-\cos(120°+x) = 0$

61. $\dfrac{\cos(\alpha-\beta)}{\cos\alpha\sin\beta} = \tan\alpha+\cot\beta$

62. $\dfrac{\sin(s+t)}{\cos s\cos t} = \tan s+\tan t$

63. $\dfrac{\sin(x-y)}{\sin(x+y)} = \dfrac{\tan x-\tan y}{\tan x+\tan y}$

64. $\dfrac{\sin(x+y)}{\cos(x-y)} = \dfrac{\cot x+\cot y}{1+\cot x\cot y}$

65. $\dfrac{\sin(s-t)}{\sin t}+\dfrac{\cos(s-t)}{\cos t} = \dfrac{\sin s}{\sin t\cos t}$

66. $\dfrac{\tan(\alpha+\beta)-\tan\beta}{1+\tan(\alpha+\beta)\tan\beta} = \tan\alpha$

Find the exact value of each of the following. (See the explanation preceding Exercise 73 in Section 5.3.)

67. $\sin 165°$ **68.** $\tan 165°$ **69.** $\tan 255°$ **70.** $\sin 255°$ **71.** $\sin 285°$

72. $\tan 285°$ **73.** $\sin 11\pi/12$ **74.** $\tan 11\pi/12$ **75.** $\tan(-13\pi/12)$ **76.** $\sin(-13\pi/12)$

77. Derive the identity for $\tan(A-B)$ using the identity for $\tan(A+B)$, and the fact that $A-B = A+(-B)$.

78. Derive the identity for $\tan(A-B)$ using the identities for $\sin(A-B)$ and $\cos(A-B)$, and the fact that $\tan(A-B) = \dfrac{\sin(A-B)}{\cos(A-B)}$.

Derive a formula for each of the following.

79. $\sin(A+B+C)$

80. $\cos(A+B+C)$

81. Let $f(x) = \sin x$. Show that $\dfrac{f(x+h)-f(x)}{h} = \sin x\left(\dfrac{\cos h-1}{h}\right)+\cos x\left(\dfrac{\sin h}{h}\right)$.

82. The slope of a line is defined as the ratio of the vertical change and the horizontal change. As shown in the sketch below, the tangent of the *angle of inclination* θ is given by the ratio of the side opposite and the side adjacent. This ratio is the same as that used in finding the slope, *m*, so that $m = \tan \theta$. In the figure on the right, let the two lines have angles of inclination α and β, and slopes m_1 and m_2, respectively. Let θ be the smallest positive angle between the lines. Show that

$$\tan \theta = \frac{m_2 - m_1}{1 + m_1 m_2}.$$

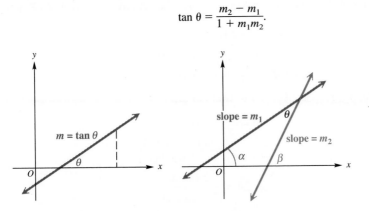

Use the results from Exercise 82 to find the angle between the following pairs of lines. Use a calculator and round to the nearest tenth of a degree.

83. $x + y = 9,\quad 2x + y = -1$

84. $5x - 2y + 4 = 0,\quad 3x + 5y = 6$

5.5 ——— DOUBLE-ANGLE IDENTITIES

Some special cases of the identities for the sum of two angles are used often enough to be expressed as separate identities. These are the identities that result from the addition identities when $A = B$, so that $A + B = 2A$. These identities, called the **double-angle identities,** are derived in this section.

In the identity $\cos (A + B) = \cos A \cos B - \sin A \sin B$, let $B = A$ to derive an expression for $\cos 2A$.

$$\cos 2A = \cos (A + A)$$
$$= \cos A \cos A - \sin A \sin A$$
$$\cos 2A = \cos^2 A - \sin^2 A$$

Two other useful forms of this identity can be obtained by substituting either $\cos^2 A = 1 - \sin^2 A$ or $\sin^2 A = 1 - \cos^2 A$. Replace $\cos^2 A$ with $1 - \sin^2 A$ to get

$$\cos 2A = \cos^2 A - \sin^2 A$$
$$= (1 - \sin^2 A) - \sin^2 A$$
$$\cos 2A = 1 - 2\sin^2 A,$$

and replace $\sin^2 A$ with $1 - \cos^2 A$ to get

$$\cos 2A = \cos^2 A - (1 - \cos^2 A)$$
$$= \cos^2 A - 1 + \cos^2 A$$
$$\cos 2A = 2 \cos^2 A - 1.$$

Find $\sin 2A$ with the identity $\sin (A + B) = \sin A \cos B + \cos A \sin B$, letting $B = A$.

$$\sin 2A = \sin (A + A)$$
$$= \sin A \cos A + \cos A \sin A$$
$$\sin 2A = 2 \sin A \cos A$$

Use the identity for $\tan (A + B)$ to find $\tan 2A$.

$$\tan 2A = \tan(A + A)$$
$$= \frac{\tan A + \tan A}{1 - \tan A \tan A}$$
$$\tan 2A = \frac{2 \tan A}{1 - \tan^2 A}$$

A summary of the double-angle identities follows.

DOUBLE-ANGLE IDENTITIES		
	$\cos 2A = \cos^2 A - \sin^2 A$	$\cos 2A = 1 - 2 \sin^2 A$
	$\cos 2A = 2 \cos^2 A - 1$	$\sin 2A = 2 \sin A \cos A$
		$\tan 2A = \dfrac{2 \tan A}{1 - \tan^2 A}$

[handwritten annotations: $2\text{sin}^2 = 1 - \cos 2A$; $\cos 2A + 1 = 2\cos^2$]

■ *Example 1*

USING THE DOUBLE-ANGLE IDENTITIES TO FIND $\sin 2\theta$, $\cos 2\theta$, AND $\tan 2\theta$

Given $\cos \theta = 3/5$ and $\sin \theta < 0$, find $\sin 2\theta$, $\cos 2\theta$, and $\tan 2\theta$.

In order to find $\sin 2\theta$, we must first find the value of $\sin \theta$. From the identity $\sin^2 \theta + \cos^2 \theta = 1$, we obtain

$$\sin^2 \theta + \left(\frac{3}{5}\right)^2 = 1$$

$$\sin^2 \theta = \frac{16}{25}$$

$$\sin \theta = -\frac{4}{5}. \qquad \text{Choose the negative square root, since } \sin \theta < 0.$$

Using the double-angle identity for sine, we get

$$\sin 2\theta = 2 \sin \theta \cos \theta = 2\left(-\frac{4}{5}\right)\left(\frac{3}{5}\right) = -\frac{24}{25}.$$

Now find $\cos 2\theta$, using the first form of the identity. (Any form may be used.)

$$\cos 2\theta = \cos^2 \theta - \sin^2 \theta = \frac{9}{25} - \frac{16}{25} = -\frac{7}{25}$$

The value of tan 2θ can be found in either of two ways. We can use the double-angle identity, and the fact that tan $\theta = (\sin \theta)/(\cos \theta) = (-4/5)/(3/5) = -4/3$.

$$\tan 2\theta = \frac{2 \tan \theta}{1 - \tan^2 \theta} = \frac{2\left(-\dfrac{4}{3}\right)}{1 - \dfrac{16}{9}} = \frac{-\dfrac{8}{3}}{-\dfrac{7}{9}} = \frac{24}{7}$$

As an alternative method, we can find tan 2θ by finding the quotient of sin 2θ and cos 2θ.

$$\tan 2\theta = \frac{\sin 2\theta}{\cos 2\theta} = \frac{-24/5}{-7/25} = \frac{24}{7} \quad ■$$

■ *Example 2*

FINDING FUNCTIONS
OF θ GIVEN
INFORMATION
ABOUT 2θ

Find the values of the six trigonometric functions of θ if cos $2\theta = 4/5$ and $90° < \theta < 180°$.

Use one of the double-angle identities for cosine to get a trigonometric function of θ.

$$\cos 2\theta = 1 - 2 \sin^2 \theta$$

$$\frac{4}{5} = 1 - 2 \sin^2 \theta$$

$$-\frac{1}{5} = -2 \sin^2 \theta$$

$$\frac{1}{10} = \sin^2 \theta$$

$$\sin \theta = \sqrt{\frac{1}{10}} = \frac{\sqrt{10}}{10}$$

Choose the positive square root since θ terminates in quadrant II. Values of cos θ and tan θ can now be found by using the fundamental identities or by sketching and labeling a right triangle in quadrant II. Using identities gives

$$\sin^2 \theta + \cos^2 \theta = 1$$

$$\frac{1}{10} + \cos^2 \theta = 1$$

$$\cos^2 \theta = \frac{9}{10}$$

$$\cos \theta = \frac{-3}{\sqrt{10}} = -\frac{3\sqrt{10}}{10}.$$

Choose the negative square root, since cosine is negative in quadrant II.

Verify that tan $\theta = \sin \theta/\cos \theta = -1/3$. Find the other three functions using reciprocals.

$$\csc \theta = \frac{1}{\sin \theta} = \sqrt{10}, \qquad \sec \theta = \frac{1}{\cos \theta} = -\frac{\sqrt{10}}{3}, \qquad \cot \theta = \frac{1}{\tan \theta} = -3 \quad ■$$

■ *Example 3*

SIMPLIFYING
EXPRESSIONS USING
DOUBLE-ANGLE
IDENTITIES

Simplify each of the following using double-angle identities.

(a) $\cos^2 7x - \sin^2 7x$

This expression suggests one of the identities for $\cos\ 2A$: $\cos 2A = \cos^2 A - \sin^2 A$. Substituting $7x$ for A gives

$$\cos^2 7x - \sin^2 7x = \cos 2(7x) = \cos 14x.$$

(b) $\sin 15° \cos 15°$

If this expression were $2 \sin 15° \cos 15°$, then we could apply the identity for $\sin 2A$ directly, since $\sin 2A = 2 \sin A \cos A$. We can still apply the identity with $A = 15°$ by writing the multiplicative identity element 1 as $(1/2)(2)$:

$$\sin 15° \cos 15° = \left(\frac{1}{2}\right)(2) \sin 15° \cos 15° \qquad \text{Multiply by 1 in the form } \frac{1}{2}\,(2).$$

$$= \frac{1}{2}\,(2 \sin 15° \cos 15°) \qquad \text{Associative property}$$

$$= \frac{1}{2}\,(\sin 2 \cdot 15°) \qquad 2 \sin A \cos A = \sin 2A, \text{ with } A = 15°$$

$$= \frac{1}{2}\, \sin 30°$$

$$= \frac{1}{2} \cdot \frac{1}{2} \qquad \sin 30° = \frac{1}{2}$$

$$= \frac{1}{4}. \quad ■$$

Double-angle identities can be used to verify certain identities, as shown in the next example.

■ *Example 4*

VERIFYING AN
IDENTITY

Verify the identity

$$\cot x \sin 2x = 1 + \cos 2x.$$

Work on the left side.

$$\cot x \sin 2x = \frac{\cos x}{\sin x} \cdot \sin 2x \qquad \cot x = \frac{\cos x}{\sin x}$$

$$= \frac{\cos x}{\sin x}\,(2 \sin x \cos x) \qquad \sin 2x = 2 \sin x \cos x$$

$$= 2 \cos^2 x$$

$$= 1 + \cos 2x \qquad 2 \cos^2 x - 1 = \cos 2x \quad ■$$

The methods used earlier to derive the identities for double angles can also be used to find identities for expressions such as $\sin 3s$, as in the final example.

▪ **Example 5**

DERIVING A
MULTIPLE-ANGLE
IDENTITY

Write sin 3s in terms of sin s.

$$\sin 3s = \sin (2s + s)$$
$$= \sin 2s \cos s + \cos 2s \sin s$$
$$= (2 \sin s \cos s) \cos s + (\cos^2 s - \sin^2 s) \sin s$$
$$= 2 \sin s \cos^2 s + \cos^2 s \sin s - \sin^3 s$$
$$= 2 \sin s(1 - \sin^2 s) + (1 - \sin^2 s) \sin s - \sin^3 s$$
$$= 2 \sin s - 2 \sin^3 s + \sin s - \sin^3 s - \sin^3 s$$
$$= 3 \sin s - 4 \sin^3 s \quad ▪$$

5.5 EXERCISES ▪

Use the identities in this section to find values of the six trigonometric functions for each of the following. See Examples 1 and 2.

1. θ, given $\cos 2\theta = 3/5$ and θ terminates in quadrant I

2. α, given $\cos 2\alpha = 3/4$ and α terminates in quadrant III

3. x, given $\cos 2x = -5/12$ and $\pi/2 < x < \pi$ **4.** t, given $\cos 2t = 2/3$ and $\pi/2 < t < \pi$

5. 2θ, given $\sin \theta = 2/5$ and $\cos \theta < 0$ **6.** 2β, given $\cos \beta = -12/13$ and $\sin \beta > 0$

7. $2x$, given $\tan x = 2$ and $\cos x > 0$ **8.** $2x$, given $\tan x = 5/3$ and $\sin x < 0$

9. 2α, given $\sin \alpha = -\sqrt{5}/7$ and $\cos \alpha > 0$ **10.** 2α, given $\cos \alpha = \sqrt{3}/5$ and $\sin \alpha > 0$

Use an identity to write each of the following as a single trigonometric function or as a single number. See Example 3.

11. $2 \cos^2 15° - 1$ **12.** $\cos^2 15° - \sin^2 15°$ **13.** $\dfrac{2 \tan 15°}{1 - \tan^2 15°}$ **14.** $1 - 2 \sin^2 15°$

15. $2 \sin \pi/3 \cos \pi/3$ **16.** $\dfrac{2 \tan \pi/3}{1 - \tan^2 \pi/3}$ **17.** $1 - 2 \sin^2 22\frac{1}{2}°$ **18.** $2 \sin 22\frac{1}{2}° \cos 22\frac{1}{2}°$

19. $2 \cos^2 67\frac{1}{2}° - 1$ **20.** $\dfrac{2 \tan 67\frac{1}{2}°}{1 - \tan^2 67\frac{1}{2}°}$ **21.** $\sin \pi/8 \cos \pi/8$ **22.** $\cos^2 \pi/8 - 1/2$

23. $\dfrac{\tan 51°}{1 - \tan^2 51°}$ **24.** $\dfrac{\tan 34°}{2(1 - \tan^2 34°)}$ **25.** $\dfrac{1}{4} - \dfrac{1}{2} \sin^2 47.1°$ **26.** $\dfrac{1}{8} \sin 29.5° \cos 29.5°$

27. $\sin^2 2\pi/5 - \cos^2 2\pi/5$ **28.** $\dfrac{2 \tan \pi/5}{\tan^2 \pi/5 - 1}$ **29.** $2 \sin 5x \cos 5x$ **30.** $2 \cos^2 6\alpha - 1$

31. $\cos^2 2\alpha - \sin^2 2\alpha$ **32.** $\dfrac{2 \tan 3r}{1 - \tan^2 3r}$ **33.** $\dfrac{2 \tan x/9}{1 - \tan^2 x/9}$ **34.** $\dfrac{\tan 2y/5}{1 - \tan^2 2y/5}$

35. State in your own words each of the following identities. (*Example:* The sine of twice an angle is two times the sine of the angle times the cosine of the angle.)
(a) $\cos 2A$ (first form derived in this section)
(b) $\cos 2A$ (second form)   **(c)** $\cos 2A$ (third form)   **(d)** $\tan 2A$

36. Specific identities for $\sec 2A$, $\csc 2A$, and $\cot 2A$ are usually not studied in detail. Why do you think this is so? Give these identities in terms of $\cos A$, $\sin A$, and $\tan A$.

Find the exact value of each of the following in two ways: evaluate the expression directly, and use an appropriate identity. Do not use a calculator.

37. $\sin 2(45°)$ **38.** $\cos 2(45°)$ **39.** $\cos 2(60°)$ **40.** $\tan 2(60°)$

41. $\cos 2\left(\dfrac{5\pi}{3}\right)$ **42.** $\sin 2\left(\dfrac{\pi}{3}\right)$ **43.** $\tan 2\left(-\dfrac{\pi}{3}\right)$ **44.** $\cos 2\left(-\dfrac{9\pi}{4}\right)$

45. $\tan 2\left(-\dfrac{4\pi}{3}\right)$ **46.** $\tan 2\left(-\dfrac{13\pi}{6}\right)$ **47.** $\sin 2\left(-\dfrac{11\pi}{2}\right)$ **48.** $\sin 2\left(-\dfrac{17\pi}{2}\right)$

Verify each of the following identities. See Example 4.

49. $(\sin \gamma + \cos \gamma)^2 = \sin 2\gamma + 1$

50. $\cos 2s = \cos^4 s - \sin^4 s$

51. $\sec 2x = \dfrac{\sec^2 x + \sec^4 x}{2 + \sec^2 x - \sec^4 x}$

52. $\sin 2\theta = \dfrac{4 \tan \theta \cos^2 \theta - 2 \tan \theta}{1 - \tan^2 \theta}$

53. $\cot 2\beta = \dfrac{\cot^2 \beta - 1}{2 \cot \beta}$

54. $\tan 8k - \tan 8k \tan^2 4k = 2 \tan 4k$

55. $\sin 2\gamma = \dfrac{2 \tan \gamma}{1 + \tan^2 \gamma}$

56. $-\tan 2\theta = \dfrac{2 \tan \theta}{\sec^2 \theta - 2}$

57. $\cos 2y = \dfrac{2 - \sec^2 y}{\sec^2 y}$

58. $\cot s + \tan s = 2 \csc 2s$

59. $\sin 4\alpha = 4 \sin \alpha \cos \alpha \cos 2\alpha$

60. $\dfrac{1 + \cos 2x}{\sin 2x} = \cot x$

61. $\tan (\theta - 45°) + \tan (\theta + 45°) = 2 \tan 2\theta$

62. $\cot 4\theta = \dfrac{1 - \tan^2 2\theta}{2 \tan 2\theta}$

63. $\dfrac{2 \cos 2\alpha}{\sin 2\alpha} = \cot \alpha - \tan \alpha$

64. $\sin 4\gamma = 4 \sin \gamma \cos \gamma - 8 \sin^3 \gamma \cos \gamma$

65. $\sin 2\alpha \cos 2\alpha = \sin 2\alpha - 4 \sin^3 \alpha \cos \alpha$

66. $\dfrac{\sin^3 t - \cos^3 t}{\sin t - \cos t} = \dfrac{2 + \sin 2t}{2}$

67. $\cos 2x = \dfrac{1 - \tan^2 x}{1 + \tan^2 x}$

68. $\dfrac{\sin 2\theta}{\sin \theta} - \dfrac{\cos 2\theta}{\cos \theta} = \sec \theta$

69. $\tan s + \cot s = 2 \csc 2s$

70. $\dfrac{\cot \alpha - \tan \alpha}{\cot \alpha + \tan \alpha} = \cos 2\alpha$

71. $1 + \tan x \tan 2x = \sec 2x$

72. $\cot \theta \tan (\theta + \pi) - \sin (\pi - \theta) \cos \left(\dfrac{\pi}{2} - \theta\right) = \cos^2 \theta$

Express each of the following as a trigonometric function of x. See Example 5.

73. $\tan^2 2x$ **74.** $\cos^2 2x$ **75.** $\cos 3x$ **76.** $\sin 4x$

77. $\tan 3x$ **78.** $\cos 4x$ **79.** $\tan 4x$ **80.** $\sin 5x$

Use a fundamental identity to simplify each of the following.

81. $\sin^2 2x + \cos^2 2x$ **82.** $1 + \tan^2 4\alpha$

83. $\cot^2 3r + 1$ **84.** $\sin^2 11\alpha + \cos^2 11\alpha$

If an object is dropped in a vaccum, then the distance, d, the object falls in t seconds is given by

$$d = \frac{1}{2} gt^2,$$

where g is the acceleration due to gravity. At any particular point on the earth's surface, the value of g is a constant, roughly 978 cm per sec per sec. A more exact value of g at any point on the earth's surface is given by

$$g = 978.0524(1 + .005297 \sin^2 \phi - .0000059 \sin^2 2\phi) - .000094h$$

in cm per second per second, where ϕ is the latitude of the point and h is the altitude of the point in feet. Find g, rounding to the nearest thousandth, given the following.

85. $\phi = 47° \, 12'$, $h = 387.0$ ft   **86.** $\phi = 68°47'$, $h = 1145$ ft

5.6 ——— HALF-ANGLE IDENTITIES

From the alternative forms of the identity for cos 2A, we can derive three additional identities for sin A/2, cos A/2, and tan A/2. These are known as **half-angle identities.**

To derive the identity for sin A/2, start with the following double-angle identity for cosine.

$$\cos 2x = 1 - 2 \sin^2 x$$

Then solve for sin x.

$$2 \sin^2 x = 1 - \cos 2x$$

$$\sin x = \pm \sqrt{\frac{1 - \cos 2x}{2}}$$

Now let $2x = A$, so that $x = A/2$, and substitute into this last expression.

$$\sin \frac{A}{2} = \pm \sqrt{\frac{1 - \cos A}{2}}$$

The $\pm$ sign in the identity above indicates that, in practice, the appropriate sign is chosen depending upon the quadrant of A/2. For example, if A/2 is a third quadrant angle, we choose the negative sign since the sine function is negative there.

The identity for cos A/2 is derived in a very similar way, starting with the double-angle identity cos 2x = 2 cos² x − 1. Solve for cos x.

$$\cos 2x + 1 = 2 \cos^2 x$$

$$\cos x = \pm \sqrt{\frac{1 + \cos 2x}{2}}$$

Replacing x with A/2 gives

$$\cos \frac{A}{2} = \pm \sqrt{\frac{1 + \cos A}{2}}.$$

The $\pm$ sign is used as described earlier.

Finally, an identity for tan $A/2$ comes from the half-angle identities for sine and cosine.

$$\tan \frac{A}{2} = \frac{\pm \sqrt{\dfrac{1 - \cos A}{2}}}{\pm \sqrt{\dfrac{1 + \cos A}{2}}}$$

or $\tan \dfrac{A}{2} = \pm \sqrt{\dfrac{1 - \cos A}{1 + \cos A}}$ ± chosen depending upon quadrant of $A/2$

An alternative identity for tan $A/2$ can be derived using the fact that tan $A/2 = (\sin A/2)/(\cos A/2)$.

$$\tan \frac{A}{2} = \frac{\sin \dfrac{A}{2}}{\cos \dfrac{A}{2}}$$

$$= \frac{2 \sin \dfrac{A}{2} \cos \dfrac{A}{2}}{2 \cos^2 \dfrac{A}{2}} \qquad \text{Multiply by 2 } \cos \frac{A}{2} \text{ in numerator and denominator.}$$

$$= \frac{\sin 2 \left(\dfrac{A}{2}\right)}{1 + \cos 2 \left(\dfrac{A}{2}\right)} \qquad \text{Use double-angle identities.}$$

$$\tan \frac{A}{2} = \frac{\sin A}{1 + \cos A}$$

From this identity for tan $A/2$, we can also derive

$$\tan \frac{A}{2} = \frac{1 - \cos A}{\sin A}.$$

See Exercise 53. These last two identities for tan $A/2$ do not require a sign choice, as required in the first one.

HALF-ANGLE IDENTITIES	$\cos \dfrac{A}{2} = \pm \sqrt{\dfrac{1 + \cos A}{2}}$ $\sin \dfrac{A}{2} = \pm \sqrt{\dfrac{1 - \cos A}{2}}$
	$\tan \dfrac{A}{2} = \pm \sqrt{\dfrac{1 - \cos A}{1 + \cos A}}$ $\tan \dfrac{A}{2} = \dfrac{\sin A}{1 + \cos A}$
	$\tan \dfrac{A}{2} = \dfrac{1 - \cos A}{\sin A}$

As mentioned earlier, the plus or minus sign is selected according to the quadrant in which $A/2$ terminates. For example, if A represents an angle of $324°$, then $A/2 = 162°$, which lies in quadrant II. In quadrant II, $\cos A/2$ and $\tan A/2$ are negative, while $\sin A/2$ is positive.

■ **Example 1**

USING A
HALF-ANGLE
IDENTITY TO FIND
AN EXACT VALUE

Find the exact value of $\cos 15°$ using the half-angle identity for cosine.

$$\cos 15° = \cos \frac{1}{2}(30°)$$

$$= \sqrt{\frac{1 + \cos 30°}{2}}$$

Choose the positive square root.

$$= \sqrt{\frac{1 + \frac{\sqrt{3}}{2}}{2}}$$

$$= \sqrt{\frac{\left(1 + \frac{\sqrt{3}}{2}\right) \cdot 2}{2 \cdot 2}}$$

$$= \frac{\sqrt{2 + \sqrt{3}}}{2} \quad ■$$

Compare the value of $\cos 15°$ obtained in Example 1 to the value obtained in Example 1 of Section 5.3, where we used the identity for the cosine of the difference of two angles. Although the expressions look completely different, they are indeed equal, as suggested by a calculator approximation for both, .96592583.

■ **Example 2**

USING A
HALF-ANGLE
IDENTITY TO FIND
AN EXACT VALUE

Find the exact value of $\tan 22.5°$ using the identity $\tan A/2 = \dfrac{\sin A}{1 + \cos A}$.

Since $22.5° = (1/2)(45°)$, replacing A with $45°$ gives

$$\tan 22.5° = \tan \frac{45°}{2} = \frac{\sin 45°}{1 + \cos 45°} = \frac{\frac{\sqrt{2}}{2}}{1 + \frac{\sqrt{2}}{2}}.$$

Now multiply numerator and denominator by 2. Then rationalize the denominator.

$$\tan 22.5° = \frac{\sqrt{2}}{2 + \sqrt{2}} = \frac{\sqrt{2}}{2 + \sqrt{2}} \cdot \frac{2 - \sqrt{2}}{2 - \sqrt{2}}$$

$$= \frac{2\sqrt{2} - 2}{2} = \sqrt{2} - 1 \quad ■$$

■ Example 3

FINDING FUNCTIONS OF A/2 GIVEN INFORMATION ABOUT A

Given $\cos s = 2/3$, with $3\pi/2 < s < 2\pi$, find $\cos s/2$, $\sin s/2$ and $\tan s/2$.

Since

$$\frac{3\pi}{2} < s < 2\pi,$$

dividing through by 2 gives

$$\frac{3\pi}{4} < \frac{s}{2} < \pi,$$

showing that $s/2$ terminates in quadrant II. In this quadrant the value of $\cos s/2$ is negative and the value of $\sin s/2$ is positive. Use the appropriate half-angle identities to get

$$\sin \frac{s}{2} = \sqrt{\frac{1 - \frac{2}{3}}{2}} = \sqrt{\frac{1}{6}} = \frac{\sqrt{6}}{6};$$

and

$$\cos \frac{s}{2} = -\sqrt{\frac{1 + \frac{2}{3}}{2}} = -\sqrt{\frac{5}{6}} = -\frac{\sqrt{30}}{6}.$$

Also,

$$\tan \frac{s}{2} = \frac{\sin \frac{s}{2}}{\cos \frac{s}{2}} = \frac{\frac{\sqrt{6}}{6}}{-\frac{\sqrt{30}}{6}} = -\frac{\sqrt{5}}{5}.$$

Notice that it is not necessary to use a half-angle identity for $\tan s/2$ once we find $\sin s/2$ and $\cos s/2$. However, using this identity would provide an excellent check. ■

■ Example 4

SIMPLIFYING EXPRESSIONS USING THE HALF-ANGLE IDENTITIES

Simplify each of the following using half-angle identities.

(a) $\pm\sqrt{\frac{1 + \cos 12x}{2}}$

Start with the identity for $\cos A/2$,

$$\cos \frac{A}{2} = \pm\sqrt{\frac{1 + \cos A}{2}},$$

and replace A with $12x$ to get

$$\pm\sqrt{\frac{1 + \cos 12x}{2}} = \cos \frac{12x}{2} = \cos 6x.$$

(b) $\dfrac{1 - \cos 5\alpha}{\sin 5\alpha}$

Use the third identity for $\tan A/2$ given earlier to get

$$\frac{1 - \cos 5\alpha}{\sin 5\alpha} = \tan \frac{5\alpha}{2}. \quad ■$$

■ *Example 5*

VERIFYING AN
IDENTITY

Verify the identity

$$\left(\sin\frac{x}{2} + \cos\frac{x}{2}\right)^2 = 1 + \sin x.$$

Work on the left.

$$\left(\sin\frac{x}{2} + \cos\frac{x}{2}\right)^2$$

$$= \sin^2\frac{x}{2} + 2\sin\frac{x}{2}\cos\frac{x}{2} + \cos^2\frac{x}{2} \qquad (a+b)^2 = a^2 + 2ab + b^2$$

$$= 1 + 2\sin\frac{x}{2}\cos\frac{x}{2} \qquad\qquad \sin^2\frac{x}{2} + \cos^2\frac{x}{2} = 1$$

$$= 1 + \sin 2\left(\frac{x}{2}\right) \qquad\qquad 2\sin\frac{x}{2}\cos\frac{x}{2} = \sin 2\left(\frac{x}{2}\right)$$

$$= 1 + \sin x \quad ■$$

5.6 EXERCISES ■

Use the half-angle identities of this section to find each of the following. See Examples 1 and 2.

1. $\sin 15°$ **2.** $\tan 15°$ **3.** $\cos \pi/8$ **4.** $\tan(-\pi/8)$ **5.** $\tan 67.5°$ **6.** $\cos 67.5°$

7. $\sin 67.5°$ **8.** $\sin 195°$ **9.** $\cos 195°$ **10.** $\tan 195°$ **11.** $\cos 165°$ **12.** $\sin 165°$

13. Explain how you could use an identity of this section to find the exact value of $\sin 7.5°$. (*Hint:* $7.5 = (1/2)(1/2)(30)$.)

14. The identity

$$\tan\frac{A}{2} = \pm\sqrt{\frac{(1-\cos A)}{(1+\cos A)}}$$

can be used to find $\tan 22.5° = \sqrt{3 - 2\sqrt{2}}$, and the identity

$$\tan\frac{A}{2} = \frac{\sin A}{(1+\cos A)}$$

can be used to get $\tan 22.5° = \sqrt{2} - 1$. Show that these answers are the same, without using a calculator. (*Hint:* If $a > 0$ and $b > 0$ and $a^2 = b^2$, then $a = b$.)

Find each of the following. See Example 3.

15. $\cos \theta/2$, given $\cos\theta = 1/4$, with $0 < \theta < \pi/2$

16. $\sin \theta/2$, given $\cos\theta = -5/8$, with $\pi/2 < \theta < \pi$

17. $\tan \theta/2$, given $\sin\theta = 3/5$, with $90° < \theta < 180°$

18. $\cos \theta/2$, given $\sin\theta = -1/5$, with $180° < \theta < 270°$

19. $\sin \alpha/2$, given $\tan\alpha = 2$, with $0 < \alpha < \pi/2$

20. $\cos \alpha/2$, given $\cot\alpha = -3$, with $\pi/2 < \alpha < \pi$

21. $\tan \beta/2$, given $\tan\beta = \sqrt{7}/3$, with $180° < \beta < 270°$

22. $\cot \beta/2$, given $\tan\beta = -\sqrt{5}/2$, with $90° < \beta < 180°$

23. $\sin\theta$, given $\cos 2\theta = 3/5$ and θ terminates in quadrant I

24. $\cos\theta$ given $\cos 2\theta = 1/2$ and θ terminates in quadrant II

25. $\cos x$, given $\cos 2x = -5/12$ and $\pi/2 < x < \pi$

26. $\sin x$, given $\cos 2x = 2/3$ and $\pi < x < 3\pi/2$

Use an identity to write each of the following as a single trigonometric function. See Example 4.

27. $\sqrt{\dfrac{1 - \cos 40°}{2}}$

28. $\sqrt{\dfrac{1 + \cos 76°}{2}}$

29. $\sqrt{\dfrac{1 - \cos 147°}{1 + \cos 147°}}$

30. $\sqrt{\dfrac{1 + \cos 165°}{1 - \cos 165°}}$

31. $\dfrac{1 - \cos 59.74°}{\sin 59.74°}$

32. $\dfrac{\sin 158.2°}{1 + \cos 158.2°}$

33. $\pm\sqrt{\dfrac{1 + \cos 18x}{2}}$

34. $\pm\sqrt{\dfrac{1 + \cos 20\alpha}{2}}$

35. $\pm\sqrt{\dfrac{1 - \cos 8\theta}{1 + \cos 8\theta}}$

36. $\pm\sqrt{\dfrac{1 - \cos 5A}{1 + \cos 5A}}$

37. $\pm\sqrt{\dfrac{1 + \cos x/4}{2}}$

38. $\pm\sqrt{\dfrac{1 - \cos 3\theta/5}{2}}$

Verify that each of the following equations is an identity. See Example 5.

39. $\sec^2 \dfrac{x}{2} = \dfrac{2}{1 + \cos x}$

40. $\cot^2 \dfrac{x}{2} = \dfrac{(1 + \cos x)^2}{\sin^2 x}$

41. $\sin^2 \dfrac{x}{2} = \dfrac{\tan x - \sin x}{2 \tan x}$

42. $\dfrac{\sin 2x}{2 \sin x} = \cos^2 \dfrac{x}{2} - \sin^2 \dfrac{x}{2}$

43. $\dfrac{2}{1 + \cos x} - \tan^2 \dfrac{x}{2} = 1$

44. $\tan \dfrac{\theta}{2} = \dfrac{\sin \theta}{1 + \cos \theta}$

45. $\tan \dfrac{\alpha}{2} = \dfrac{1 - \cos \alpha}{\sin \alpha}$

46. $\tan \dfrac{\gamma}{2} = \csc \gamma - \cot \gamma$

47. $\dfrac{\tan \dfrac{x}{2} + \cot \dfrac{x}{2}}{\cot \dfrac{x}{2} - \tan \dfrac{x}{2}} = \sec x$

48. $1 - \tan^2 \dfrac{\theta}{2} = \dfrac{2 \cos \theta}{1 + \cos \theta}$

49. $\cos x = \dfrac{1 - \tan^2 \dfrac{x}{2}}{1 + \tan^2 \dfrac{x}{2}}$

50. $\dfrac{\sin 2\alpha - 2 \sin \alpha}{2 \sin \alpha + \sin 2\alpha} = -\tan^2 \dfrac{\alpha}{2}$

51. $8 \sin^2 \dfrac{\gamma}{2} \cos^2 \dfrac{\gamma}{2} = 1 - \cos 2\gamma$

52. $\cos^2 \dfrac{x}{2} = \dfrac{1 + \sec x}{2 \sec x}$

53. In the text we derived the identity

$$\tan \dfrac{A}{2} = \dfrac{\sin A}{1 + \cos A}.$$

Multiply both the numerator and denominator of the right side by $1 - \cos A$ to obtain the equivalent form

$$\tan \dfrac{A}{2} = \dfrac{1 - \cos A}{\sin A}.$$

54. The identity $\tan \dfrac{A}{2} = \dfrac{\sin A}{1 + \cos A}$ can be derived alternatively as follows.

(a) Start with $\tan A/2 = \pm\sqrt{(1 - \cos A)/(1 + \cos A)}$, and multiply numerator and denominator by $\sqrt{1 + \cos A}$ to show that

$$\tan \dfrac{A}{2} = \pm\left|\dfrac{\sin A}{1 + \cos A}\right|.$$

(b) Show that $1 + \cos A \geq 0$, giving

$$\tan \dfrac{A}{2} = \dfrac{\pm|\sin A|}{1 + \cos A}.$$

(c) By considering the quadrant in which A lies, show that $\tan A/2$ and $\sin A$ have the same sign, with

$$\tan \dfrac{A}{2} = \dfrac{\sin A}{1 + \cos A}.$$

An airplane flying faster than sound sends out sound waves that form a cone, as shown in the figure. The cone intersects the ground to form a hyperbola. As this hyperbola passes over a particular point on the ground, a sonic boom is heard at that point. If α is the angle at the vertex of the cone, then

$$\sin \frac{\alpha}{2} = \frac{1}{m},$$

where m is the Mach number for the speed of the plane. (We assume m > 1.) The Mach number is the ratio of the speed of the plane and the speed of sound. Thus, a speed of Mach 1.4 means that the plane is flying at 1.4 times the speed of sound. Find α or m, as necessary, for each of the following.

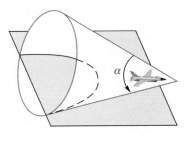

55. $m = 3/2$ **56.** $m = 5/4$ **57.** $m = 2$ **58.** $m = 5/2$ **59.** $\alpha = 30°$ **60.** $\alpha = 60°$

5.7 — SUM AND PRODUCT IDENTITIES

While the identities in this section are used less frequently than those in the preceding sections, they are important in further work in mathematics. One group of identities can be used to rewrite a product of two functions as a sum or difference. The other group can be used to rewrite a sum or difference of two functions as a product. Some of these identities can also be used to rewrite an expression involving both sine and cosine functions as an expression with only one of these functions. In Sections 6.2 and 6.3 on conditional equations, the need for this kind of change will become clear.

All of the identities in this section result from the identities for the sine and cosine of the sum or difference of two angles. For example, adding the two identities for sin $(A + B)$ and sin $(A - B)$ gives

$$\sin (A + B) = \sin A \cos B + \cos A \sin B$$
$$\underline{\sin (A - B) = \sin A \cos B - \cos A \sin B}$$
$$\sin (A + B) + \sin (A - B) = 2 \sin A \cos B$$

Dividing by 2 gives

$$\sin A \cos B = \frac{1}{2} [\sin (A + B) + \sin (A - B)].$$

Subtracting sin $(A - B)$ from sin $(A + B)$ gives a similar identity,

$$\cos A \sin B = \frac{1}{2} [\sin (A + B) - \sin (A - B)].$$

These two product-to-sum identities are summarized below.

PRODUCT-TO-SUM IDENTITIES	$\sin A \cos B = \frac{1}{2} [\sin (A + B) + \sin (A - B)]$
	$\cos A \sin B = \frac{1}{2} [\sin (A + B) - \sin (A - B)]$

The identities for cos $(A + B)$ and cos $(A - B)$ can be used in a similar manner to get the following.

PRODUCT-TO-SUM **IDENTITIES**	$\cos A \cos B = \dfrac{1}{2} [\cos (A + B) + \cos (A - B)]$
	$\sin A \sin B = \dfrac{1}{2} [\cos (A - B) - \cos (A + B)]$

■ *Example 1*
USING A
PRODUCT-TO-SUM
IDENTITY

Rewrite cos 2θ sin θ as the sum or difference of two functions.
Use the identity for cos A sin B.

$$\cos 2\theta \sin \theta = \frac{1}{2} [\sin (2\theta + \theta) - \sin (2\theta - \theta)]$$

$$= \frac{1}{2} (\sin 3\theta - \sin \theta)$$

$$= \frac{1}{2} \sin 3\theta - \frac{1}{2} \sin \theta \quad ■$$

CAUTION	Note in Example 1 that sin 3θ − sin θ is *not* equal to sin 2θ.

■ *Example 2*
USING A
PRODUCT-TO-SUM
IDENTITY

Use an identity to evaluate cos 15° cos 45°.
Use the identity for cos A cos B.

$$\cos 15° \cos 45° = \frac{1}{2} [\cos (15° + 45°) + \cos (15° - 45°)]$$

$$= \frac{1}{2} [\cos 60° + \cos (-30°)]$$

$$= \frac{1}{2} (\cos 60° + \cos 30°)$$

$$= \frac{1}{2} \left(\frac{1}{2} + \frac{\sqrt{3}}{2} \right) = \frac{1 + \sqrt{3}}{4} \quad ■$$

Now these new identities can be used to obtain identities that are useful in calculus when a sum of trigonometric functions must be written as a product. To begin, let $A + B = x$ and let $A - B = y$. Then adding x and y gives $x + y = (A + B) + (A - B) = 2A$, or

$$A = \frac{x + y}{2}.$$

Finding $x - y$ gives $x - y = (A + B) - (A - B) = 2B$, or

$$B = \frac{x - y}{2}.$$

With these results, the identity

$$\sin A \cos B = \frac{1}{2}[\sin (A + B) + \sin (A - B)]$$

becomes

$$\sin \left(\frac{x + y}{2}\right) \cos \left(\frac{x - y}{2}\right) = \frac{1}{2}(\sin x + \sin y),$$

or

$$\sin x + \sin y = 2 \sin \left(\frac{x + y}{2}\right) \cos \left(\frac{x - y}{2}\right).$$

Other identities for $\sin x - \sin y$, $\cos x + \cos y$, and $\cos x - \cos y$ can be derived in a similar way. These sum-to-product identities are summarized below. In the summary, we use A and B rather than x and y to be consistent with other identities presented in this chapter. The variables that we use can be chosen arbitrarily.

SUM-TO-PRODUCT IDENTITIES

$$\sin A + \sin B = 2 \sin \left(\frac{A + B}{2}\right) \cos \left(\frac{A - B}{2}\right)$$

$$\sin A - \sin B = 2 \cos \left(\frac{A + B}{2}\right) \sin \left(\frac{A - B}{2}\right)$$

$$\cos A + \cos B = 2 \cos \left(\frac{A + B}{2}\right) \cos \left(\frac{A - B}{2}\right)$$

$$\cos A - \cos B = -2 \sin \left(\frac{A + B}{2}\right) \sin \left(\frac{A - B}{2}\right)$$

■ *Example 3*
USING A SUM-TO-PRODUCT IDENTITY

Write $\sin 2\gamma - \sin 4\gamma$ as a product of two functions.
Use the identity for $\sin A - \sin B$.

$$\sin 2\gamma - \sin 4\gamma = 2 \cos \left(\frac{2\gamma + 4\gamma}{2}\right) \sin \left(\frac{2\gamma - 4\gamma}{2}\right)$$

$$= 2 \cos \frac{6\gamma}{2} \sin \frac{-2\gamma}{2}$$

$$= 2 \cos 3\gamma \sin (-\gamma)$$

$$= -2 \cos 3\gamma \sin \gamma \quad ■$$

■ *Example 4*
VERIFYING AN
IDENTITY

Verify the identity

$$\frac{\sin 3s + \sin s}{\cos s + \cos 3s} = \tan 2s.$$

Work as follows.

$$\frac{\sin 3s + \sin s}{\cos s + \cos 3s} = \frac{2 \sin \left(\dfrac{\overset{A}{3s} + \overset{B}{s}}{2}\right) \cos \left(\dfrac{\overset{A}{3s} - \overset{B}{s}}{2}\right)}{2 \cos \left(\dfrac{s + 3s}{2}\right) \cos \left(\dfrac{s - 3s}{2}\right)}$$

Use the sum-to-product identities.

$$= \frac{\sin 2s \cos s}{\cos 2s \cos (-s)}$$

$$= \frac{\sin 2s \cos s}{\cos 2s \cos s} \qquad \cos (-s) = \cos s$$

$$= \frac{\sin 2s}{\cos 2s}$$

$$= \tan 2s \qquad \frac{\sin A}{\cos A} = \tan A \quad ■$$

5.7 EXERCISES ■

Rewrite each of the following as a sum or difference of trigonometric functions. See Examples 1 and 2.

1. $\cos 45° \sin 25°$ **2.** $2 \sin 74° \cos 114°$ **3.** $3 \cos 5x \cos 3x$ **4.** $2 \sin 2x \sin 4x$

5. $\sin (-\theta) \sin (-3\theta)$ **6.** $4 \cos 8\alpha \sin (-4\alpha)$ **7.** $-8 \cos 4y \cos 5y$ **8.** $2 \sin 3k \sin 14k$

Rewrite each of the following as a product of trigonometric functions. See Example 3.

9. $\sin 60° - \sin 30°$ **10.** $\sin 28° + \sin (-18°)$ **11.** $\cos 42° + \cos 148°$ **12.** $\cos 2x - \cos 8x$

13. $\sin 12\beta - \sin 3\beta$ **14.** $\cos 5x + \cos 10x$ **15.** $-3 \sin 2x + 3 \sin 5x$ **16.** $-\cos 8s + \cos 14s$

Verify that each of the following is an identity. See Example 4.

17. $\tan x = \dfrac{\sin 3x - \sin x}{\cos 3x + \cos x}$

18. $\dfrac{\sin 5t + \sin 3t}{\cos 3t - \cos 5t} = \cot t$

19. $\dfrac{\cot 2\theta}{\tan 3\theta} = \dfrac{\cos 5\theta + \cos \theta}{\cos \theta - \cos 5\theta}$

20. $\dfrac{\cos \alpha + \cos \beta}{\cos \alpha - \cos \beta} = -\cot \left(\dfrac{\alpha + \beta}{2}\right) \cot \left(\dfrac{\alpha - \beta}{2}\right)$

21. $\dfrac{1}{\tan 2s} = \dfrac{\sin 3s - \sin s}{\cos s - \cos 3s}$

22. $\dfrac{\sin^2 5\alpha - 2 \sin 5\alpha \sin 3\alpha + \sin^2 3\alpha}{\sin^2 5\alpha - \sin^2 3\alpha} = \dfrac{\tan \alpha}{\tan 4\alpha}$

23. $\sin 6\theta \cos 4\theta - \sin 3\theta \cos 7\theta = \sin 3\theta \cos \theta$

24. $\sin 8\beta \sin 4\beta + \cos 10\beta \cos 2\beta = \cos 6\beta \cos 2\beta$

25. $\sin^2 u - \sin^2 v = \sin (u + v) \sin (u - v)$

26. $\cos^2 u - \cos^2 v = -\sin (u + v) \sin (u - v)$

27. Show that the double-angle identity for sine can be considered a special case of the identity $\sin s \cos t = (1/2)[\sin (s + t) + \sin (s - t)]$.

28. Show that the double-angle identity $\cos 2s = 2 \cos^2 s - 1$ is a special case of the identity $\cos s \cos t = (1/2)[\cos (s + t) + \cos (s - t)]$.

29. The product-to-sum identity

$$\sin A \cos B = \frac{1}{2} [\sin (A + B) + \sin (A - B)]$$

can be stated in words as follows: The product of the sine of one angle and the cosine of another angle is equal to one-half the sum of the sine of the sum of the angles and the sine of the difference of the angles. State the other three product-to-sum identities in a similar way.

30. The sum-to-product identity

$$\sin A + \sin B = 2 \sin \left(\frac{A + B}{2}\right) \cos \left(\frac{A - B}{2}\right)$$

can be stated in words as follows: The sum of the sines of two angles is equal to twice the product of the sine of half the sum of the angles and the cosine of half the difference of the angles. State the other three sum-to-product identities in a similar way.

CHAPTER 5   SUMMARY ■

SECTION	KEY IDEAS
5.1 Fundamental Identities	**Reciprocal Identities** $$\cot \theta = \frac{1}{\tan \theta} \qquad \sec \theta = \frac{1}{\cos \theta} \qquad \csc \theta = \frac{1}{\sin \theta}$$ **Quotient Identities** $$\tan \theta = \frac{\sin \theta}{\cos \theta} \qquad \cot \theta = \frac{\cos \theta}{\sin \theta}$$ **Pythagorean Identities** $$\sin^2 \theta + \cos^2 \theta = 1 \qquad \tan^2 \theta + 1 = \sec^2 \theta \qquad 1 + \cot^2 \theta = \csc^2 \theta$$ **Negative-angle Identities** $$\sin (-\theta) = -\sin \theta \qquad \cos (-\theta) = \cos \theta \qquad \tan (-\theta) = -\tan \theta$$
5.3, 5.4 Sum and Difference Identities for Cosine **Sum and Difference Identities for Sine and Tangent**	**Sum and Difference Identities** $$\cos (A - B) = \cos A \cos B + \sin A \sin B$$ $$\cos (A + B) = \cos A \cos B - \sin A \sin B$$ $$\sin (A + B) = \sin A \cos B + \cos A \sin B$$ $$\sin (A - B) = \sin A \cos B - \cos A \sin B$$ $$\tan (A + B) = \frac{\tan A + \tan B}{1 - \tan A \tan B}$$ $$\tan (A - B) = \frac{\tan A - \tan B}{1 + \tan A \tan B}$$

SECTION	KEY IDEAS
5.5 Double-Angle Identities	**Double-Angle Identities** $$\cos 2A = \cos^2 A - \sin^2 A$$ $$\cos 2A = 1 - 2 \sin^2 A$$ $$\cos 2A = 2 \cos^2 A - 1$$ $$\sin 2A = 2 \sin A \cos A$$ $$\tan 2A = \frac{2 \tan A}{1 - \tan^2 A}$$
5.6 Half-Angle Identities	**Half-Angle Identities** $$\sin \frac{A}{2} = \pm \sqrt{\frac{1 - \cos A}{2}} \qquad \tan \frac{A}{2} = \frac{1 - \cos A}{\sin A}$$ $$\cos \frac{A}{2} = \pm \sqrt{\frac{1 + \cos A}{2}} \qquad \tan \frac{A}{2} = \frac{\sin A}{1 + \cos A}$$ $$\tan \frac{A}{2} = \pm \sqrt{\frac{1 - \cos A}{1 + \cos A}}$$ (The sign is chosen based on the quadrant of $A/2$.)
5.7 Sum and Product Identities	**Sum and Product Identities** $$\sin A \cos B = \frac{1}{2} [\sin (A + B) + \sin (A - B)]$$ $$\cos A \sin B = \frac{1}{2} [\sin (A + B) - \sin (A - B)]$$ $$\cos A \cos B = \frac{1}{2} [\cos (A + B) + \cos (A - B)]$$ $$\sin A \sin B = \frac{1}{2} [\cos (A - B) - \cos (A + B)]$$ $$\sin A + \sin B = 2 \sin \left(\frac{A + B}{2}\right) \cos \left(\frac{A - B}{2}\right)$$ $$\sin A - \sin B = 2 \cos \left(\frac{A + B}{2}\right) \sin \left(\frac{A - B}{2}\right)$$ $$\cos A + \cos B = 2 \cos \left(\frac{A + B}{2}\right) \cos \left(\frac{A - B}{2}\right)$$ $$\cos A - \cos B = -2 \sin \left(\frac{A + B}{2}\right) \sin \left(\frac{A - B}{2}\right)$$

1. Use the trigonometric identities to find the remaining five trigonometric functions of x, given that $\cos x = 3/5$ and x is in quadrant IV.

2. Given $\tan x = -5/4$, where $\pi/2 < x < \pi$, use the trigonometric identities to find the other trigonometric functions of x.

3. Find the exact values of $\sin x$, $\cos x$, and $\tan x$, for $x = \pi/12$, using
   (a) difference identities;   **(b)** half-angle identities.

4. Find the exact values of the six trigonometric functions of $165°$ using the method of Exercises 73–78 in Section 5.3.

For each item in Column I, give the letter of the item in Column II that completes an identity.

Column I	Column II
5. $\cos 210°$	**(a)** $\sin(-35°)$
6. $\sin 35°$	**(b)** $\cos 55°$
7. $\tan(-35°)$	**(c)** $\sqrt{\dfrac{1 + \cos 150°}{2}}$
8. $-\sin 35°$	**(d)** $2 \sin 150° \cos 150°$
9. $\cos 35°$	**(e)** $\cos 150° \cos 60° - \sin 150° \sin 60°$
10. $\cos 75°$	**(f)** $\cot(-35°)$
11. $\sin 75°$	**(g)** $\cos^2 150° - \sin^2 150°$
12. $\sin 300°$	**(h)** $\sin 15° \cos 60° + \cos 15° \sin 60°$
13. $\cos 300°$	**(i)** $\cos(-35°)$
	(j) $\cot 125°$

For each item in Column I, give the letter of the item in Column II that completes an identity.

Column I	Column II
14. $\sec x$	**(a)** $\dfrac{1}{\sin x}$
15. $\csc x$	**(b)** $\dfrac{1}{\cos x}$
16. $\tan x$	**(c)** $\dfrac{\sin x}{\cos x}$
17. $\cot x$	**(d)** $\dfrac{1}{\cot^2 x}$
18. $\sin^2 x$	**(e)** $\dfrac{1}{\cos^2 x}$
19. $\tan^2 x + 1$	**(f)** $\dfrac{\cos x}{\sin x}$
20. $\tan^2 x$	**(g)** $\dfrac{1}{\sin^2 x}$
	(h) $1 - \cos^2 x$

Use identities to express each of the following in terms of sin θ *and* cos θ, *and simplify.*

21. $\sec^2 \theta - \tan^2 \theta$

22. $\dfrac{\cot \theta}{\sec \theta}$

23. $\tan^2 \theta \, (1 + \cot^2 \theta)$

24. $\csc \theta + \cot \theta$

25. $\csc^2 \theta + \sec^2 \theta$

26. $\tan \theta - \sec \theta \csc \theta$

For each of the following find sin (x + y), cos (x − y), tan (x + y), *and the quadrant of x + y.*

27. $\sin x = -1/4$, $\cos y = -4/5$,   *x* and *y* in quadrant III

28. $\sin y = -2/3$, $\cos x = -1/5$,   *x* in quadrant II, *y* in quadrant III

29. $\sin x = 1/10$, $\cos y = 4/5$,   *x* in quadrant I, *y* in quadrant IV

30. $\cos x = 2/9$, $\sin y = -1/2$,   *x* in quadrant IV, *y* in quadrant III

Find sine and cosine of each of the following.

31. θ, given $\cos 2\theta = -3/4$, $90° < 2\theta < 180°$

32. *B*, given $\cos 2B = 1/8$,   *B* in quadrant IV

33. 2*x*, given $\tan x = 3$, $\sin x < 0$

34. 2*y*, given $\sec y = -5/3$, $\sin y > 0$

Find each of the following.

35. cos θ/2, given $\cos \theta = -1/2$,   with $90° < \theta < 180°$

36. sin *A*/2, given $\cos A = -3/4$,   with $90° < A < 180°$

37. tan *x*, given $\tan 2x = 2$, $\pi < x < 3\pi/2$

38. sin *y*, given $\cos 2y = -1/3$, $\pi/2 < y < \pi$

Verify that each of the following equations is an identity.

39. $\sin^2 x - \sin^2 y = \cos^2 y - \cos^2 x$

40. $2 \cos^3 x - \cos x = \dfrac{\cos^2 x - \sin^2 x}{\sec x}$

41. $-\cot \dfrac{x}{2} = \dfrac{\sin 2x + \sin x}{\cos 2x - \cos x}$

42. $\dfrac{\sin^2 x}{2 - 2 \cos x} = \cos^2 \dfrac{x}{2}$

43. $\dfrac{\sin 2x}{\sin x} = \dfrac{2}{\sec x}$

44. $2 \cos A - \sec A = \cos A - \dfrac{\tan A}{\csc A}$

45. $\dfrac{2 \tan B}{\sin 2B} = \sec^2 B$

46. $\tan \beta = \dfrac{1 - \cos 2\beta}{\sin 2\beta}$

47. $1 + \tan^2 \alpha = 2 \tan \alpha \csc 2\alpha$

48. $-\dfrac{\sin (A - B)}{\sin (A + B)} = \dfrac{\cot A - \cot B}{\cot A + \cot B}$

49. $\dfrac{\sin t}{1 - \cos t} = \cot \dfrac{t}{2}$

50. $2 \cos (A + B) \sin (A + B) = \sin 2A \cos 2B + \sin 2B \cos 2A$

51. $\dfrac{2 \cot x}{\tan 2x} = \csc^2 x - 2$

52. $\sin t = \dfrac{\cos t \sin 2t}{1 + \cos 2t}$

53. $\tan \theta \sin 2\theta = 2 - 2 \cos^2 \theta$

54. $\csc A \sin 2A - \sec A = \cos 2A \sec A$

55. $2 \tan x \csc 2x - \tan^2 x = 1$

56. $2 \cos^2 \theta - 1 = \dfrac{1 - \tan^2 \theta}{1 + \tan^2 \theta}$

57. $\tan \theta \cos^2 \theta = \dfrac{2 \tan \theta \cos^2 \theta - \tan \theta}{1 - \tan^2 \theta}$

58. $-\cot \dfrac{x}{2} = \dfrac{\sin 2x + \sin x}{\cos 2x - \cos x}$

59. $2 \cos^3 x - \cos x = \dfrac{\cos^2 x - \sin^2 x}{\sec x}$

60. $\sin^3 \theta = \sin \theta - \cos^2 \theta \sin \theta$

61. $\cos^4 \theta = \dfrac{3}{8} + \dfrac{1}{2} \cos 2\theta + \dfrac{1}{8} \cos 4\theta$

62. $\tan \dfrac{7}{2} x = \dfrac{2 \tan \dfrac{7}{4} x}{1 - \tan^2 \dfrac{7}{4} x}$

63. $\sec^2 \alpha - 1 = \dfrac{\sec 2\alpha - 1}{\sec 2\alpha + 1}$

64. $\dfrac{\sin 3t + \sin 2t}{\sin 3t - \sin 2t} = \dfrac{\tan \dfrac{5t}{2}}{\tan \dfrac{t}{2}}$

65. $\tan 4\theta = \dfrac{2 \tan 2\theta}{2 - \sec^2 2\theta}$

66. $\sin 2\alpha = \dfrac{2(\sin \alpha - \sin^3 \alpha)}{\cos \alpha}$

67. $2 \cos^2 \dfrac{x}{2} \tan x = \tan x + \sin x$

68. $\csc \theta - \cot \theta = \tan \dfrac{\theta}{2}$

69. $\tan \left(\dfrac{x}{2} + \dfrac{\pi}{4} \right) = \sec x + \tan x$

70. $\dfrac{1}{2} \cot \dfrac{x}{2} - \dfrac{1}{2} \tan \dfrac{x}{2} = \cot x$

71. Exact values of the trigonometric functions of 15° can be found by the following method, an alternative to the use of the half-angle formulas. Start with a right triangle ABC having a 60° angle at A and a 30° angle at B. Let the hypotenuse of this triangle have length 2. Extend side BC and draw a semicircle with diameter along BC extended, center at B, and radius AB. Draw segment AE. (See the figure.) Since any angle inscribed in a semicircle is a right angle, triangle AED is a right triangle.

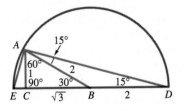

Prove each of the following statements.
(a) Triangle ABD is isosceles.
(b) Angle ABD is 150°.
(c) Angle DAB and angle ADB are each 15°.
(d) DC has length $2 + \sqrt{3}$.
(e) Since AC has length 1, the length of AD is $AD = \sqrt{1^2 + (2 + \sqrt{3})^2}$. Reduce this to $\sqrt{8 + 4\sqrt{3}}$, and show that this result equals $\sqrt{6} + \sqrt{2}$.

72. Use angle ADB of triangle ADE and find $\cos 15°$.

73. Show that AE has length $\sqrt{6} - \sqrt{2}$. Then find $\sin 15°$.

74. Use triangle ACE and find $\tan 15°$.

■ THE GRAPHING CALCULATOR ■

When we verify a trigonometric identity, we are essentially showing that two trigonometric functions have the same function values for all numbers in their domains. In Section 5.2 we discussed how trigonometric identities can be verified by using algebraic manipulations and trigonometric substitutions.

A graphing calculator can help us determine whether a given equation is an identity—that is, whether two functions are identical. Consider the equation

$$\tan^2 x(1 + \cot^2 x) = \frac{1}{1 - \sin^2 x},$$

which was verified as an identity in Example 2 of Section 5.2. We can also use a popular model graphing calculator, the TI-81 by Texas Instruments, to verify this identity. First, be sure that the calculator is set in the radian and sequential graphing modes, using the [MODE] key. Then, use the [ZOOM] key to set the limits for x and y based on the "Trig" domain and range. Now enter the function on the left side of the equation as Y_1, using the [Y=] key. Care should be taken here to use parentheses as necessary. One way of entering the function will result in the following display.

$$Y_1 = (\tan x)_\wedge 2 * \left(1 + \left(\frac{1}{\tan x}\right)_\wedge 2\right)$$

Notice how cot x is entered as 1/tan x. The function on the right side of the equation can now be entered as Y_2, resulting in the following display.

$$Y_2 = 1/(1 - (\sin x)_\wedge 2)$$

Now press the [GRAPH] key. The calculator will begin by graphing Y_1. See the figure.

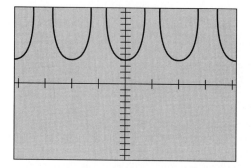

Notice that a small square appears in the upper right corner of the screen, indicating that graphing is taking place. After Y_1 is graphed, the graph of Y_2 will follow. Since this equation is indeed an identity, the graph of Y_2 will coincide with that of Y_1, as indicated by the prolonged presence of the small square. Had this not been an identity, the graph of Y_2 would have been different from that of Y_1.

You may wish to experiment with the following identities, using your graphing calculator as described above. They are identities that appear in the exercises for Section 5.2. For simplicity, we will use the variable x in all of them.

1. $\dfrac{\cot x}{\csc x} = \cos x$

2. $\cos^2 x (\tan^2 x + 1) = 1$

3. $\dfrac{\sin^2 x}{\cos x} = \sec x - \cos x$

4. $\cot x + \tan x = \sec x \csc x$

5. $\dfrac{\cos x}{\sin x \cot x} = 1$

6. $\dfrac{\cos x + 1}{\tan^2 x} = \dfrac{\cos x}{\sec x - 1}$

7. $\dfrac{\tan^2 x - 1}{\sec^2 x} = \dfrac{\tan x - \cot x}{\tan x + \cot x}$

8. $\sin x(1 + \tan x) + \cos x(1 + \cot x) = \sec x + \csc x$

9. $\dfrac{1}{\cos^2 x} + \dfrac{1}{\sin^2 x} = \dfrac{1}{\cos^2 x - \cos^4 x}$

10. $\dfrac{\sec x}{1 + \tan x} = \dfrac{\csc x}{1 + \cot x}$

INVERSE TRIGONOMETRIC FUNCTIONS AND TRIGONOMETRIC EQUATIONS

In many applications of trigonometry, we are given the trigonometric function value of an angle or number, and must find the value of the angle or number. We saw this first in Section 2.3, and in this chapter we will study the *inverse trigonometric (or circular) functions* in detail. In Chapter 5 we studied trigonometric equations that were identities; in this chapter we will learn how to solve *conditional* trigonometric equations involving trigonometric functions or inverse trigonometric functions.

6.1 ———— INVERSE TRIGONOMETRIC FUNCTIONS

This section begins with a review of the basic concepts of inverse functions, a topic usually studied in intermediate or college algebra courses. Recall from Section 1.1 that for a function f, every element x in the domain corresponds to one and only one element y, or $f(x)$, in the range. This means that if the point (a, b) lies on the graph of f, then there is no other point on the graph that has a as a first coordinate. However, there may be other points having b as a second coordinate, since the definition of a function allows range elements to be used more than once.

If a function is defined so that each range element is used only once, then it is called a **one-to-one function.** For example, the function $f(x) = x^3$ is a one-to-one function, because every real number has exactly one real cube root. On the other hand, $f(x) = x^2$ is not one-to-one, because, for example, $f(2) = 4$ and $f(-2) = 4$. We can find two domain elements, 2 and -2, that correspond to the range element 4.

There is a graphical test, called the horizontal line test, that will help determine whether a function is one-to-one.

HORIZONTAL LINE TEST	Any horizontal line will intersect the graph of a one-to-one function in at most one point. If it is possible to draw a horizontal line that intersects the graph of a function in more than one point, then the function is not one-to-one.

Figure 6.1(a) shows the graph of $f(x) = x^3$. Since any horizontal line will intersect the graph in exactly one point, it passes the horizontal line test. Figure 6.1(b) shows the graph of $f(x) = x^2$. Since many horizontal lines will intersect the graph in two points, it is not a one-to-one function.

By interchanging the components of the ordered pairs of a one-to-one function f, we obtain a set of ordered pairs that satisfies the definition of function. This new function is called the **inverse function,** or **inverse,** of f, and is symbolized by f^{-1}.

INVERSE FUNCTION	The inverse function of the one-to-one function f is defined as $$f^{-1} = \{(y, x) \mid (x, y) \text{ belongs to } f\}.$$

CAUTION In this context, the -1 is not an exponent. That is,

$$f^{-1}(x) \neq \frac{1}{f(x)}.$$

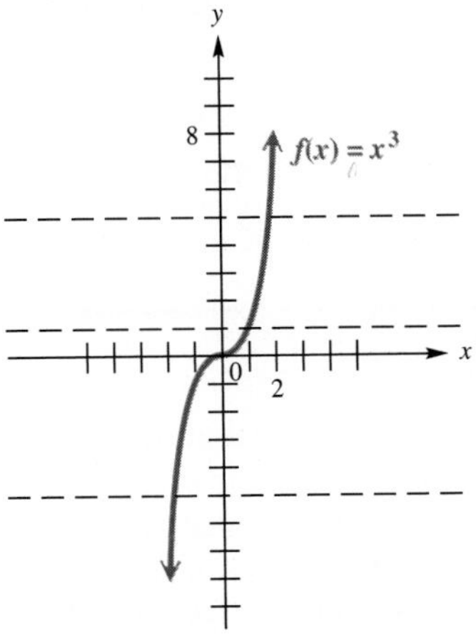

$f(x) = x^3$ is a one-to-one function. It satisfies the conditions of the horizontal line test.

(a)

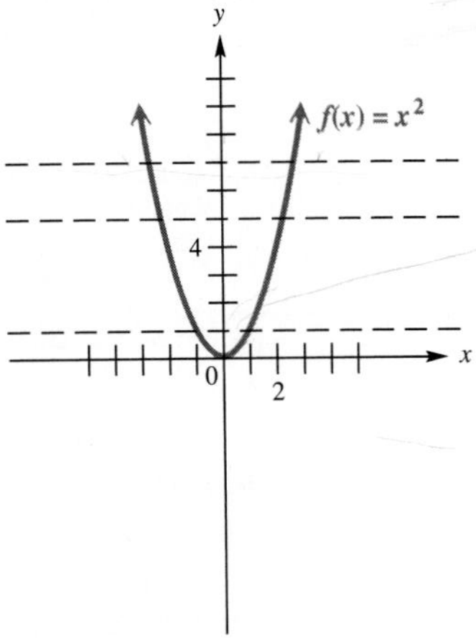

$f(x) = x^2$ is not one-to-one. It does not satisfy the conditions of the horizontal line test.

(b)

■ **FIGURE 6.1**

Based on the definition of an inverse function, we can make the following general statements about inverses.

1. If the point (a, b) lies on the graph of the one-to-one function f, then the point (b, a) lies on the graph of f^{-1}.

2. The domain of f is equal to the range of f^{-1}, and the range of f is equal to the domain of f^{-1}.

3. For all x in the domain of f, $f^{-1}[f(x)] = x$, and for all x in the domain of f^{-1}, $f[f^{-1}(x)] = x$.

4. Because the point (b, a) is a reflection of the point (a, b) about the line $y = x$, the graph of f^{-1} is a reflection of the graph of f about this line. See Figure 6.2.

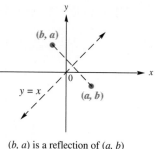

(b, a) is a reflection of (a, b) about the line $y = x$.

(a)

The graph of f^{-1} is a "mirror image" of the graph of f about the line $y = x$.

(b)

■ **FIGURE 6.2**

As we shall see later in this section, there may be a reason for restricting the domain of a function that is not one-to-one to one that is. In such a restriction, we will also require that the range be unchanged. For example, we saw that $f(x) = x^2$, with its natural domain $(-\infty, \infty)$, is not one-to-one. However, if we restrict its domain to be the set of non-negative numbers $[0, \infty)$, we obtain a new function that is one-to-one, and has the same range as before. See Figure 6.3.

We could also have chosen to restrict the domain to $(-\infty, 0]$ so that a one-to-one function was obtained. For important functions, such as the trigonometric functions, such choices are usually made based on general agreement by mathematicians.

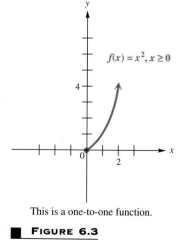

This is a one-to-one function.

■ **FIGURE 6.3**

Now let us consider the function $y = \sin x$. From Figure 6.4 and the horizontal line test, it is clear that $y = \sin x$ is not a one-to-one function. By suitably restricting the domain of the sine function, however, a one-to-one function can be defined. It is common to restrict the domain of $y = \sin x$ to the interval $[-\pi/2, \pi/2]$, which gives the part of the graph shown in color in Figure 6.4. We will call this function with restricted domain $y = \text{Sin } x$ (with a capital S) to distinguish it from $y = \sin x$ (with a lowercase s), which has the real numbers as domain. As Figure 6.4 shows, the range of both $y = \sin x$ and

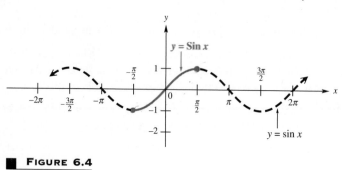

■ **FIGURE 6.4**

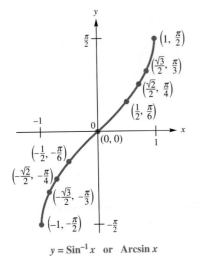

$y = \text{Sin}^{-1} x$ or $\text{Arcsin } x$

$y = \text{Sin } x$ is $[-1, 1]$. Reflecting the graph of $y = \text{Sin } x$ about the line $y = x$ gives the graph of the inverse function, shown in Figure 6.5. Some key points are labelled on the graph.

The equation of the inverse of $y = \text{Sin } x$ is found by first exchanging x and y to get $x = \text{Sin } y$. This equation then is solved for y by writing $y = \text{Sin}^{-1} x$ (read "inverse sine of x". Note that $\text{Sin}^{-1} x$ does not mean $1/\text{Sin } x$.) As Figure 6.5 shows, the domain of $y = \text{Sin}^{-1} x$ is $[-1, 1]$, while the restricted domain of $y = \text{Sin } x$, $[-\pi/2, \pi/2]$, is the range of $y = \text{Sin}^{-1} x$. An alternative notation for $\text{Sin}^{-1} x$ is **Arcsin x.**

SIN⁻¹ X OR ARCSIN X	$y = \text{Sin}^{-1} x$ or $y = \text{Arcsin } x$ means $x = \text{Sin } y$, for y in $[-\pi/2, \pi/2]$.

Thus, we may think of $y = \text{Sin}^{-1} x$ or $y = \text{Arcsin } x$ as "y is the number in $[-\pi/2, \pi/2]$ whose sine is x." These two types of notation will be used in the rest of this book.

■ **Example 1**

FINDING INVERSE SINE VALUES

Find y in each of the following.

(a) $y = \text{Arcsin } \dfrac{1}{2}$

The graph of the function $y = \text{Arcsin } x$ (Figure 6.5) shows that the point $(1/2, \pi/6)$ lies on the graph. Therefore, $\text{Arcsin } (1/2) = \pi/6$. Alternatively, we may think of $y = \text{Arcsin } (1/2)$ as

$$y \text{ is the number in } \left[-\frac{\pi}{2}, \frac{\pi}{2}\right] \text{ whose sine is } \frac{1}{2}.$$

Rewrite the equation as $\text{Sin } y = 1/2$. Since $\sin \pi/6 = 1/2$ and $\pi/6$ is in the range of the Arcsin function, $y = \pi/6$.

(b) $y = \text{Sin}^{-1} (-1)$

Writing the alternative equation, $\text{Sin } y = -1$, shows that $y = -\pi/2$. This can be verified by noticing that the point $(-1, -\pi/2)$ is on the graph of $y = \text{Sin}^{-1} x$. ■

CAUTION

In Example 1(b), it is tempting to give the value of $\text{Sin}^{-1}(-1)$ as $3\pi/2$, since $\sin(3\pi/2) = -1$. Notice, however, that $3\pi/2$ is not in the range of the inverse sine function. Be certain (in dealing with *all* inverse trigonometric functions) that the number given for an inverse function value is in the range of the particular inverse function being considered.

The function $y = \text{Cos}^{-1} x$ (or $y = \text{Arccos } x$) is defined by choosing a function $y = \text{Cos } x$ with the restricted domain $[0, \pi]$. This domain becomes the range of $y = \text{Cos}^{-1} x$. The range of $y = \text{Cos } x$, the interval $[-1, 1]$ becomes the domain of $y = \text{Cos}^{-1}x$. The graph of $y = \text{Cos}^{-1} x$ is shown in Figure 6.6. Again, some key points are shown.

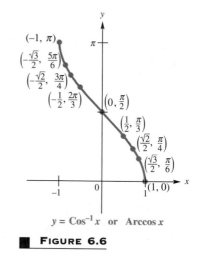

$y = \text{Cos}^{-1}x$ or Arccos x

■ **FIGURE 6.6**

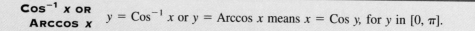

COS⁻¹ X OR ARCCOS X $y = \text{Cos}^{-1} x$ or $y = \text{Arccos } x$ means $x = \text{Cos } y$, for y in $[0, \pi]$.

■ *Example 2*
FINDING INVERSE COSINE VALUES

Find y in each of the following.

(a) $y = \text{Arccos } 1$

Since the point $(1, 0)$ lies on the graph of $y = \text{Arccos } x$, the value of y is 0. Alternatively, we may think of $y = \text{Arccos } 1$ as "y is the number in $[0, \pi]$ whose cosine is 1," or $\cos y = 1$. Then $y = 0$, since $\cos 0 = 1$ and 0 is in the range of the Arccos function.

(b) $y = \text{Cos}^{-1}\left(-\dfrac{\sqrt{2}}{2}\right)$

We must find the value of y that satisfies $\text{Cos } y = -\sqrt{2}/2$, where y is in the interval $[0, \pi]$, the range of the function $y = \text{Cos}^{-1} x$. The only value for y that satisfies these conditions is $3\pi/4$. This can be verified from the graph in Figure 6.6. ■

The inverse tangent function is defined below. Its graph is given in Figure 6.7.

TAN^{-1} X OR ARCTAN X	$y = \text{Tan}^{-1} x$ or $y = \text{Arctan } x$ means $x = \text{Tan } y$, for y in $(-\pi/2, \pi/2)$.

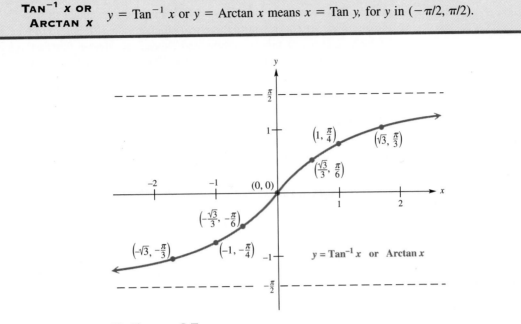

FIGURE 6.7

From the graph we can see that Arctan $0 = 0$, Arctan $1 = \pi/4$, Arctan $(-\sqrt{3}) = -\pi/3$, and so on.

We have defined the inverse sine, cosine, and tangent functions with suitable restrictions on the domains. The other three inverse trigonometric functions are similarly defined.* The six inverse trigonometric functions with their domains and ranges are given in the table. This information, particularly the range for each function, should be learned. (The graphs of the last three inverse trigonometric functions are left for the exercises.)

INVERSE TRIGONOMETRIC FUNCTIONS	Function	Domain	Range	Quadrants of the Unit Circle Range Values Come From
	$y = \text{Sin}^{-1} x$	$[-1, 1]$	$[-\pi/2, \pi/2]$	I and IV
	$y = \text{Cos}^{-1} x$	$[-1, 1]$	$[0, \pi]$	I and II
	$y = \text{Tan}^{-1} x$	$(-\infty, \infty)$	$(-\pi/2, \pi/2)$	I and IV
	$y = \text{Cot}^{-1} x$	$(-\infty, \infty)$	$(0, \pi)$	I and II
	$y = \text{Sec}^{-1} x$	$(-\infty, -1] \cup [1, \infty)$	$[0, \pi], y \neq \pi/2$	I and II
	$y = \text{Csc}^{-1} x$	$(-\infty, -1] \cup [1, \infty)$	$[-\pi/2, \pi/2], y \neq 0$	I and IV

*Sec^{-1} and Csc^{-1} are sometimes defined with different ranges.

The inverse trigonometric functions are formally defined with real number ranges. However, there are times when it may be convenient to find the degree-measured angles equivalent to these real number values. It is also often convenient to think in terms of the unit circle, and choose the inverse function values based on the quadrants given in the table above. The next example uses these ideas.

■ *Example 3*
FINDING INVERSE VALUES (DEGREE-MEASURED ANGLES)

Find the *degree measure* of θ in each of the following.

(a) θ, if θ = Arctan 1

Here θ must be in $(-90°, 90°)$, but since $1 > 0$, θ must be in quadrant I. The alternative statement, Tan θ = 1, leads to θ = 45°.

(b) θ, if θ = Sec^{-1} 2

Write the equation as Sec θ = 2. For Sec^{-1} x, θ is in quadrant I or II. Since 2 is positive, θ is in quadrant I and θ = 60°, since Sec 60° = 2. Note that 60° (the degree equivalent of $\pi/3$) is in the range of the inverse secant function. ■

The inverse trigonometric function keys on a calculator give results in the proper quadrant, for the Sin^{-1}, Cos^{-1}, and Tan^{-1} functions, according to the definitions of these functions. For example, on a calculator, in degrees,

$$\text{Sin}^{-1}.5 = 30°, \qquad \text{Sin}^{-1}(-.5) = -30°,$$
$$\text{Tan}^{-1}(-1) = -45°, \qquad \text{and} \qquad \text{Cos}^{-1}(-.5) = 120°.$$

Similar results are found when the calculator is set for radian measure. This is not the case for Cot^{-1}. For example, since the sequence [1/x] [INV] [tan] is used to find Cot^{-1}, the calculator gives values of Cot^{-1} with the same range as Tan^{-1}, $(-\pi/2, \pi/2)$, which is not the correct range for Cot^{-1}. For Cot^{-1} the proper range must be considered and the results adjusted accordingly.

■ *Example 4*
FINDING AN INVERSE FUNCTION VALUE WITH A CALCULATOR

Find θ in degrees if θ = Arccot $(-.35410000)$.

Enter $-.35410000$ in the calculator and press the keys [1/x] [INV] [tan] (or an equivalent sequence) to get $-70.500946°$. The restriction on the range of Arccot means that θ must be in quadrant II, and the absolute value of the angle obtained in the display is the reference angle of θ. Therefore,

$$θ = 180° - |-70.500946°| = 109.499054°. \quad ■$$

■ *Example 5*
FINDING FUNCTION VALUES WITHOUT A CALCULATOR

Evaluate each of the following without a calculator.

(a) $\sin\left(\text{Tan}^{-1}\dfrac{3}{2}\right)$

Let

$$θ = \text{Tan}^{-1}\frac{3}{2}, \text{ so that Tan } θ = \frac{3}{2}.$$

Since Tan^{-1} is defined only in quadrants I and IV and since 3/2 is positive, θ is in quadrant I. Sketch θ in quadrant I, and label a triangle as shown in Figure 6.8. The hypotenuse is $\sqrt{13}$ and the value of sine is the quotient of the side opposite and the hypotenuse, so

$$\sin\left(\text{Tan}^{-1}\frac{3}{2}\right) = \sin\theta = \frac{3}{\sqrt{13}} = \frac{3\sqrt{13}}{13}.$$

To check this result on a calculator, enter 3/2 as 1.5. Then find Tan^{-1} 1.5, and finally find $\sin(\text{Tan}^{-1}$ 1.5). Store this result and calculate $3\sqrt{13}/13$, which should agree with the result for $\sin(\text{Tan}^{-1}$ 1.5). Since the values are only approximations, this check does not *prove* that the result is correct, but it is highly suggestive that it is correct.

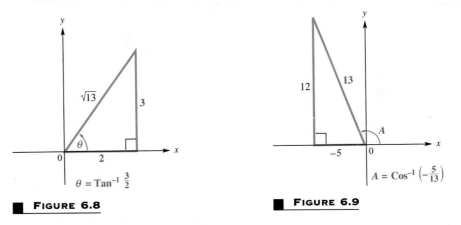

$\theta = \text{Tan}^{-1}\frac{3}{2}$

■ **FIGURE 6.8**

$A = \text{Cos}^{-1}\left(-\frac{5}{13}\right)$

■ **FIGURE 6.9**

(b) $\tan\left(\text{Cos}^{-1}\left(-\frac{5}{13}\right)\right)$

Let $A = \text{Cos}^{-1}(-5/13)$. Then $\text{Cos } A = -5/13$. Since $\text{Cos}^{-1} x$ for a negative value of x is in quadrant II, sketch A in quadrant II, as shown in Figure 6.9.

From the triangle in Figure 6.9,

$$\tan\left(\text{Cos}^{-1}\left(-\frac{5}{13}\right)\right) = \tan A = -\frac{12}{5}. \quad ■$$

■ *Example 6*

FINDING FUNCTION
VALUES USING SUM
AND DOUBLE-ANGLE
FORMULAS

Evaluate the following without using a calculator.

(a) $\cos\left(\text{Arctan }\sqrt{3} + \text{Arcsin }\frac{1}{3}\right)$

Let $A = \text{Arctan }\sqrt{3}$ and $B = \text{Arcsin } 1/3$ so that $\text{Tan } A = \sqrt{3}$ and $\text{Sin } B = 1/3$. Sketch both A and B in quadrant I, as shown in Figure 6.10.

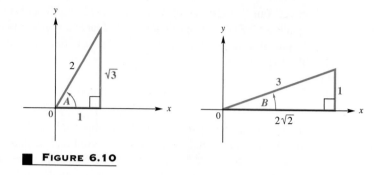

■ FIGURE 6.10

Now use the identity for cos $(A + B)$.

$$\cos(A + B) = \cos A \cos B - \sin A \sin B$$

$$\cos\left(\text{Arctan } \sqrt{3} + \text{Arcsin } \frac{1}{3}\right) = \cos(\text{Arctan } \sqrt{3})\cos\left(\text{Arcsin } \frac{1}{3}\right)$$

$$-\sin(\text{Arctan } \sqrt{3})\sin\left(\text{Arcsin } \frac{1}{3}\right) \quad \textbf{(1)}$$

From the sketch in Figure 6.10,

$$\cos(\text{Arctan } \sqrt{3}) = \cos A = \frac{1}{2}, \qquad \cos\left(\text{Arcsin } \frac{1}{3}\right) = \cos B = \frac{2\sqrt{2}}{3},$$

$$\sin(\text{Arctan } \sqrt{3}) = \sin A = \frac{\sqrt{3}}{2}, \qquad \sin\left(\text{Arcsin } \frac{1}{3}\right) = \sin B = \frac{1}{3}.$$

Substitute these values into equation (1) to get

$$\cos\left(\text{Arctan } \sqrt{3} + \text{Arcsin } \frac{1}{3}\right) = \frac{1}{2}\cdot\frac{2\sqrt{2}}{3} - \frac{\sqrt{3}}{2}\cdot\frac{1}{3}$$

$$= \frac{2\sqrt{2}}{6} - \frac{\sqrt{3}}{6}$$

$$= \frac{2\sqrt{2} - \sqrt{3}}{6}.$$

(b) $\tan\left(2 \text{ Arcsin } \frac{2}{5}\right)$

Let Arcsin $(2/5) = B$. Then, from the identity for the tangent of the double angle,

$$\tan\left(2 \text{ Arcsin } \frac{2}{5}\right) = \tan(2B)$$

$$= \frac{2\tan B}{1 - \tan^2 B}.$$

Since Arcsin (2/5) = *B*, sin *B* = 2/5. Sketch a triangle in quadrant I, find the length of the third side, and then find tan *B*. From the triangle in Figure 6.11, tan *B* = 2/√21, and

$$\tan\left(2\,\text{Arcsin}\,\frac{2}{5}\right) = \frac{2\left(\dfrac{2}{\sqrt{21}}\right)}{1 - \left(\dfrac{2}{\sqrt{21}}\right)^2} = \frac{\dfrac{4}{\sqrt{21}}}{1 - \dfrac{4}{21}}$$

$$= \frac{\dfrac{4}{\sqrt{21}}}{\dfrac{17}{21}} = \frac{4\sqrt{21}}{17}. \blacksquare$$

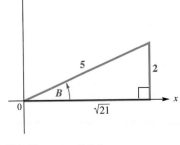

FIGURE 6.11

■ **Example 7**
WRITING A
FUNCTION VALUE IN
TERMS OF *u*

Write sin (Tan⁻¹ *u*) as an expression in *u*.
    Let θ = Tan⁻¹ *u*, so that Tan θ = *u*.
Here *u* may be positive or negative. Since
−π/2 < Tan⁻¹ *u* < π/2, sketch θ in quadrants I and IV and label two triangles as shown in Figure 6.12. Since sine is given by the quotient of the side opposite and the hypotenuse,

$$\sin(\text{Tan}^{-1} u) = \sin\theta = \frac{u}{\sqrt{u^2 + 1}}$$

$$= \frac{u\sqrt{u^2 + 1}}{u^2 + 1}.$$

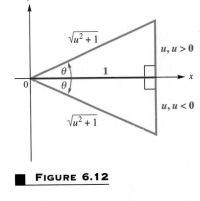

FIGURE 6.12

The result is positive when *u* is positive and negative when *u* is negative. ■

6.1 EXERCISES ■

For each of the following, give the exact real number value of y. Do not use a calculator. See Examples 1 and 2.

1. $y = \text{Arcsin}\left(-\dfrac{1}{2}\right)$
2. $y = \text{Arccos}\,\dfrac{\sqrt{3}}{2}$
3. $y = \text{Tan}^{-1}\,1$
4. $y = \text{Cot}^{-1}(-1)$

5. $y = \text{Arcsin}\,0$
6. $y = \text{Arccos}\,(-1)$
7. $y = \text{Cos}^{-1}\,\dfrac{1}{2}$
8. $y = \text{Sin}^{-1}\left(-\dfrac{\sqrt{3}}{2}\right)$

9. $y = \text{Sec}^{-1}(-\sqrt{2})$
10. $y = \text{Csc}^{-1}(-2)$
11. $y = \text{Arccot}\,(-\sqrt{3})$
12. $y = \text{Arccos}\,0$

For each of the following, give the degree measure of θ. Do not use a calculator. See Example 3.

13. $\theta = \text{Arctan}\,(-1)$
14. $\theta = \text{Arccos}\left(-\dfrac{1}{2}\right)$
15. $\theta = \text{Arcsin}\left(-\dfrac{\sqrt{3}}{2}\right)$
16. $\theta = \text{Arcsin}\left(-\dfrac{\sqrt{2}}{2}\right)$

17. $\theta = \text{Cot}^{-1}\left(-\dfrac{\sqrt{3}}{3}\right)$
18. $\theta = \text{Sec}^{-1}(-2)$
19. $\theta = \text{Csc}^{-1}(-2)$
20. $\theta = \text{Csc}^{-1}(-1)$

Use a calculator to give the real number value of each of the following.

21. Arctan 1.1111111

22. Arcsin .81926439

23. $Cot^{-1}(-.92170128)$

24. $Sec^{-1}(-1.2871684)$

25. Arcsin .92837781

26. Arccos .44624593

Use a calculator to give the value of each of the following in decimal degrees. See Example 4.

27. $Sin^{-1}(-.13349122)$

28. $Cos^{-1}(-.13348816)$

29. Arccos $(-.39876459)$

30. Arcsin .77900016

31. Csc^{-1} 1.9422833

32. Cot^{-1} 1.7670492

Graph each of the following as defined in the text, and give the domain and range.

33. $y = Cot^{-1} x$

34. $y = Arccsc\ x$

35. $y = Arcsec\ x$

36. The following expressions were used by the mathematicians who computed the value of π to 100,000 decimal places. Use a calculator to verify that each is (approximately) correct.

(a) $\pi = 16\ Tan^{-1}\dfrac{1}{5} - 4\ Tan^{-1}\dfrac{1}{239}$

(b) $\pi = 24\ Tan^{-1}\dfrac{1}{8} + 8\ Tan^{-1}\dfrac{1}{57} + 4\ Tan^{-1}\dfrac{1}{239}$

(c) $\pi = 48\ Tan^{-1}\dfrac{1}{18} + 32\ Tan^{-1}\dfrac{1}{57} - 20\ Tan^{-1}\dfrac{1}{239}$

37. Enter 1.003 in your calculator, and press the keys for inverse sine. The response will indicate that something is wrong. What is wrong?

38. Enter 1.003 in your calculator, and press the keys for inverse tangent. This time, unlike in Exercise 37, you get an answer. What is different?

39. Enter 1.74283 in your calculator (set for radians), and press the sine key. Then press the keys for inverse sine. You get 1.398763 instead of 1.74283. What happened?

40. Explain why $\sin[Sin^{-1} x]$ will always equal x, but $Sin^{-1}(\sin x)$ will not necessarily equal x.

Give the value of each of the following without using a calculator. See Examples 5 and 6.

41. $\tan\left(Arccos\dfrac{3}{4}\right)$

42. $\sin\left(Arccos\dfrac{1}{4}\right)$

43. $\cos(Tan^{-1}(-2))$

44. $\sec\left(Sin^{-1}\left(-\dfrac{1}{5}\right)\right)$

45. $\cot\left(Arcsin\left(-\dfrac{2}{3}\right)\right)$

46. $\cos\left(Arctan\dfrac{8}{3}\right)$

47. $\sec\left(Arccot\dfrac{3}{5}\right)$

48. $\cos\left(Arcsin\dfrac{12}{13}\right)$

49. $\cos\left(Arccos\dfrac{1}{2}\right)$

50. $\sin\left(Arcsin\dfrac{\sqrt{3}}{2}\right)$

51. $\tan(Tan^{-1}(-1))$

52. $\cot(Cot^{-1}(-\sqrt{3}))$

53. $\sec(Sec^{-1} 2)$

54. $\csc(Csc^{-1}\sqrt{2})$

55. $Arccos\left(\cos\dfrac{\pi}{4}\right)$

56. $Arctan\left(\tan\left(-\dfrac{\pi}{4}\right)\right)$

57. $Arcsin\left(\sin\dfrac{\pi}{3}\right)$

58. $Arccos(\cos 0)$

59. $\sin\left(2\ Tan^{-1}\dfrac{12}{5}\right)$

60. $\cos\left(2\ Sin^{-1}\dfrac{1}{4}\right)$

61. $\cos\left(2\ Arctan\dfrac{4}{3}\right)$

62. $\tan\left(2\ Cos^{-1}\dfrac{1}{4}\right)$

63. $\sin\left(2\ Cos^{-1}\dfrac{1}{5}\right)$

64. $\cos(2\ Arctan(-2))$

65. $\tan\left(2\ Arcsin\left(-\dfrac{3}{5}\right)\right)$

66. $\sin\left(2\ Arccos\dfrac{2}{9}\right)$

67. $\sin\left(Sin^{-1}\dfrac{1}{2} + Tan^{-1}(-3)\right)$

68. $\cos\left(Tan^{-1}\dfrac{5}{12} - Cot^{-1}\dfrac{4}{3}\right)$

69. $\cos\left(Arcsin\dfrac{3}{5} + Arccos\dfrac{5}{13}\right)$

70. $\tan\left(Arccos\dfrac{\sqrt{3}}{2} - Arcsin\left(-\dfrac{3}{5}\right)\right)$

Use a calculator to find each of the following. Give answers as real numbers.

71. $\cos (\text{Tan}^{-1}.5)$

72. $\sin (\text{Cos}^{-1}.25)$

73. $\tan (\text{Arcsin} .12251014)$

74. $\cot (\text{Arccos} .58236841)$

Write each of the following as an expression in u. See Example 7.

75. $\sin (\text{Arccos } u)$

76. $\tan (\text{Arccos } u)$

77. $\sec (\text{Cot}^{-1} u)$

78. $\csc (\text{Sec}^{-1} u)$

79. $\cot (\text{Arcsin } u)$

80. $\cos (\text{Arcsin } u)$

81. $\sin \left(\text{Sec}^{-1} \dfrac{u}{2} \right)$

82. $\cos \left(\text{Tan}^{-1} \dfrac{3}{u} \right)$

83. $\tan \left(\text{Arcsin} \dfrac{u}{\sqrt{u^2 + 2}} \right)$

84. $\cos \left(\text{Arccos} \dfrac{u}{\sqrt{u^2 + 5}} \right)$

85. $\sec \left(\text{Arccot} \dfrac{\sqrt{4 - u^2}}{u} \right)$

86. $\csc \left(\text{Arctan} \dfrac{\sqrt{9 - u^2}}{u} \right)$

87. Suppose that an airplane flying faster than sound goes directly over you. Assume that the plane is flying level. At the instant that you feel the sonic boom from the plane, the angle of elevation to the plane is given by

$$\alpha = 2 \, \text{Arcsin} \, \frac{1}{m},$$

where m is the Mach number of the plane's speed. (See the exercises at the end of Section 5.6.) Find α to the nearest degree for each of the following values of m.

(a) $m = 1.2$ **(b)** $m = 1.5$
(c) $m = 2$ **(d)** $m = 2.5$

88. A painting 1 m high and 3 m from the floor will cut off an angle θ to an observer, where

$$\theta = \text{Tan}^{-1} \left(\frac{x}{x^2 + 2} \right).$$

Assume that the observer is x m from the wall where the painting is displayed and that the eyes of the observer are 2 m above the ground. See the figure. Find the value of θ for the following values of x. Round to the nearest degree.

(a) 1
(b) 2
(c) 3
(d) Derive the formula given above. (*Hint:* Use the identity for $\tan (\theta + \alpha)$. Use right triangles.)

89. The following calculator trick will not work on all calculators. However, if you have a Texas Instruments or a Sharp scientific calculator, it will work. The calculator must be in the *degree* mode.

(a) Enter the year of your birth (all four digits).
(b) Subtract the number of years that have elapsed since 1980. For example, if it is 1993, subtract 13.
(c) Find the sine of the display.
(d) Find the inverse sine of the new display.
The result should be your age when you celebrate your birthday this year.

90. Explain why the procedure in Exercise 89 works as it does.

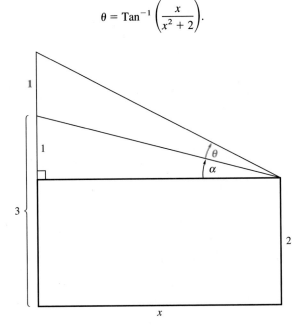

6.2 ———— # TRIGONOMETRIC EQUATIONS

In Chapter 5 we studied trigonometric equations that were identities. We now consider trigonometric equations that are **conditional**; that is, ones that are satisfied by some values but not others.

Conditional equations with trigonometric (or circular) functions can usually be solved by using algebraic methods and trigonometric identities. For example, suppose that we wish to find the solutions of the equation

$$2 \sin \theta + 1 = 0$$

for all θ in the interval $[0°, 360°)$. We use the same method here as we would in solving the algebraic equation $2y + 1 = 0$. Subtract 1 from both sides, and divide by 2.

$$2 \sin \theta + 1 = 0$$
$$2 \sin \theta = -1$$
$$\sin \theta = -\frac{1}{2}$$

Since $\sin \theta = -1/2$, we know that θ must be in either quadrant III or IV, since the sine function is negative in these two quadrants. Furthermore, the reference angle must be 30°, since $\sin 30° = 1/2$. The sketch in Figure 6.13 shows the two possible values of θ.

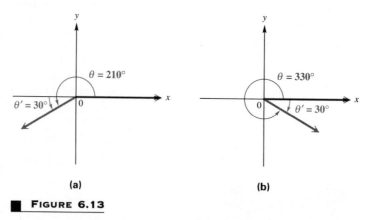

(a)   **(b)**

■ **FIGURE 6.13**

Since we are seeking the solutions in the interval $[0°, 360°)$, we find in Figure 6.13(a) the solution 210° and in Figure 6.13(b) the solution 330°.

In some cases we are required to find *all* solutions of conditional trigonometric equations. All solutions of the equation $2 \sin \theta + 1 = 0$ would be written as

$$\theta = 210° + 360° \cdot n \quad \text{or} \quad \theta = 330° + 360° \cdot n,$$

where n is any integer. We add integer multiples of 360° to obtain all angles coterminal with 210° and 330°. If we had been required to solve this equation for real

numbers (or angles in radians) in the interval $[0, 2\pi)$, the two solutions would be
$7\pi/6$ and $11\pi/6$, while all solutions would be written as

$$\theta = \frac{7\pi}{6} + 2n\pi \quad \text{or} \quad \theta = \frac{11\pi}{6} + 2n\pi,$$

where n is any integer.

In the examples in this section, we will find solutions in the intervals $[0°, 360°)$ or $[0, 2\pi)$. Remember that *all* solutions can be found using the methods described above.

▪ *Example 1*

SOLVING A
TRIGONOMETRIC
EQUATION

Solve $2\cos^2\theta - \cos\theta - 1 = 0$ in the interval $[0°, 360°)$.

We solve this equation by factoring. (It is quadratic in the term $\cos\theta$.)

$$2\cos^2\theta - \cos\theta - 1 = 0$$
$$(2\cos\theta + 1)(\cos\theta - 1) = 0$$

$$2\cos\theta + 1 = 0 \quad \text{or} \quad \cos\theta - 1 = 0$$

$$\cos\theta = -\frac{1}{2} \qquad\qquad \cos\theta = 1$$

In the first case, we have $\cos\theta = -1/2$, indicating that θ must be in either quadrant II or III, with a reference angle of 60°. Using a sketch similar to the ones in Figure 6.13 would indicate that two solutions are 120° and 240°. The second case, $\cos\theta = 1$, has the quadrantal angle 0° as its only solution in the interval. (We do not include 360° since it is not in the stated interval.) Therefore, the solutions of this equation are 0°, 120°, and 240°. ▪

NOTE

Using the concept of the inverse cosine function in Example 1, we would obtain Arccos $(-1/2) = 120°$ and Arccos $1 = 0°$. Note that the third solution, 240°, would have to be found by inspection. The method of using reference angles provides an advantage in this respect.

▪ *Example 2*

SOLVING A
TRIGONOMETRIC
EQUATION

Solve $\sin x \tan x = \sin x$ in the interval $[0°, 360°)$.

Subtract $\sin x$ from both sides, then factor on the left.

$$\sin x \tan x = \sin x$$
$$\sin x \tan x - \sin x = 0$$
$$\sin x(\tan x - 1) = 0$$

Now set each factor equal to 0.

$$\sin x = 0 \quad \text{or} \quad \tan x - 1 = 0$$
$$\tan x = 1$$

$x = 0°$ or $x = 180°$ $\qquad$ $x = 45°$ or $x = 225°$ ▪

CAUTION There are four solutions for Example 2. Trying to solve the equation by dividing both sides by sin x would give just tan $x = 1$, which would give $x = 45°$ or $x = 225°$. The other two solutions would not appear. The missing solutions are the ones that make the divisor, sin x, equal 0. For this reason, it is best to avoid dividing by a variable expression.

Recall from algebra that squaring both sides of an equation, such as $\sqrt{x+4} = x + 2$, will yield all solutions, but may also give extraneous values. (In this equation, 0 is a solution, while -3 is extraneous. Verify this.) The same situation may occur when trigonometric equations are solved in this manner, as shown in the next example.

■ *Example 3*
SOLVING A
TRIGONOMETRIC
EQUATION

Solve $\tan x + \sqrt{3} = \sec x$ in the interval $[0, 2\pi)$.

Since the tangent and secant functions are related by the identity $1 + \tan^2 x = \sec^2 x$, one method of solving this equation is to square both sides, and express $\sec^2 x$ in terms of $\tan^2 x$.

$$\tan x + \sqrt{3} = \sec x$$
$$\tan^2 x + 2\sqrt{3}\tan x + 3 = \sec^2 x$$
$$\tan^2 x + 2\sqrt{3}\tan x + 3 = 1 + \tan^2 x$$
$$2\sqrt{3}\tan x = -2$$
$$\tan x = -\frac{1}{\sqrt{3}} = -\frac{\sqrt{3}}{3}$$

The possible solutions in the given interval are $5\pi/6$ and $11\pi/6$. Now check the possible solutions. Try $5\pi/6$ first.

Left side: $\tan x + \sqrt{3} = \tan\dfrac{5\pi}{6} + \sqrt{3} = -\dfrac{\sqrt{3}}{3} + \sqrt{3} = \dfrac{2\sqrt{3}}{3}$

Right side: $\sec x = \sec\dfrac{5\pi}{6} = \dfrac{-2\sqrt{3}}{3}$

The check shows that $5\pi/6$ is not a solution. Now check $11\pi/6$.

Left side: $\tan\dfrac{11\pi}{6} + \sqrt{3} = -\dfrac{\sqrt{3}}{3} + \sqrt{3} = \dfrac{2\sqrt{3}}{3}$

Right side: $\sec\dfrac{11\pi}{6} = \dfrac{2\sqrt{3}}{3}$

This solution satisfies the equation, so $11\pi/6$ is the only solution of the given equation.   ■

In some cases trigonometric equations require a calculator to obtain approximate solutions, as in the next example.

▪ *Example 4*

SOLVING A
TRIGONOMETRIC
EQUATION USING A
CALCULATOR

Solve $\tan^2 x + \tan x - 2 = 0$ in the interval $[0, 2\pi)$.

Like Example 1, this equation is quadratic in form and may be solved for $\tan x$ by factoring.

$$\tan^2 x + \tan x - 2 = 0$$
$$(\tan x - 1)(\tan x + 2) = 0.$$

Set each factor equal to 0.

$$\tan x - 1 = 0 \quad \text{or} \quad \tan x + 2 = 0$$
$$\tan x = 1 \quad \text{or} \quad \tan x = -2$$

The solutions for $\tan x = 1$ in the interval $[0, 2\pi)$ are $x = \pi/4$ or $5\pi/4$. To solve $\tan x = -2$ in the interval, use a calculator set in the *radian* mode. We find that $\text{Tan}^{-1}(-2) \approx -1.1071487$. This is a quadrant IV number, based on the range of the inverse tangent function. However, since we want solutions in the interval $[0, 2\pi)$, we must first add π to -1.1071487, and then add 2π:

$$x \approx -1.1071487 + \pi \approx 2.03444394$$
$$x \approx -1.1071487 + 2\pi \approx 5.1760367.$$

The solutions in the required interval are

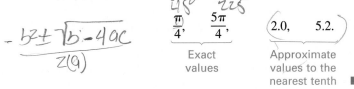

$$\frac{\pi}{4}, \quad \frac{5\pi}{4}, \qquad 2.0, \quad 5.2.$$

Exact   Approximate
values   values to the
         nearest tenth ▪

When a trigonometric equation that is quadratic in form cannot be factored, the quadratic formula can be used to solve the equation.

▪ *Example 5*

SOLVING A
TRIGONOMETRIC
EQUATION WITH THE
QUADRATIC
FORMULA

Solve $\cot^2 x + 3 \cot x = 1$ in $[0°, 360°)$.

Write the equation in quadratic form, with 0 on one side.

$$\cot^2 x + 3 \cot x - 1 = 0$$

Since this equation cannot be solved by factoring, use the quadratic formula, with $a = 1$, $b = 3$, $c = -1$, and $\cot x$ as the variable.

$$\cot x = \frac{-3 \pm \sqrt{9 + 4}}{2} = \frac{-3 \pm \sqrt{13}}{2} \approx \frac{-3 \pm 3.6055513}{2}$$

$$\cot x \approx .30277564 \quad \text{or} \quad \cot x \approx -3.3027756$$

$$x \approx 73.2°, 253.2°, 163.2°, 343.2°$$

The final answers were obtained using a calculator set in the degree mode. ▪

The methods for solving trigonometric equations illustrated in the examples can be summarized as follows.

SOLVING TRIGONOMETRIC EQUATIONS	1. If only one trigonometric function is present, first solve the equation for that function.
	2. If more than one trigonometric function is present, rearrange the equation so that one side equals 0. Then try to factor and set each factor equal to 0 to solve.
	3. If Step 2 does not work, try using identities to change the form of the equation. It may be helpful to square both sides of the equation first. If this is done, check for extraneous solutions.
	4. If the equation is quadratic in form, but not factorable, use the quadratic formula.

④

$2 \sec x + 1 = \sec x + 3$
$-\sec x - 1 \quad -\sec x - 1$
$\sec x = 2$

$x \in I$
$x = \frac{\pi}{3}$

① $2 \cot x + 1 + 1 = 0$
$2 \cot x + 2 = 0$
$2 (\cot x + 1) = 0$
$2 \neq 0 \qquad \cot x + 1 = 0$
$\qquad -1 \quad -1$
$\qquad \cot x = -1$

6.2 EXERCISES ■

Solve each of the following equations for solutions in the interval $[0, 2\pi)$. *See Examples 3 and 4.*

1. $2 \cot x + 1 = -1$ **2.** $\sin x + 2 = 3$ **3.** $2 \sin x + 3 = 4$

4. $2 \sec x + 1 = \sec x + 3$ **5.** $\tan^2 x - 3 = 0$ **6.** $\sec^2 x - 2 = -1$

7. $(\cot x - \sqrt{3})(2 \sin x + \sqrt{3}) = 0$ **8.** $(\tan x - 1)(\cos x - 1) = 0$ **9.** $(\cot x - 1)(\sqrt{3} \cot x + 1) = 0$

10. $(\csc x + 2)(\csc x - \sqrt{2}) = 0$ **11.** $\cos^2 x + 2 \cos x + 1 = 0$ **12.** $2 \cos^2 x - \sqrt{3} \cos x = 0$

13. $-2 \sin^2 x = 3 \sin x + 1$ **14.** $\cos^2 x - \sin^2 x = 0$

Solve each of the following equations for solutions in the interval $[0°, 360°)$. *See Examples 1 and 2.*

15. $2 \sin \theta - 1 = \csc \theta$ **16.** $\tan \theta + 1 = \sqrt{3} + \sqrt{3} \cot \theta$ **17.** $\tan \theta - \cot \theta = 0$

18. $\cos^2 \theta = \sin^2 \theta + 1$ **19.** $\csc^2 \theta - 2 \cot \theta = 0$ **20.** $\tan^3 \theta = 3 \tan \theta$

21. $2 \cos^4 \theta = \cos^2 \theta$ **22.** $\sin^2 \theta \cos \theta = \cos \theta$ **23.** $2 \tan^2 \theta \sin \theta - \tan^2 \theta = 0$

24. $\sin^2 \theta \cos^2 \theta = 0$ **25.** $\sec^2 \theta \tan \theta = 2 \tan \theta$ **26.** $4(1 + \sin \theta) = \dfrac{3}{1 - \sin \theta}$

27. $\sin \theta + \cos \theta = 1$ **28.** $\sec \theta - \tan \theta = 1$

Solve each of the following equations for solutions in the interval $[0°, 360°)$. *Use a calculator and express approximate solutions to the nearest tenth of a degree. In Exercises 35–42, you will need to use the quadratic formula. See Examples 4 and 5.*

29. $3 \sin^2 x - \sin x = 2$ **30.** $\dfrac{2 \tan x}{3 - \tan^2 x} = 1$ **31.** $\sec^2 \theta = 2 \tan \theta + 4$

32. $5 \sec^2 \theta = 6 \sec \theta$ **33.** $3 \cot^2 \theta = \cot \theta$ **34.** $8 \cos \theta = \cot \theta$

35. $9 \sin^2 x - 6 \sin x = 1$ **36.** $4 \cos^2 x + 4 \cos x = 1$ **37.** $\tan^2 x + 4 \tan x + 2 = 0$

38. $3 \cot^2 x - 3 \cot x - 1 = 0$ **39.** $\sin^2 x - 2 \sin x + 3 = 0$ **40.** $2 \cos^2 x + 2 \cos x - 1 = 0$

41. $\cot x + 2 \csc x = 3$ **42.** $2 \sin x = 1 - 2 \cos x$

43. Refer to Example 1. The solutions in the interval $[0°, 360°)$ are $0°$, $120°$, and $240°$. See the discussion at the beginning of this section, and express *all* solutions of this equation (in degrees).

44. Refer to Example 4. The solutions in the interval $[0, 2\pi)$ are $\pi/4$, $5\pi/4$, 2.0, and 5.2. See the discussion at the beginning of this section and express *all* solutions of this equation (in real numbers).

45. What is wrong with the following solution for all x in the interval $[0, 2\pi)$? Solve $\sin^2 x - \sin x = 0$.

$$\sin^2 x - \sin x = 0$$
$$\sin x - 1 = 0 \qquad \text{Divide by } \sin x.$$
$$\sin x = 1 \qquad \text{Add 1.}$$
$$x = \frac{\pi}{2}$$

46. What is wrong with the following solution for all θ in the interval $[0°, 360°)$? Solve $\tan^2 \theta - 1 = 0$.

$$\tan^2 \theta - 1 = 0$$
$$\tan^2 \theta = 1 \qquad \text{Add 1.}$$
$$\tan \theta = 1 \qquad \text{Take square root on each side.}$$
$$\theta = 45°, 225°$$

47. In an electric circuit, let V represent the electromotive force in volts at t seconds. Assume $V = \cos 2\pi t$. Find the smallest positive value of t where $0 \le t \le 1/2$ for each of the following values of V.
   (a) $V = 0$ **(b)** $V = .5$ **(c)** $V = .25$

48. A coil of wire rotating in a magnetic field induces a voltage given by

$$e = 20 \sin \left(\frac{\pi t}{4} - \frac{\pi}{2} \right),$$

where t is time in seconds. Find the smallest positive time to produce the following voltages.
   (a) 0   **(b)** $10\sqrt{3}$

49. The equation

$$.342D \cos \theta + h \cos^2 \theta = \frac{16D^2}{V_0^2}$$

is used in reconstructing accidents in which a vehicle vaults into the air after hitting an obstruction. V_0 is the velocity in feet per second of the vehicle when it hits, D is the distance (in feet) from the obstruction to the landing point, and h is the difference in height (in feet) between the landing point and the takeoff point. Angle θ is the takeoff angle, the angle between the horizontal and the path of the vehicle. Find θ to the nearest degree if $V_0 = 60$, $D = 80$, and $h = 2$.

50. If $0 < k < 1$, how many solutions does the equation $\sin^2 \theta = k$ have in the interval $[0, 360°)$?

6.3 ——— **TRIGONOMETRIC EQUATIONS WITH MULTIPLE ANGLES**

In this section we discuss trigonometric equations that involve functions of half angles and multiples of angles.

▪ *Example 1*

SOLVING AN
EQUATION WITH A
DOUBLE ANGLE

Solve $\cos 2x = \cos x$ in the interval $[0, 2\pi)$.

First change $\cos 2x$ to a trigonometric function of x. Use the identity $\cos 2x = 2 \cos^2 x - 1$ so that the equation involves only the cosine of x. Then use the methods of the previous section.

$$\cos 2x = \cos x$$

$$2\cos^2 x - 1 = \cos x$$

$$2\cos^2 x - \cos x - 1 = 0$$

$$(2\cos x + 1)(\cos x - 1) = 0$$

$$2\cos x + 1 = 0 \qquad \text{or} \qquad \cos x - 1 = 0$$

$$\cos x = -\frac{1}{2} \qquad \text{or} \qquad \cos x = 1$$

In the required interval,

$$x = \frac{2\pi}{3} \qquad \text{or} \qquad \frac{4\pi}{3} \qquad \text{or} \qquad x = 0.$$

The solutions are 0, $2\pi/3$, and $4\pi/3$. ■

CAUTION

In Example 1 it is important to notice that $\cos 2x$ cannot be changed to $\cos x$ by dividing by 2, since 2 is not a factor of the numerator.

$$\frac{\cos 2x}{2} \neq \cos x$$

The only way to change $\cos 2x$ to a trigonometric function of x is by using one of the identities for $\cos 2x$.

Conditional trigonometric equations in which a half angle or multiple angle is involved often require an additional step to solve. This step involves adjusting the interval of solution to fit the requirements of the half or multiple angle. This is shown in the following examples.

■ *Example 2*
SOLVING AN
EQUATION WITH A
HALF ANGLE

Solve $2\sin\frac{x}{2} = 1$ in the interval [0°, 360°).

As a compound inequality, the interval [0°, 360°) is written

$$0° \le x < 360°.$$

Dividing both sides by 2 gives

$$0° \le \frac{x}{2} < 180°.$$

To find all values of $x/2$ in the interval 0° to 180°, begin by solving for the trigonometric function.

$$2\sin\frac{x}{2} = 1$$

$$\sin\frac{x}{2} = \frac{1}{2} \qquad \text{Divide by 2.}$$

Both $\sin 30° = 1/2$ and $\sin 150° = 1/2$ and $30°$ and $150°$ are in the given interval for $x/2$, so

$$\frac{x}{2} = 30° \quad \text{or} \quad \frac{x}{2} = 150°$$

$$x = 60° \quad \text{or} \quad x = 300°. \qquad \text{Multiply by 2.}$$

The solutions in the given interval are $60°$ and $300°$. ■

■ *Example 3*

SOLVING AN
EQUATION WITH A
MULTIPLE ANGLE

Solve $4 \sin x \cos x = \sqrt{3}$ in the interval $[0°, 360°)$.
   The identity $2 \sin x \cos x = \sin 2x$ is useful here.

$$4 \sin x \cos x = \sqrt{3}$$

$$2(2 \sin x \cos x) = \sqrt{3} \qquad 4 = 2 \cdot 2$$

$$2 \sin 2x = \sqrt{3} \qquad 2 \sin x \cos x = \sin 2x$$

$$\sin 2x = \frac{\sqrt{3}}{2} \qquad \text{Divide by 2.}$$

From the given domain $0° \le x < 360°$, the domain for $2x$ is $0° \le 2x < 720°$. Now list all solutions in this interval:

$$2x = 60°, 120°, 420°, 480°$$

or $\qquad\qquad x = 30°, 60°, 210°, 240°. \qquad \text{Divide by 2.}$

The last two solutions for $2x$ were found by adding $360°$ to $60°$ and $120°$, respectively. ■

■ *Example 4*

SOLVING AN
EQUATION WITH A
MULTIPLE ANGLE

Solve $\tan 3x + \sec 3x = 2$ in the interval $[0, 2\pi)$.
   Since the tangent and secant functions are related by the identity $1 + \tan^2 \theta = \sec^2 \theta$, one way to begin is to express everything in terms of the secant. This may be done by subtracting $\sec 3x$ from both sides and then squaring.

$$\tan 3x + \sec 3x = 2$$

$$\tan 3x = 2 - \sec 3x \qquad\qquad \text{Subtract sec 3x.}$$

$$\tan^2 3x = 4 - 4 \sec 3x + \sec^2 3x \qquad \begin{array}{l}\text{Square both sides;}\\ (a - b)^2 = a^2 - 2ab + b^2\end{array}$$

$$\sec^2 3x - 1 = 4 - 4 \sec 3x + \sec^2 3x \qquad \text{Replace } \tan^2 3x \text{ with } \sec^2 3x - 1.$$

$$0 = 5 - 4 \sec 3x$$

$$4 \sec 3x = 5$$

$$\sec 3x = \frac{5}{4}$$

$$\frac{1}{\cos 3x} = \frac{5}{4} \qquad\qquad \sec \theta = \frac{1}{\cos \theta}$$

$$\cos 3x = \frac{4}{5} \qquad\qquad \text{Use reciprocals.}$$

Multiply the inequality $0 \le x < 2\pi$ by 3 to find the interval for $3x$: $[0, 6\pi)$. Using a calculator and knowing that cosine is positive in quadrants I and IV, we get

$$3x \approx .64350111, 5.6396842, 6.9266864, 11.922870, 13.209872, 18.206055.$$

Dividing by 3 gives

$$x \approx .21450037, 1.8798947, 2.3088955, 3.9742898, 4.4032906, 6.0686849.$$

Since both sides of the equation were squared, each of these proposed solutions must be checked. It can be verified by substitution in the given equation that the solutions are .21450037, 2.3088955, and 4.4032906. ■

6.3 EXERCISES ■

Solve each of the following equations for solutions in the interval $[0, 2\pi)$. See Examples 1–4.

1. $\cos 2x = \dfrac{\sqrt{3}}{2}$ **2.** $\cos 2x = -\dfrac{1}{2}$ **3.** $\sin 3x = -1$ **4.** $\sin 3x = 0$

5. $3 \tan 3x = \sqrt{3}$ **6.** $\cot 3x = \sqrt{3}$ **7.** $\sqrt{2} \cos 2x = -1$ **8.** $2\sqrt{3} \sin 2x = \sqrt{3}$

9. $\sin \dfrac{x}{2} = \sqrt{2} - \sin \dfrac{x}{2}$ **10.** $\sin x = \sin 2x$ **11.** $\tan 4x = 0$ **12.** $\cos 2x - \cos x = 0$

13. $\sec^4 2x = 4$ **14.** $\sin^2 \dfrac{x}{2} - 1 = 0$ **15.** $\sin \dfrac{x}{2} = \cos \dfrac{x}{2}$ **16.** $\sec \dfrac{x}{2} = \cos \dfrac{x}{2}$

17. $\cos 2x + \cos x = 0$ **18.** $\sin x \cos x = \dfrac{1}{4}$

Solve each of the following equations for solutions in the interval $[0°, 360°)$. When an approximation is appropriate, use a calculator as necessary to find solutions to the nearest tenth of a degree. See Examples 1–4.

19. $\sqrt{2} \sin 3\theta - 1 = 0$ **20.** $-2 \cos 2\theta = \sqrt{3}$ **21.** $\cos \dfrac{\theta}{2} = 1$

22. $\sin \dfrac{\theta}{2} = 1$ **23.** $2\sqrt{3} \sin \dfrac{\theta}{2} = 3$ **24.** $2\sqrt{3} \cos \dfrac{\theta}{2} = -3$

25. $2 \sin \theta = 2 \cos 2\theta$ **26.** $\cos \theta - 1 = \cos 2\theta$ **27.** $1 - \sin \theta = \cos 2\theta$

28. $\sin 2\theta = 2 \cos^2 \theta$ **29.** $\csc^2 \dfrac{\theta}{2} = 2 \sec \theta$ **30.** $\cos \theta = \sin^2 \dfrac{\theta}{2}$

31. $2 - \sin 2\theta = 4 \sin 2\theta$ **32.** $4 \cos 2\theta = 8 \sin \theta \cos \theta$ **33.** $2 \cos^2 2\theta = 1 - \cos 2\theta$

34. $\sin \theta = \cos \dfrac{\theta}{2}$ **35.** $\tan 3\theta + \sec 3\theta = 1$ **36.** $\cot \dfrac{\theta}{2} = \csc \dfrac{\theta}{2} + 1$

For the following equations, use the sum and product identities from Section 5.7. Give all solutions in the interval $[0, 2\pi)$.

37. $\sin x + \sin 3x = \cos x$ **38.** $\cos 4x - \cos 2x = \sin x$ **39.** $\sin 3x - \sin x = 0$

40. $\cos 2x + \cos x = 0$ **41.** $\sin 4x + \sin 2x = 2 \cos x$ **42.** $\cos 5x + \cos 3x = 2 \cos 4x$

43. What is wrong with the following solution? Solve $2 \sin (1/2)x = -1$ in the interval $[0°, 360°)$.

$$2 \sin \frac{1}{2}x = -1$$

$$\sin \frac{1}{2}x = -\frac{1}{2}$$

$$\frac{1}{2}x = 210° \quad \text{or} \quad \frac{1}{2}x = 330°$$

$$x = 420° \qquad\qquad x = 660°$$

The solutions are 420° and 660°.

44. What is wrong with the following solution? Solve $\tan 2\theta = 2$ in the interval $[0, 2\pi)$.

$$\tan 2\theta = 2$$

$$\frac{\tan 2\theta}{2} = \frac{2}{2}$$

$$\tan \theta = 1$$

$$\theta = \frac{\pi}{4} \quad \text{or} \quad \theta = \frac{5\pi}{4}$$

The solutions are $\pi/4$ and $5\pi/4$.

45. The seasonal variation in the length of daylight can be represented by a sine function. For example, the daily number of hours of daylight in New Orleans is given by

$$h = \frac{35}{3} + \frac{7}{3} \sin \frac{2\pi x}{365},$$

where x is the number of days after March 21 (disregarding leap year).*
(a) On what date will there be about 14 hours of daylight?
(b) What date has the least number of hours of daylight?
(c) When will there be about 10 hours of daylight?

46. The British nautical mile is defined as the length of a minute of arc of a meridian. Since the earth is flat at its poles, the nautical mile, in feet, is given by

$$L = 6{,}077 - 31 \cos 2\theta,$$

where θ is the latitude in degrees. (See the figure.)*
(a) Find the latitude(s) at which the nautical mile is 6,074 feet.

A nautical mile is the length on any of these meridians cut by a central angle of measure 1 minute.

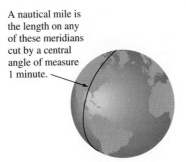

(b) At what latitude(s) is the nautical mile 6,108 feet?
(c) In the United States the nautical mile is defined everywhere as 6,080.2 feet. At what latitude(s) does this agree with the British nautical mile?

The study of alternating electric current requires the solutions of equations of the form $i = I_{max} \sin 2\pi f t$, *for time t in seconds, where i is instantaneous current in amperes,* I_{max} *is maximum current in amperes, and f is the number of cycles per second.† Find the smallest positive value of t, given the following data.*

47. $i = 40, I_{max} = 100, f = 60$

48. $i = 50, I_{max} = 100, f = 120$

49. $i = I_{max}, f = 60$

50. $i = \frac{1}{2} I_{max}, f = 60$

*From *A Sourcebook of Applications of School Mathematics* by Donald Bushaw et al. Copyright © 1980 by The Mathematical Association of America. Reprinted by permission. The material was prepared with the support of National Science Foundation Grant No. SED72-01123 A05. However, any opinions, findings, conclusions, or recommendations expressed herein are those of the authors and do not necessarily reflect the views of NSF.

†Ralph H. Hannon, *Basic Technical Mathematics with Calculus* (Philadelphia: W. B. Saunders Co., 1978), pp. 300–302.

6.4 ————— **INVERSE TRIGONOMETRIC EQUATIONS**

Section 6.1 introduced the inverse trigonometric functions. Recall, for example, that $x = \text{Sin } y$ means the same thing as $y = \text{Arcsin } x$ or $y = \text{Sin}^{-1} x$. Sometimes the solution of a trigonometric equation with more than one variable requires inverse trigonometric functions, as shown in the following examples.

■ *Example 1*

SOLVING AN
EQUATION FOR A
VARIABLE USING
INVERSE NOTATION

Solve $y = 3 \cos 2x$ for x.

We want $\cos 2x$ alone on one side of the equation so we can solve for $2x$ and then for x. First, divide both sides of the equation by 3.

$$y = 3 \cos 2x$$

$$\frac{y}{3} = \cos 2x$$

Now write the statement in the alternate form

$$2x = \text{Arccos } \frac{y}{3}.$$

Finally, multiply both sides by 1/2.

$$x = \frac{1}{2} \text{Arccos } \frac{y}{3} \quad ■$$

The next examples show how to solve equations involving inverse trigonometric functions.

■ *Example 2*

SOLVING AN
EQUATION
INVOLVING AN
INVERSE
TRIGONOMETRIC
FUNCTION

Solve $2 \text{ Arcsin } x = \pi$.

First solve for Arcsin x.

$$2 \text{ Arcsin } x = \pi$$

$$\text{Arcsin } x = \frac{\pi}{2} \quad \text{Divide by 2.}$$

Use the definition of Arcsin x to get

$$x = \text{Sin } \frac{\pi}{2}$$

or $x = 1$.

Verify that the solution satisfies the given equation. ■

■ *Example 3*

SOLVING AN
EQUATION
INVOLVING INVERSE
TRIGONOMETRIC
FUNCTIONS

Solve $\text{Cos}^{-1} x = \text{Sin}^{-1} \frac{1}{2}$.

Let $\text{Sin}^{-1}(1/2) = u$. Then $\text{Sin } u = 1/2$ and the equation becomes

$$\text{Cos}^{-1} x = u,$$

for u in quadrant I. This can be written as

$$\text{Cos } u = x.$$

Sketch a triangle and label it using the facts that u is in quadrant I and $\sin u = 1/2$. See Figure 6.14. Since $x = \text{Cos } u$,

$$x = \frac{\sqrt{3}}{2}. \quad ■$$

Some equations with inverse trigonometric functions require the use of identities to solve.

■ *Example 4*

SOLVING AN
INVERSE
TRIGONOMETRIC
EQUATION USING
AN IDENTITY

Solve $\text{Arcsin } x - \text{Arccos } x = \pi/6$.
    Begin by adding $\text{Arccos } x$ to both sides of the equation so that one inverse function is alone on one side of the equation.

$$\text{Arcsin } x - \text{Arccos } x = \frac{\pi}{6}$$

$$\text{Arcsin } x = \text{Arccos } x + \frac{\pi}{6} \qquad (1)$$

Use the definition of Arcsin to write this statement as

$$\sin \left(\text{Arccos } x + \frac{\pi}{6} \right) = x.$$

Let $u = \text{Arccos } x$, so $0 \le u \le \pi$ by definition. Then

$$\sin \left(u + \frac{\pi}{6} \right) = x. \qquad (2)$$

Using the identity for $\sin (A + B)$,

$$\sin \left(u + \frac{\pi}{6} \right) = \sin u \cos \frac{\pi}{6} + \cos u \sin \frac{\pi}{6}.$$

Substitute this result into equation (2) to get

$$\sin u \cos \frac{\pi}{6} + \cos u \sin \frac{\pi}{6} = x. \qquad (3)$$

From equation (1) and by the definition of the Arcsin function,

$$-\frac{\pi}{2} \le \text{Arccos } x + \frac{\pi}{6} \le \frac{\pi}{2}.$$

Subtract $\pi/6$ from each expression to get

$$-\frac{2\pi}{3} \le \text{Arccos } x \le \frac{\pi}{3}.$$

Since $0 \leq \text{Arccos } x \leq \pi$, it follows that here we must have $0 \leq \text{Arccos } x \leq \pi/3$. Thus $x > 0$, and we can sketch the triangle in Figure 6.15. From this triangle we find that $\sin u = \sqrt{1 - x^2}$. Now substitute into equation (3) using $\sin u = \sqrt{1 - x^2}$, $\sin \pi/6 = 1/2$, $\cos \pi/6 = \sqrt{3}/2$, and $\cos u = x$.

$$(\sqrt{1 - x^2})\frac{\sqrt{3}}{2} + x \cdot \frac{1}{2} = x$$

$$(\sqrt{1 - x^2})\sqrt{3} + x = 2x$$

$$(\sqrt{3})\sqrt{1 - x^2} = x$$

Squaring both sides gives

$$3(1 - x^2) = x^2$$

$$3 - 3x^2 = x^2$$

$$3 = 4x^2$$

$$x = \sqrt{\frac{3}{4}} \qquad \text{Choose the positive square root because } x > 0.$$

$$= \frac{\sqrt{3}}{2}.$$

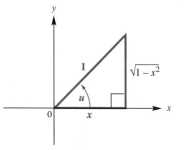

FIGURE 6.15

To check, replace x with $\sqrt{3}/2$ in the original equation:

$$\text{Arcsin } \frac{\sqrt{3}}{2} - \text{Arccos } \frac{\sqrt{3}}{2} = \frac{\pi}{3} - \frac{\pi}{6} = \frac{\pi}{6},$$

as required. The solution is $\sqrt{3}/2$. ■

6.4 EXERCISES ■

Solve each of the following equations for x. See Example 1.

1. $y = 5 \cos x$

2. $4y = \sin x$

3. $2y = \cot 3x$

4. $6y = \frac{1}{2} \sec x$

5. $y = 3 \tan 2x$

6. $y = 3 \sin \frac{x}{2}$

7. $y = 6 \cos \frac{x}{4}$

8. $y = -\sin \frac{x}{3}$

9. $y = -2 \cos 5x$

10. $y = 3 \cot 5x$

11. $y = \cos (x + 3)$

12. $y = \tan (2x - 1)$

13. $y = \sin x - 2$

14. $y = \cot x + 1$

15. $y = 2 \sin x - 4$

16. $y = 4 + 3 \cos x$

17. Refer to Exercise 13. A student attempting to solve this problem wrote as the first step
$$y = \sin (x - 2),$$
inserting parentheses as shown. Explain why this is incorrect.

18. Explain why the equation
$$\text{Sin}^{-1} x = \text{Cos}^{-1} 2$$
cannot have a solution. (No work needs to be shown here.)

Solve each of the following equations. See Examples 2 and 3.

19. $\frac{4}{3} \text{Sin}^{-1} \frac{y}{4} = \pi$

20. $4\pi + 4 \text{Tan}^{-1} y = \pi$

21. $2 \text{Arccos} \left(\frac{y - \pi}{3} \right) = 2\pi$

22. $\text{Arccos}\left(y - \dfrac{\pi}{3}\right) = \dfrac{\pi}{6}$

23. $\text{Arcsin } x = \text{Arctan } \dfrac{3}{4}$

24. $\text{Arctan } x = \text{Arccos } \dfrac{5}{13}$

25. $\text{Cos}^{-1} x = \text{Sin}^{-1} \dfrac{3}{5}$

26. $\text{Cot}^{-1} x = \text{Tan}^{-1} \dfrac{4}{3}$

Solve each of the following equations. See Example 4.

27. $\text{Sin}^{-1} x - \text{Tan}^{-1} 1 = -\dfrac{\pi}{4}$

28. $\text{Sin}^{-1} x + \text{Tan}^{-1} \sqrt{3} = \dfrac{2\pi}{3}$

29. $\text{Arccos } x + 2 \text{ Arcsin } \dfrac{\sqrt{3}}{2} = \pi$

30. $\text{Arccos } x + 2 \text{ Arcsin } \dfrac{\sqrt{3}}{2} = \dfrac{\pi}{3}$

31. $\text{Arcsin } 2x + \text{Arccos } x = \dfrac{\pi}{6}$

32. $\text{Arcsin } 2x + \text{Arcsin } x = \dfrac{\pi}{2}$

33. $\text{Cos}^{-1} x + \text{Tan}^{-1} x = \dfrac{\pi}{2}$

34. $\text{Tan}^{-1} x + \text{Cos}^{-1} x = \dfrac{\pi}{2}$

35. Solve $d = 550 + 450 \cos \dfrac{\pi}{50} t$ for t in terms of d.

36. Solve $d = 40 + 60 \cos \dfrac{\pi}{6} (t - 2)$ for t in terms of d.

37. In the study of alternating electric current, instantaneous voltage is given by

$$e = E_{\max} \sin 2\pi ft,$$

where f is the number of cycles per second, $E_{\max}$ is the maximum voltage, and t is time in seconds.
(a) Solve the equation for t.
(b) Find the smallest positive value of t if $E_{\max} = 12$, $e = 5$, and $f = 100$. Use a calculator.

38. When a large-view camera is used to take a picture of an object that is not parallel to the film, the lens board should be tilted so that the planes containing the subject, the lens board, and the film intersect in a line (see the figure). This gives the best "depth of field."*

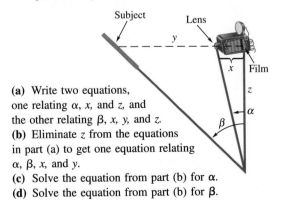

(a) Write two equations, one relating α, x, and z, and the other relating β, x, y, and z.
(b) Eliminate z from the equations in part (a) to get one equation relating α, β, x, and y.
(c) Solve the equation from part (b) for α.
(d) Solve the equation from part (b) for β.

39. In many computer languages, such as BASIC and FORTRAN, only the Arctan function is available. To use the other inverse trigonometric functions, it is necessary to express them in terms of Arctangent. This can be done as follows.
(a) Let $u = \text{Arcsin } x$. Solve the equation for x in terms of u.
(b) Use the result of part (a) to label the three sides of the triangle in the figure in terms of x.

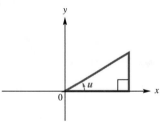

(c) Use the triangle from part (b) to write an equation for $\tan u$ in terms of x.
(d) Solve the equation from part (c) for u.

40. In the exercises for Section 4.1 we found the equation

$$y = \dfrac{1}{3} \sin \dfrac{4\pi t}{3},$$

where t is time (in seconds) and y is the angle formed by a rhythmically moving arm.
(a) Solve the equation for t.
(b) At what time(s) does the arm form an angle of .3 radian?

*From *A Sourcebook of Applications of School Mathematics* by Donald Bushaw et al. Copyright © 1980 by The Mathematical Association of America. Reprinted by permission. The material was prepared with the support of National Science Foundation Grant No. SED72-01123 A05. However, any opinions, findings, conclusions, or recommendations expressed herein are those of the authors and do not necessarily reflect the views of NSF.

CHAPTER 6 SUMMARY ■

SECTION	KEY IDEAS
6.1 Inverse Trigonometric Functions	**Inverse Trigonometric Functions**

Quadrants of the Unit Circle Range Values Come From

Function	Domain	Range	Quadrants of the Unit Circle Range Values Come From
$y = \text{Sin}^{-1} x$	$[-1, 1]$	$[-\pi/2, \pi/2]$	I and IV
$y = \text{Cos}^{-1} x$	$[-1, 1]$	$[0, \pi]$	I and II
$y = \text{Tan}^{-1} x$	$(-\infty, \infty)$	$(-\pi/2, \pi/2)$	I and IV
$y = \text{Cot}^{-1} x$	$(-\infty, \infty)$	$(0, \pi)$	I and II
$y = \text{Sec}^{-1} x$	$(-\infty, -1] \cup [1, \infty)$	$[0, \pi], y \neq \pi/2$	I and II
$y = \text{Csc}^{-1} x$	$(-\infty, -1] \cup [1, \infty)$	$[-\pi/2, \pi/2], y \neq 0$	I and IV

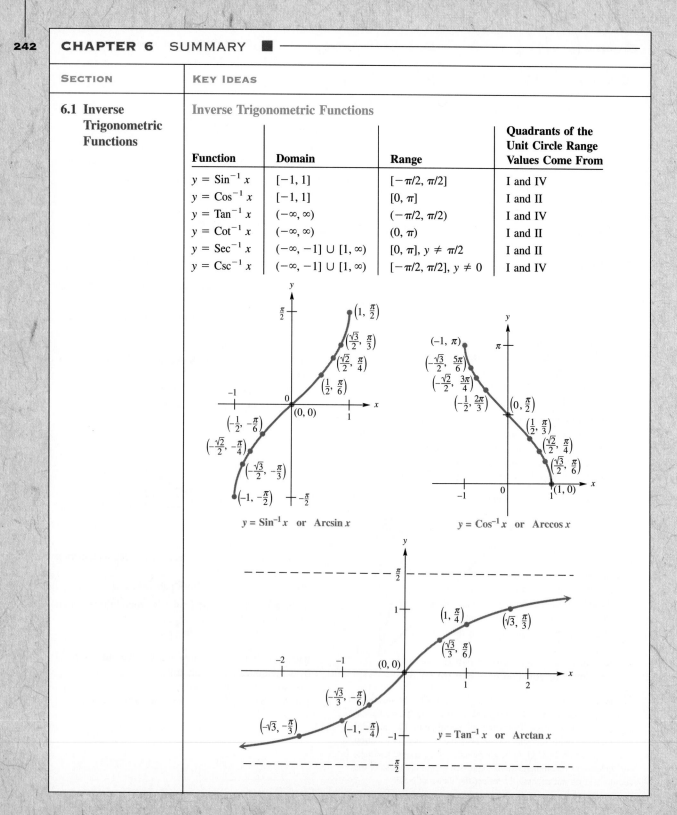

$y = \text{Sin}^{-1}x$ or Arcsin x

$y = \text{Cos}^{-1}x$ or Arccos x

$y = \text{Tan}^{-1} x$ or Arctan x

SECTION	KEY IDEAS
6.2 Trigonometric Equations	**Solving Trigonometric Equations** **1.** If only one trigonometric function is present, first solve the equation for that function. **2.** If more than one trigonometric function is present, rearrange the equation so that one side equals 0. Then try to factor and set each factor equal to 0 to solve. **3.** If Step 2 does not work, try using identities to change the form of the equation. It may be helpful to square both sides of the equation first. If this is done, check for extraneous solutions. **4.** If the equation is quadratic in form, but not factorable, use the quadratic formula.

CHAPTER 6　REVIEW EXERCISES ■

For each of the following, give the exact real number value of y. Do not use a calculator.

1. $y = \text{Sin}^{-1} \dfrac{\sqrt{2}}{2}$

2. $y = \text{Arccos}\left(-\dfrac{1}{2}\right)$

3. $y = \text{Tan}^{-1}(-\sqrt{3})$

4. $y = \text{Arcsin}(-1)$

5. $y = \text{Cos}^{-1}\left(-\dfrac{\sqrt{2}}{2}\right)$

6. $y = \text{Arctan}\dfrac{\sqrt{3}}{3}$

7. $y = \text{Sec}^{-1}(-2)$

8. $y = \text{Arccsc}\dfrac{2\sqrt{3}}{3}$

9. $y = \text{Arccot}(-1)$

For each of the following, give the degree measure of θ. Do not use a calculator.

10. $\theta = \text{Arccos}\dfrac{1}{2}$

11. $\theta = \text{Arcsin}\left(-\dfrac{\sqrt{3}}{2}\right)$

12. $\theta = \text{Tan}^{-1} 0$

Use a calculator to give the degree measure of θ.

13. $\theta = \text{Arctan } 1.7804675$

14. $\theta = \text{Sin}^{-1}(-.66045320)$

15. $\theta = \text{Cos}^{-1}.80396577$

16. $\theta = \text{Cot}^{-1} 4.5046388$

17. $\theta = \text{Arcsec } 3.4723155$

18. $\theta = \text{Csc}^{-1} 7.4890096$

19. Explain why $\text{Sin}^{-1} 3$ cannot be defined.

20. $\text{Arcsin}(\sin 5\pi/6) \neq 5\pi/6$. Explain why this is so.

21. What is the domain of the Arccot function?

22. What is the range of the Arcsec function as defined in this text?

Find each of the following without using a calculator.

23. $\sin\left(\text{Sin}^{-1}\dfrac{1}{2}\right)$

24. $\tan\left(\text{Tan}^{-1}\dfrac{2}{3}\right)$

25. $\cos(\text{Arccos}(-1))$

26. $\sin\left(\text{Arcsin}\left(-\dfrac{\sqrt{3}}{2}\right)\right)$

27. $\text{Arccos}\left(\cos\dfrac{3\pi}{4}\right)$

28. $\text{Arcsec}(\sec \pi)$

29. $\text{Tan}^{-1}\left(\tan\dfrac{\pi}{4}\right)$

30. $\text{Cos}^{-1}(\cos 0)$

31. $\sin\left(\text{Arccos}\dfrac{3}{4}\right)$

32. $\cos(\text{Arctan } 3)$

33. $\cos(\mathrm{Csc}^{-1}(-2))$

34. $\sec\left(2\,\mathrm{Sin}^{-1}\left(-\dfrac{1}{3}\right)\right)$

35. $\tan\left(\mathrm{Arcsin}\,\dfrac{3}{5}+\mathrm{Arccos}\,\dfrac{5}{7}\right)$

Write each of the following as an expression in u.

36. $\sin(\mathrm{Tan}^{-1}u)$

37. $\cos\left(\mathrm{Arctan}\,\dfrac{u}{\sqrt{1-u^2}}\right)$

38. $\tan\left(\mathrm{Arcsec}\,\dfrac{\sqrt{u^2+1}}{u}\right)$

Graph each of the following, and give the domain and range.

39. $y=\mathrm{Sin}^{-1}x$

40. $y=\mathrm{Cos}^{-1}x$

41. $y=\mathrm{Arccot}\,x$

Solve each of the following equations for solutions in the interval $[0, 2\pi)$. Use a calculator in Exercises 43 and 44.

42. $\sin^2 x=1$

43. $2\tan x-1=0$

44. $3\sin^2 x-5\sin x+2=0$

45. $\tan x=\cot x$

46. $\sec^4 2x=4$

47. $\tan^2 2x-1=0$

48. $\sec\dfrac{x}{2}=\cos\dfrac{x}{2}$

49. $\cos 2x+\cos x=0$

50. $4\sin x\cos x=\sqrt{3}$

Solve each of the following equations for solutions in the interval $[0°, 360°)$. Use a calculator and when appropriate, express solutions to the nearest tenth of a degree.

51. $\sin^2\theta+3\sin\theta+2=0$

52. $2\tan^2\theta=\tan\theta+1$

53. $\sin 2\theta=\cos 2\theta+1$

54. $2\sin 2\theta=1$

55. $3\cos^2\theta+2\cos\theta-1=0$

56. $5\cot^2\theta-\cot\theta-2=0$

57. $\sin 2\theta+\sin 4\theta=0$

58. $\cos\theta-\cos 2\theta=2\cos\theta$

Solve each equation for x.

59. $4y=2\sin x$

60. $y=3\cos\dfrac{x}{2}$

61. $2y=\tan(3x+2)$

62. $5y=4\sin x-3$

63. $\dfrac{4}{3}\,\mathrm{Arctan}\,\dfrac{x}{2}=\pi$

64. $\mathrm{Arccos}\,x=\mathrm{Arcsin}\,\dfrac{2}{7}$

65. $\mathrm{Arccos}\,x+\mathrm{Arctan}\,1=\dfrac{11\pi}{12}$

66. $\mathrm{Arccot}\,x=\mathrm{Arcsin}\left(\dfrac{-\sqrt{2}}{2}\right)+\dfrac{3\pi}{4}$

67. Recall Snell's law from Exercises 61–64 of Section 2.3:

$$\frac{c_1}{c_2}=\frac{\sin\theta_1}{\sin\theta_2},$$

where c_1 is the speed of light in one medium, c_2 is the speed of light in a second medium, and θ_1 and θ_2 are the angles shown in the figure. Suppose that a light is shining up through water into the air as in the figure.

As θ_1 increases, θ_2 approaches 90°, at which point no light will emerge from the water. Assume the ratio c_1/c_2 in this case is .752. For what value of θ_1 does $\theta_2=90°$? This value of θ_1 is called the *critical angle* for water.

68. Refer to Exercise 67. What happens when θ_1 is greater than the critical angle?

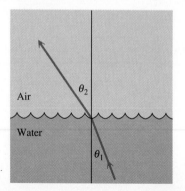

■ THE GRAPHING CALCULATOR ■

In Sections 6.2 and 6.3 we discussed methods of solving trigonometric equations using concepts of equation solving from algebra, along with ideas from trigonometry, such as using identities for substitution and reference angles to determine solutions. Conditional trigonometric equations can also be solved by using a graphing calculator. For example, suppose that we wish to solve the equation

$$\tan x + \sqrt{3} = \sec x$$

in the interval $[0, 2\pi)$. This equation was solved in Example 3 of Section 6.2, and we found the exact solution $11\pi/6$. This number is approximately equal to 5.7595865.

Recall from algebra that the solutions of the equation $f(x) = 0$ are the x-intercepts of the graph of $y = f(x)$. We must first write the given equation in the form $f(x)$, or y, equal to 0:

$$\tan x + \sqrt{3} - \sec x = 0.$$

Here we have $y = \tan x + \sqrt{3} - \sec x$. Now use the RANGE function of the calculator to get the following maximum and minimum values:

x min $= 0$

x max $= 6.3$ (This is just a bit larger than 2π.)

y min $= -4$

y max $= 4$ (These are arbitrarily chosen.)

With the TI-81 calculator set in the radian mode, enter the function

$$y_1 = \tan x + \sqrt{3} - \left(\frac{1}{\cos x}\right).$$

Notice that we must enter $\sec x$ as $1/\cos x$. Now, use the GRAPH function to get the following display.

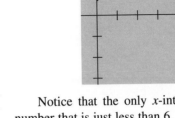

Notice that the only x-intercept for $[0, 2\pi)$ is a number that is just less than 6. Using the TRACE feature of the calculator, we find that two points on the graph are

$$(5.7031579, -.1186889)$$

and $(5.7694737, .01960658).$

Since the value of y changes sign between the two x-values in these points, by the intermediate value theorem from algebra, there must be an x-value between them for which $y = 0$. Using the ZOOM feature of the calculator, we can get better and better approximations of the solution. Of course, the calculator will not display the exact value $11\pi/6$, but it will give us 5.7595865 if we zoom in a sufficient number of times. (Notice that the extraneous value $5\pi/6$ obtained in Example 3 in Section 6.2 does not appear.)

You may wish to experiment with your graphing calculator by solving the following equations taken from the exercises in Sections 6.2 and 6.3. In each case, find the approximate solutions in the interval $[0, 2\pi)$. They are all taken from odd-numbered exercises so that you can verify your approximate solutions by comparing them to the exact solutions given in the answer section.

1. $3 \sin^2 x - \sin x = 2$ (Section 6.2, Exercise 29)

2. $\sec^2 x = 2 \tan x + 4$ (Section 6.2, Exercise 31)

3. $9 \sin^2 x - 6 \sin x = 1$ (Section 6.2, Exercise 35)

4. $\sin^2 x - 2 \sin x + 3 = 0$ (Section 6.2, Exercise 39)

5. $\cot x + 2 \csc x = 3$ (Section 6.2, Exercise 41)

6. $\cos 2x = \dfrac{\sqrt{3}}{2}$ (Section 6.3, Exercise 1)

7. $\sin \dfrac{x}{2} = \sqrt{2} - \sin \dfrac{x}{2}$ (Section 6.3, Exercise 9)

8. $\cos 2x + \cos x = 0$ (Section 6.3, Exercise 17)

9. $\sin x + \sin 3x = \cos x$ (Section 6.3, Exercise 37)

10. $\sin 4x + \sin 2x = 2 \cos x$ (Section 6.3, Exercise 41)

APPLICATIONS OF TRIGONOMETRY AND VECTORS

Until now, our applied work with trigonometry has been limited to right triangles. However, the concepts developed in the earlier chapters can be extended so that our work can apply to *all* triangles. Every triangle has three sides and three angles. In this chapter we show that if any three of the six measures of a triangle (provided at least one measure is a side) are known, then the other three measures can be found. This process is called solving a triangle. In the latter part of the chapter this knowledge is used to solve problems involving vectors.

7.1 ———— OBLIQUE TRIANGLES AND THE LAW OF SINES

Recall from geometry the following axioms that allow us to prove that two triangles are congruent (that is, their corresponding sides and angles are equal).

CONGRUENCE AXIOMS	Side-Angle-Side (SAS)	If two sides and the included angle of one triangle are equal, respectively, to two sides and the included angle of a second triangle, then the triangles are congruent.
	Angle-Side-Angle (ASA)	If two angles and the included side of one triangle are equal, respectively, to two angles and the included side of a second triangle, then the triangles are congruent.
	Side-Side-Side (SSS)	If three sides of one triangle are equal, respectively, to three sides of a second triangle, then the triangles are congruent.

Throughout this chapter keep in mind that whenever any of the groups of data described above are given, the triangle is uniquely determined; that is, all other data in the triangle are given by one and only one set of measures.

A triangle that is not a right triangle is called an **oblique triangle.** The measures of the three sides and the three angles of a triangle can be found if at least one side and any other two measures are known. There are four possible cases.

SOLVING OBLIQUE TRIANGLES	**1.** One side and two angles are known.
	2. Two sides and one angle not included between the two sides are known. This case may lead to more than one triangle.
	3. Two sides and the angle included between the two sides are known.
	4. Three sides are known.

NOTE	If we know three angles of a triangle, we cannot find unique side lengths, since AAA assures us only of similarity, not congruence. For example, there are infinitely many triangles ABC with $A = 35°$, $B = 65°$, and $C = 80°$.

The first two cases require the *law of sines,* which is discussed in this section and the next. The last two cases require the *law of cosines,* which is discussed in Section 7.3.

To derive the law of sines, start with an oblique triangle, such as the acute triangle in Figure 7.1(a) or the obtuse triangle in Figure 7.1(b). (Recall: These terms were defined in Section 1.3.) The following discussion applies to both triangles. First, construct the perpendicular from B to side AC. Let h be the length of this perpendicular. Then c is the hypotenuse of right triangle ADB, and a is the hypotenuse of right triangle BDC. By results from Chapter 2,

$$\text{in triangle } ADB, \qquad \sin A = \frac{h}{c} \quad \text{or} \quad h = c \sin A,$$

$$\text{in triangle } BDC, \qquad \sin C = \frac{h}{a} \quad \text{or} \quad h = a \sin C.$$

Since $h = c \sin A$ and $h = a \sin C$,

$$a \sin C = c \sin A,$$

or, upon dividing both sides by $\sin A \sin C$,

$$\frac{a}{\sin A} = \frac{c}{\sin C}.$$

In a similar way, by constructing the perpendiculars from other vertices, it can be shown that

$$\frac{a}{\sin A} = \frac{b}{\sin B} \qquad \text{and} \qquad \frac{b}{\sin B} = \frac{c}{\sin C}.$$

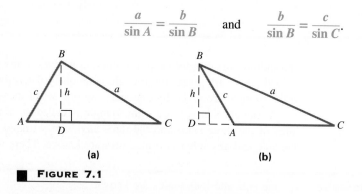

(a)   (b)

■ **FIGURE 7.1**

This discussion proves the following theorem.

LAW OF SINES   In any triangle ABC, with sides a, b, and c,

$$\frac{a}{\sin A} = \frac{b}{\sin B}, \qquad \frac{a}{\sin A} = \frac{c}{\sin C}, \qquad \text{and} \qquad \frac{b}{\sin B} = \frac{c}{\sin C}.$$

This can be written in compact form as

$$\frac{a}{\sin A} = \frac{b}{\sin B} = \frac{c}{\sin C}.$$

Sometimes an alternate form of the law of sines,

$$\frac{\sin A}{a} = \frac{\sin B}{b} = \frac{\sin C}{c},$$

is convenient to use.

If two angles and the side opposite one of the angles are known, the law of sines can be used directly to solve for the side opposite the other known angle. The triangle can then be solved completely, as shown in the first example.

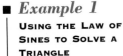

■ Example 1

USING THE LAW OF SINES TO SOLVE A TRIANGLE

Solve triangle ABC if $A = 32.0°$, $B = 81.8°$, and $a = 42.9$ centimeters. See Figure 7.2.

Start by drawing a triangle, roughly to scale, and labelling the given parts as in Figure 7.2. Since the values of A, B, and a are known, use the part of the law of sines that involves these variables.

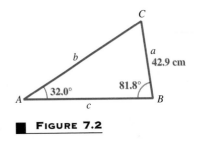

■ FIGURE 7.2

$$\frac{a}{\sin A} = \frac{b}{\sin B}$$

Substituting the known values gives

$$\frac{42.9}{\sin 32.0°} = \frac{b}{\sin 81.8°}.$$

Multiply both sides of the equation by $\sin 81.8°$.

$$b = \frac{42.9 \sin 81.8°}{\sin 32.0°}$$

When using a calculator to find b, keep intermediate answers in the calculator until the final result is found. Then round to the proper number of significant digits. In this case, find $\sin 81.8°$, and then multiply that number by 42.9. Keep the result in the calculator while you find $\sin 32.0°$, and then divide. Since the given information is accurate to three significant digits, round the value of b to get

$$b = 80.1 \text{ centimeters.}$$

Find C from the fact that the sum of the angles of any triangle is 180°.

$$A + B + C = 180°$$
$$C = 180° - A - B$$
$$C = 180° - 32.0° - 81.8°$$
$$= 66.2°$$

Now use the law of sines again to find c. (Why does the Pythagorean theorem not apply?)

$$\frac{a}{\sin A} = \frac{c}{\sin C}$$

$$\frac{42.9}{\sin 32.0°} = \frac{c}{\sin 66.2°}$$

$$c = \frac{42.9 \sin 66.2°}{\sin 32.0°}$$

$$c = 74.1 \text{ centimeters} \quad \blacksquare$$

| CAUTION | In applications of oblique triangles, such as the one in Example 2, a correctly labelled sketch is essential in order to set up the correct equation. |

■ *Example 2*

USING THE LAW OF SINES IN AN APPLICATION

Tri Nguyen wishes to measure the distance across the Big Muddy River. See Figure 7.3. He finds that $C = 112° 53'$, $A = 31° 06'$, and $b = 347.6$ feet. Find the required distance.

To use the law of sines, one side and the angle opposite it must be known. Since the only side whose length is given is b, angle B must be found before the law of sines can be used.

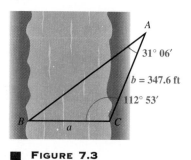

■ **FIGURE 7.3**

$$B = 180° - A - C$$
$$= 180° - 31° 06' - 112° 53' = 36° 01'$$

Now the required distance a can be found. Use the form of the law of sines involving A, B, and b.

$$\frac{a}{\sin A} = \frac{b}{\sin B}$$

Substitute the known values.

$$\frac{a}{\sin 31° 06'} = \frac{347.6}{\sin 36° 01'}$$

$$a = \frac{347.6 \sin 31° 06'}{\sin 36° 01'}$$

$$a = 305.3 \text{ feet} \qquad \text{Use a calculator.}$$

It may be necessary to convert to decimal degrees on a calculator before the final calculation can be made. ■

The next example involves the use of bearing, first discussed in Section 2.5.

■ *Example 3*
USING THE LAW OF
SINES IN AN
APPLICATION

Two tracking stations are on an east-west line 110 miles apart. A forest fire is located on a bearing of N 42° E from the western station and a bearing of N 15° E from the eastern station. How far is the fire from the western station?

Figure 7.4 shows the two stations at points A and B and the fire at point C. Angle *BAC* = 90° − 42° = 48°, the obtuse angle at B equals 90° + 15° = 105°, and the third angle, C, equals 180° − 105° − 48° = 27°. Using the law of sines to find side b gives

$$\frac{b}{\sin 105°} = \frac{110}{\sin 27°}$$

$$b = 234,$$

or 230 miles (to two significant digits). ■

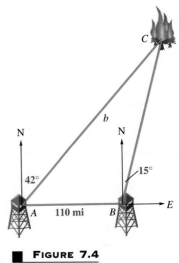

■ FIGURE 7.4

AREA   The method used to derive the law of sines can also be used to derive a useful formula to find the area of a triangle. A familiar formula for the area of a triangle is $K = (1/2)bh$, where K represents the area, b the base, and h the height. This formula cannot always be used easily, since in practice h is often unknown. To find a more useful formula, refer to acute triangle *ABC* in Figure 7.5(a) or obtuse triangle *ABC* in Figure 7.5(b).

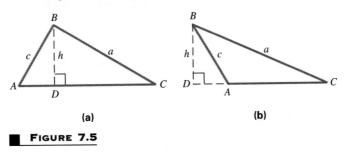

(a)   (b)

■ FIGURE 7.5

A perpendicular has been drawn from B to the base of the triangle (or the extension of the base). This perpendicular forms two right triangles. Using triangle *ABD*,

$$\sin A = \frac{h}{c},$$

or $$h = c \sin A.$$

Substituting into the formula $K = (1/2)bh$,

$$K = \frac{1}{2} b(c \sin A)$$

or
$$K = \frac{1}{2} bc \sin A.$$

Any other pair of sides and the angle between them could have been used. This result is summarized in the next theorem.

AREA OF A TRIANGLE

In any triangle ABC, the area K is given by any of the following formulas:

$$K = \frac{1}{2} bc \sin A, \qquad K = \frac{1}{2} ab \sin C, \qquad K = \frac{1}{2} ac \sin B.$$

That is, the area is given by half the product of the lengths of two sides and the sine of the angle included between them.

■ *Example 4*

FINDING THE AREA OF A TRIANGLE

USING $K = \frac{1}{2} ab \sin C$

Find the area of triangle ABC if $A = 24° \ 40'$, $b = 27.3$ centimeters, and $C = 52° \ 40'$.

Before we can use the formula given above, we must use the law of sines to find either a or c. Since the sum of the measures of the angles of any triangle is 180°,

$$B = 180° - 24° \ 40' - 52° \ 40' = 102° \ 40'.$$

Now use the form of the law of sines that relates a, b, A, and B to find a.

$$\frac{a}{\sin A} = \frac{b}{\sin B}$$

$$\frac{a}{\sin 24° \ 40'} = \frac{27.3}{\sin 102° \ 40'}$$

Solve for a to verify that $a = 11.7$ centimeters. Now find the area.

$$K = \frac{1}{2} ab \sin C = \frac{1}{2} (11.7)(27.3) \sin 52° \ 40' = 127$$

The area of triangle ABC is 127 square centimeters to three significant digits. ■

NOTE

Whenever possible, it is a good idea to use given values in solving triangles or finding areas rather than values obtained in intermediate steps. This avoids possible rounding errors.

7.1 800 1- 17; 20

7.2:

7.1 EXERCISES ■ ─────────────────

In this exercise set, a calculator will be necessary to solve the triangles.

Solve each of the following triangles. See Example 1.

1. $A = 37°, B = 48°, c = 18$ m

2. $B = 52°, C = 29°, a = 43$ cm

3. $A = 46° 30', B = 52° 50', b = 87.3$ mm

4. $A = 59° 30', B = 48° 20', b = 32.9$ m

5. $A = 27.2°, C = 115.5°, c = 76.0$ ft

6. $B = 124.1°, C = 18.7°, c = 94.6$ m

7. $A = 68.41°, B = 54.23°, a = 12.75$ ft

8. $C = 74.08°, B = 69.38°, c = 45.38$ m

9. $A = 87.2°, b = 75.9$ yd, $C = 74.3°$

10. $B = 38° 40', a = 19.7$ cm, $C = 91° 40'$

11. $B = 20° 50', C = 103° 10', AC = 132$ ft

12. $A = 35.3°, B = 52.8°, AC = 675$ ft

13. $A = 39.70°, C = 30.35°, b = 39.74$ m

14. $C = 71.83°, B = 42.57°, a = 2.614$ cm

15. $B = 42.88°, C = 102.40°, b = 3974$ ft

16. $A = 18.75°, B = 51.53°, c = 2798$ yd

17. $A = 39° 54', a = 268.7$ m, $B = 42° 32'$

18. $C = 79° 18', c = 39.81$ mm, $A = 32° 57'$

19. Explain why the law of sines cannot be used to solve a triangle if we are given the lengths of the three sides of a triangle.

20. In Example 1, we ask the question "Why does the Pythagorean theorem not apply?" Answer this question.

21. State the law of sines in your own words.

22. Susan Katz, a perceptive trigonometry student, makes the statement "If we know *any* two angles and one side of a triangle, then the triangle is uniquely determined." Is this a valid statement? Explain, referring to the congruence axioms given in this section.

Solve each of the following problems. See Examples 2 and 3.

23. To find the distance AB across a river, a distance $BC = 354$ m is laid off on one side of the river. It is found that $B = 112° 10'$ and $C = 15° 20'$. Find AB.

24. To determine the distance RS across a deep canyon, Joanna lays off a distance $TR = 582$ yd. She then finds that $T = 32° 50'$ and $R = 102° 20'$. Find RS.

25. Radio direction finders are placed at points A and B, which are 3.46 mi apart on an east-west line, with A west of B. From A the bearing of a certain radio transmitter is 47.7°, and from B the bearing is 302.5°. Find the distance of the transmitter from A.

26. A ship is sailing due north. At a certain point the bearing of a lighthouse 12.5 km distant is N 38.8° E. Later on, the captain notices that the bearing of the lighthouse has become S 44.2° E. How far did the ship travel between the two observations of the lighthouse?

27. A folding chair is to have a seat 12.0 in deep with angles as shown in the figure. How far down from the seat should the crossing legs be joined? (Find x in the figure.)

28. Mark notices that the bearing of a tree on the opposite bank of a river flowing north is 115.45°. Lisa is on the same bank as Mark, but 428.3 m away. She notices that the bearing of the tree is 45.47°. The two banks are parallel. What is the distance across the river?

29. Three gears are arranged as shown in the figure. Find angle θ.

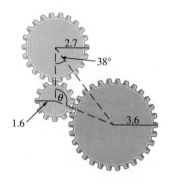

30. Three atoms with atomic radii of 2.0, 3.0, and 4.5 are arranged as in the figure. Find the distance between the centers of atoms A and C.

31. The bearing of a lighthouse from a ship was found to be N 37° E. After the ship sailed 2.5 miles due south, the new bearing was N 25° E. Find the distance between the ship and the lighthouse at each location.

32. A balloonist is directly above a straight road 1.5 miles long that joins two villages. She finds that the town closer to her is at an angle of depression of 35° and the farther town is at an angle of depression of 31°. How high above the ground is the balloon? (See the figure.)

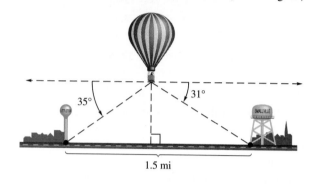

Find the area of each of the following triangles. See Example 4.

33. $A = 42.5°$, $b = 13.6$ m, $c = 10.1$ m

34. $C = 72.2°$, $b = 43.8$ ft, $a = 35.1$ ft

35. $B = 124.5°$, $a = 30.4$ cm, $c = 28.4$ cm

36. $C = 142.7°$, $a = 21.9$ km, $b = 24.6$ km

37. $A = 56.80°$, $b = 32.67$ in, $c = 52.89$ in

38. $A = 34.97°$, $b = 35.29$ m, $c = 28.67$ m

39. $A = 24° 25'$, $B = 56° 20'$, $c = 78.40$ cm

40. $B = 48° 30'$, $C = 74° 20'$, $a = 462$ km

41. A painter is going to apply a special coating to a triangular metal plate on a new building. Two sides measure 16.1 m and 15.2 m. She knows that the angle between these sides is 125°. What is the area of the surface she plans to cover with the coating?

42. A real estate agent wants to find the area of a triangular lot. A surveyor takes measurements and finds that two sides are 52.1 m and 21.3 m, and the angle between them is 42.2°. What is the area of the lot?

43. In any triangle having sides a, b, and c, it must be true that $a + b > c$. Use this fact and the law of sines to show that $\sin A + \sin B > \sin (A + B)$ for any two angles A and B of a triangle.

44. Show that the area of a triangle having sides a, b, and c and corresponding angles A, B, and C is given by

$$\frac{a^2 \sin B \sin C}{2 \sin A}.$$

7.2 THE AMBIGUOUS CASE OF THE LAW OF SINES

The law of sines can be used when given two angles and the side opposite one of these angles. Also, if two angles and the included side are known, then the third angle can be found by using the fact that the sum of the angles of a triangle is 180°, and then applying the law of sines. However, if we are given the lengths of two sides and the angle opposite one of them, it is possible that 0, 1, or 2 such triangles exist. (Recall that there is no "SSA" congruence theorem.)

To illustrate these facts, suppose that the measure of acute angle A of triangle ABC, the length of side a, and the length of side b are given. Draw angle A having a terminal side of length b. Now draw a side of length a opposite angle A. The following chart shows that there might be more than one possible outcome. This situation is called the **ambiguous case of the law of sines.**

Number of Possible Triangles	Sketch	Condition Necessary for Case to Hold
0	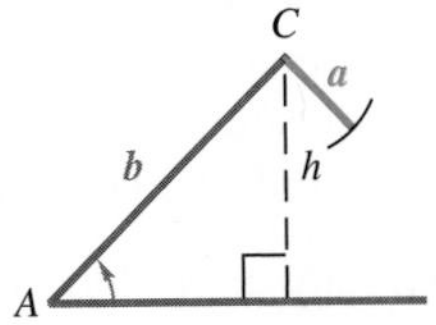	$a < h$ ($h = b \sin A$)
1	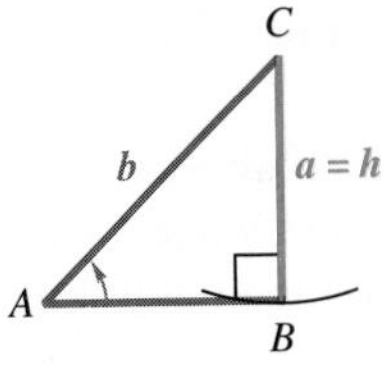	$a = h$
1	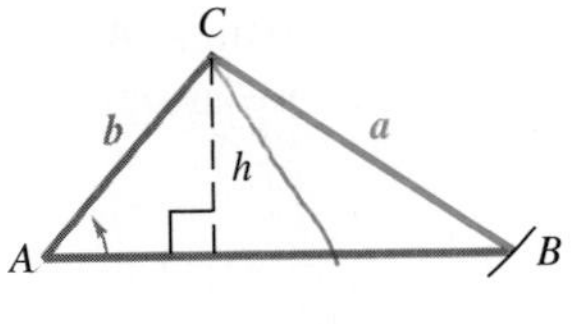	$a > b$
2	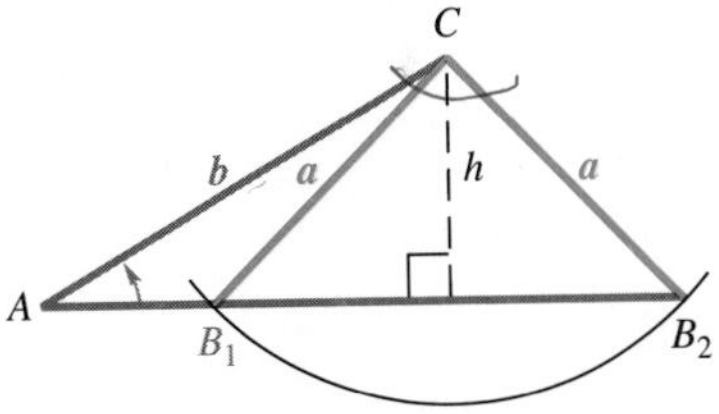	$b > a > h$

If angle A is obtuse, there are two possible outcomes, as shown in the next chart.

Number of Possible Triangles	Sketch	Condition Necessary for Case to Hold
0	 *(diagram with C, a, b, A)*	$a \leq b$
1	 *(diagram with C, b, a, A, B)*	$a > b$

As the various examples of this section will illustrate, applying the law of sines to the values of a, b, and A and some basic properties of geometry and trigonometry will allow us to determine which of these cases applies. The following basic facts should be kept in mind.

1. For any angle θ of a triangle, $0 < \sin \theta \leq 1$. If $\sin \theta = 1$, then $\theta = 90°$ and the triangle is a right triangle.
2. $\sin \theta = \sin (180° - \theta)$ (That is, supplementary angles have the same sine value.)
3. The smallest angle is opposite the shortest side, the largest angle is opposite the longest side, and the middle-valued angle is opposite the medium side (assuming the triangle is scalene).

■ **Example 1**

SOLVING A TRIANGLE USING THE LAW OF SINES (NO SUCH TRIANGLE)

Solve the triangle ABC if $B = 55° \, 40'$, $b = 8.94$ meters, and $a = 25.1$ meters.

Since we are given B, b, and a, use the law of sines to find A.

$$\frac{\sin A}{a} = \frac{\sin B}{b}$$

Substitute the given values.

$$\frac{\sin A}{25.1} = \frac{\sin 55° \, 40'}{8.94}$$

$$\sin A = \frac{25.1 \sin 55° \, 40'}{8.94}$$

$$\sin A = 2.3184379$$

Since $\sin A$ cannot be greater than 1, there can be no such angle A and thus no triangle with the given information. An attempt to sketch such a triangle leads to the situation seen in Figure 7.6. ■

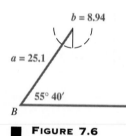

■ **FIGURE 7.6**

■ *Example 2*

SOLVING A
TRIANGLE USING
THE LAW OF SINES
(TWO TRIANGLES)

Solve triangle ABC if $A = 55° \, 20'$, $a = 22.8$ feet, and $b = 24.9$ feet.

To begin, use the law of sines to find angle B.

$$\frac{a}{\sin A} = \frac{b}{\sin B}$$

$$\frac{22.8}{\sin 55° \, 20'} = \frac{24.9}{\sin B}$$

$$\sin B = \frac{24.9 \sin 55° \, 20'}{22.8}$$

$$\sin B = .89822938$$

Since $\sin B = .89822938$, to the nearest ten minutes we have one value of B as

$$B = 64° \, 00'$$

using the inverse sine function of a calculator. However, since supplementary angles have the same sine value, another *possible* value of B is

$$B = 180° - 64° \, 00' = 116° \, 00'.$$

To see if $B = 116° \, 00'$ is a valid possibility, simply add $116° \, 00'$ to the measure of the given value of A, $55° \, 20'$. Since $116° \, 00' + 55° \, 20' = 171° \, 20'$, and this sum is less than $180°$ (the sum of the angles of a triangle), we know that it is a valid angle measure for this triangle.

To keep track of these two different values of B, let

$$B_1 = 116° \, 00' \quad \text{and} \quad B_2 = 64° \, 00'.$$

Now separately solve triangles AB_1C_1 and AB_2C_2 shown in Figure 7.7.

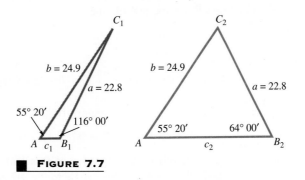

■ FIGURE 7.7

Let us begin with AB_1C_1. Find C_1 first.

$$C_1 = 180° - A - B_1 = 8° \, 40'.$$

Now, use the law of sines to find c_1.

$$\frac{a}{\sin A} = \frac{c_1}{\sin C_1}$$

$$\frac{22.8}{\sin 55° \, 20'} = \frac{c_1}{\sin 8° \, 40'}$$

$$c_1 = \frac{22.8 \sin 8° \, 40'}{\sin 55° \, 20'}$$

$$c_1 = 4.18 \text{ feet}$$

To solve triangle AB_2C_2, first find C_2.

$$C_2 = 180° - A - B_2 = 60° \, 40'$$

By the law of sines,

$$\frac{22.8}{\sin 55° \, 20'} = \frac{c_2}{\sin 60° \, 40'}$$

$$c_2 = \frac{22.8 \sin 60° \, 40'}{\sin 55° \, 20'}$$

$$c_2 = 24.2 \text{ feet.} \quad ■$$

CAUTION | When solving a triangle using the type of data given in Example 2, do not forget to find the possible obtuse angle. The inverse sine function of the calculator will not give it directly. As we shall see in the next example, it is possible that the obtuse angle will not be a valid measure.

■ *Example 3*
SOLVING A
TRIANGLE USING
THE LAW OF SINES
(ONE TRIANGLE)

Solve triangle ABC given $A = 43.5°$, $a = 10.7$ inches, and $b = 7.2$ inches.
To find angle B use the law of sines.

$$\frac{\sin B}{7.2} = \frac{\sin 43.5°}{10.7}$$

$$\sin B = \frac{7.2 \sin 43.5°}{10.7} = .46319186$$

The inverse sine function of the calculator gives us

$$B = 27.6°$$

as the acute angle. The other possible value of B is $180° - 27.6° = 152.4°$. However, when we add this possible obtuse angle to the given angle $A = 43.5°$, we get $152.4° + 43.5° = 195.9°$, which is greater than $180°$. So there can be only one triangle. Then angle $C = 180° - 27.6° - 43.5° = 108.9°$, and side c can be found with the law of sines.

$$\frac{c}{\sin 108.9°} = \frac{10.7}{\sin 43.5°}$$

$$c = \frac{10.7 \sin 108.9°}{\sin 43.5°}$$

$$c = 14.7 \text{ inches} \quad ■$$

■ *Example 4*

**ANALYZING DATA
INVOLVING AN
OBTUSE ANGLE**

Without using the law of sines, explain why the data

$$A = 104°, \; a = 26.8 \text{ meters}, \; b = 31.3 \text{ meters}$$

cannot be valid for a triangle *ABC*.

Since *A* is an obtuse angle, the largest side of the triangle must be *a*, the side opposite *A*. However, we are given $b > a$, which is impossible if *A* is obtuse. Therefore, no such triangle *ABC* exists. ■

odd 1–21

7.2 EXERCISES ■

A calculator will be needed throughout this exercise set.

Find the unknown angles in triangle ABC for each of the triangles that exists. See Examples 1–3.

1. $A = 29.7°$, $b = 41.5$ ft, $a = 27.2$ ft

2. $B = 48.2°$, $a = 890$ cm, $b = 697$ cm

3. $C = 41° 20'$, $b = 25.9$ m, $c = 38.4$ m

4. $B = 48° 50'$, $a = 3850$ in, $b = 4730$ in

5. $B = 74.3°$, $a = 859$ m, $b = 783$ m

6. $C = 82.2°$, $a = 10.9$ km, $c = 7.62$ km

7. $A = 142.13°$, $b = 5.432$ ft, $a = 7.297$ ft

8. $B = 113.72°$, $a = 189.6$ yd, $b = 243.8$ yd

9. $C = 129° 18'$, $a = 372.9$ cm, $c = 416.7$ cm

10. $A = 132° 07'$, $b = 7.481$ mi, $a = 8.219$ mi

Solve each of the triangles that exists. See Examples 1–3.

11. $A = 42.5°$, $a = 15.6$ ft, $b = 8.14$ ft

12. $C = 52.3°$, $a = 32.5$ yd, $c = 59.8$ yd

13. $B = 72.2°$, $b = 78.3$ m, $c = 145$ m

14. $C = 68.5°$, $c = 258$ cm, $b = 386$ cm

15. $A = 38° 40'$, $a = 9.72$ km, $b = 11.8$ km

16. $C = 29° 50'$, $a = 8.61$ m, $c = 5.21$ m

17. $B = 32° 50'$, $a = 7540$ cm, $b = 5180$ cm

18. $C = 22° 50'$, $b = 159$ mm, $c = 132$ mm

19. $A = 96.80°$, $b = 3.589$ ft, $a = 5.818$ ft

20. $C = 88.70°$, $b = 56.87$ yd, $c = 112.4$ yd

21. $B = 39.68°$, $a = 29.81$ m, $b = 23.76$ m

22. $A = 51.20°$, $c = 7986$ cm, $a = 7208$ cm

23. Apply the law of sines to the following: $a = \sqrt{5}$, $c = 2\sqrt{5}$, $A = 30°$. What is the value of sin *C*? What is the measure of *C*? Based on its angle measures, what kind of triangle is triangle *ABC*?

24. In your own words, explain the condition that must exist to determine that there is no triangle satisfying the given values of *a*, *b*, and *B*, once the value of sin *B* is found.

25. Without using the law of sines, explain why no triangle *ABC* exists satisfying $A = 103° 20'$, $a = 14.6$ ft, $b = 20.4$ ft.

26. Apply the law of sines to the data given in Example 4. Describe in your own words what happens when you try to find the measure of angle *B* using a calculator.

27. A surveyor reported the following data about a piece of property: "The property is triangular in shape, with dimensions as shown in the figure." Use the law of sines to see whether such a piece of property could exist.

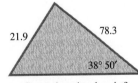

Can such a triangle exist?

28. The surveyor tries again: "A second triangular piece of property has dimensions as shown." This time it turns out that the surveyor did not consider every possible case. Use the law of sines to show why.

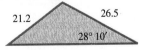

Use the law of sines to prove that each of the following statements is true for any triangle ABC, with corresponding sides a, b, and c.

29. $\dfrac{a + b}{b} = \dfrac{\sin A + \sin B}{\sin B}$

30. $\dfrac{a - b}{a + b} = \dfrac{\sin A - \sin B}{\sin A + \sin B}$

260 7.3 ─────── # THE LAW OF COSINES

Recall from Section 7.1 that if we are given two sides and the included angle or three sides of a triangle, a unique triangle is formed. These are the SAS and SSS cases, respectively. In both cases, however, we cannot begin the solution of the triangle by using the law of sines. Both of these cases require the use of the law of cosines, introduced in this section.

It will be helpful to remember the following property of triangles when applying the law of cosines.

> In any triangle, the sum of the lengths of any two sides must be greater than the length of the remaining side.

For example, it would be impossible to construct a triangle with sides of lengths 3, 4, and 10. See Figure 7.8.

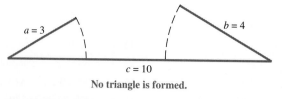

No triangle is formed.

■ **FIGURE 7.8**

To derive the law of cosines, let ABC be any oblique triangle. Choose a coordinate system so that vertex B is at the origin and side BC is along the positive x-axis. See Figure 7.9.

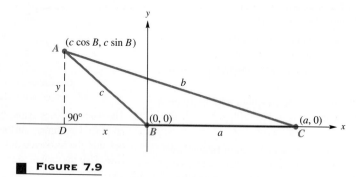

■ **FIGURE 7.9**

Let (x, y) be the coordinates of vertex A of the triangle. Verify that for angle B, whether obtuse or acute,

$$\sin B = \frac{y}{c} \quad \text{and} \quad \cos B = \frac{x}{c}.$$

(Here we assume that x is negative if B is obtuse.) From these results

$$y = c \sin B \quad \text{and} \quad x = c \cos B,$$

so that the coordinates of point A become

$$(c \cos B, c \sin B).$$

Point C has coordinates $(a, 0)$, and AC has length b. By the distance formula,

$$b = \sqrt{(c \cos B - a)^2 + (c \sin B)^2}.$$

Squaring both sides and simplifying gives

$$\begin{aligned}
b^2 &= (c \cos B - a)^2 + (c \sin B)^2 \\
&= c^2 \cos^2 B - 2ac \cos B + a^2 + c^2 \sin^2 B \\
&= a^2 + c^2(\cos^2 B + \sin^2 B) - 2ac \cos B \\
&= a^2 + c^2(1) - 2ac \cos B \\
&= a^2 + c^2 - 2ac \cos B.
\end{aligned}$$

This result is one form of the law of cosines. In the work above, we could just as easily have placed A or C at the origin. This would have given the same result, but with the variables rearranged. These various forms of the law of cosines are summarized in the following theorem.

LAW OF COSINES

In any triangle ABC, with sides a, b, and c,

$$a^2 = b^2 + c^2 - 2bc \cos A$$
$$b^2 = a^2 + c^2 - 2ac \cos B$$
$$c^2 = a^2 + b^2 - 2ab \cos C.$$

The law of cosines says that the square of a side of a triangle is equal to the sum of the squares of the other two sides, minus twice the product of the two sides and the cosine of the angle included between them.

NOTE

If we let $C = 90°$ in the third form of the law of cosines given above, we have $\cos C = \cos 90° = 0$, and the formula becomes

$$c^2 = a^2 + b^2,$$

the familiar equation of the Pythagorean theorem. Thus, the Pythagorean theorem is a special case of the law of cosines.

The first example shows how the law of cosines can be used to solve an applied problem.

262

■ *Example 1*

USING THE LAW OF COSINES IN AN APPLICATION

A surveyor wishes to find the distance between two inaccessible points A and B on opposite sides of a lake. While standing at point C, she finds that $AC = 259$ meters, $BC = 423$ meters, and angle ACB measures $132°\ 40'$. Find the distance AB. (See Figure 7.10.)

The law of cosines can be used here, since we know the lengths of two sides of the triangle and the measure of the included angle.

$$AB^2 = 259^2 + 423^2 - 2(259)(423) \cos 132°\ 40'$$
$$AB^2 = 394510.6 \qquad \text{Use a calculator.}$$
$$AB \approx 628 \qquad \text{Take the square root and round to 3 significant digits.}$$

The distance between the points is approximately 628 meters. ■

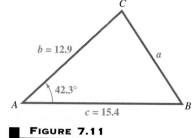

■ **FIGURE 7.10**

■ *Example 2*

USING THE LAW OF COSINES TO SOLVE A TRIANGLE

Solve triangle ABC if $A = 42.3°$, $b = 12.9$ meters, and $c = 15.4$ meters. See Figure 7.11.

■ **FIGURE 7.11**

Start by finding a with the law of cosines.

$$a^2 = b^2 + c^2 - 2bc \cos A$$
$$a^2 = 12.9^2 + 15.4^2 - 2(12.9)(15.4) \cos 42.3°$$
$$a^2 = 109.7$$
$$a = 10.5 \text{ meters}$$

We now must find the measures of angles B and C. There are several approaches that can be used at this point. Let us use the law of sines to find one of these angles. Of the two remaining angles, B must be the smaller since it is opposite the shorter of the two sides b and c. Therefore, it cannot be obtuse, and we will avoid any ambiguity when we find its sine.

$$\frac{\sin 42.3°}{10.5} = \frac{\sin B}{12.9}$$

$$\sin B = \frac{12.9 \sin 42.3°}{10.5}$$

$$B = 55.8° \qquad \text{Use the inverse sine function of a calculator.}$$

The easiest way to find C is to subtract the sum of A and B from 180°.

$$C = 180° - A - B = 81.9°. \quad ■$$

CAUTION | Had we chosen to use the law of sines to find C rather than B in Example 2, we would not have known whether C equals 81.9° or its supplement, 98.1°.

■ *Example 3*

USING THE LAW OF COSINES TO SOLVE A TRIANGLE

Solve triangle ABC if $a = 9.47$ feet, $b = 15.9$ feet, and $c = 21.1$ feet.

We are given the lengths of three sides of the triangle, so we may use the law of cosines to solve for any angle of the triangle. Let us solve for C, the largest angle, using the law of cosines. We will be able to tell if C is obtuse if $\cos C < 0$. Use the form of the law of cosines that involves C.

$$c^2 = a^2 + b^2 - 2ab \cos C,$$

or

$$\cos C = \frac{a^2 + b^2 - c^2}{2ab}.$$

Inserting the given values leads to

$$\cos C = \frac{(9.47)^2 + (15.9)^2 - (21.1)^2}{2(9.47)(15.9)}$$

$$\cos C = -.34109402. \qquad \text{Use a calculator.}$$

Using the inverse cosine function of the calculator, we get the obtuse angle C.

$$C = 109.9°$$

We can use either the law of sines or the law of cosines to find $B = 45.1°$. (Verify this.) Since $A = 180° - B - C$,

$$A = 25.0°. \quad ■$$

As shown in this section and the previous one, four possible cases can occur when solving an oblique triangle. These cases are summarized in the chart below, along with a suggested procedure for solving in each case. There are other procedures that work, but we give the one that is most efficient. In all four cases, it is assumed that the given information actually produces a triangle.

Case	Suggested Procedure for Solving
One side and two angles are known. (SAA or ASA)	1. Find the remaining angle using the angle sum formula ($A + B + C = 180°$). 2. Find the remaining sides using the law of sines.
Two sides and one angle (not included between the two sides) are known. (SSA)	*Be aware of the ambiguous case; there may be two triangles.* 1. Find an angle using the law of sines. 2. Find the remaining angle using the angle sum formula. 3. Find the remaining side using the law of sines. *If two triangles exist, repeat Steps 1, 2, and 3.*
Two sides and the included angle are known. (SAS)	1. Find the third side using the law of cosines. 2. Find the smaller of the two remaining angles using the law of sines. 3. Find the remaining angle using the angle sum formula.
Three sides are known. (SSS)	1. Find the largest angle using the law of cosines. 2. Find either remaining angle using the law of sines. 3. Find the remaining angle using the angle sum formula.

AREA The law of cosines can be used to derive a formula for the area of a triangle when only the lengths of the three sides are known. This formula is known as Heron's formula, named after the Greek mathematician Heron of Alexandria, who lived around A.D. 75. It is found in his work *Metrica,* and is now given.

HERON'S AREA FORMULA If a triangle has sides of lengths a, b, and c, and if the **semiperimeter** is

$$s = \frac{1}{2}(a + b + c),$$

then the area of the triangle is

$$K = \sqrt{s(s - a)(s - b)(s - c)}.$$

A proof of Heron's formula is suggested in Exercises 49–54.

■ *Example 4*

**FINDING AN AREA
USING HERON'S
FORMULA**

Find the area of the triangle having sides of lengths $a = 29.7$ feet, $b = 42.3$ feet, and $c = 38.4$ feet.

To use Heron's area formula, first find s.

$$s = \frac{1}{2}(a + b + c)$$

$$s = \frac{1}{2}(29.7 + 42.3 + 38.4)$$

$$= 55.2$$

The area is

$$K = \sqrt{s(s - a)(s - b)(s - c)}$$

$$K = \sqrt{55.2(55.2 - 29.7)(55.2 - 42.3)(55.2 - 38.4)}$$

$$= \sqrt{55.2(25.5)(12.9)(16.8)}$$

$$= 552 \text{ square feet.} \quad ■$$

7.3 EXERCISES ■

In this exercise set, a calculator will be necessary.

Solve each of the following problems, using the law of sines or the law of cosines. See Example 1.

1. Points A and B are on opposite sides of Lake Yankee. From a third point, C, the angle between the lines of sight to A and B is 46.3°. If AC is 350 m long and BC is 286 m long, find AB.

2. The sides of a parallelogram are 4.0 cm and 6.0 cm. One angle is 58° while another is 122°. Find the lengths of the diagonals of the parallelogram.

3. Airports A and B are 450 km apart, on an east-west line. Tom flies in a northeast direction from A to airport C. From C he flies 359 km on a bearing of 128° 40′ to B. How far is C from A?

4. Two ships leave a harbor together, traveling on courses that have an angle of 135° 40′ between them. If they each travel 402 mi, how far apart are they?

5. The layout for a child's dollhouse in her backyard shows the dimensions given in the figure. Find x.

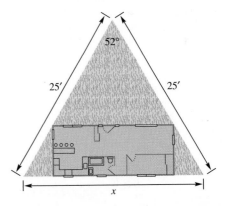

6. A hill slopes at an angle of 12.47° with the horizontal. From the base of the hill, the angle of elevation of a 459.0 ft tower at the top of the hill is 35.98°. How much rope would be required to reach from the top of the tower to the bottom of the hill?

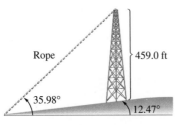

Rope
459.0 ft
35.98°
12.47°

7. A crane with a counterweight is shown in the figure. Find the horizontal distance between points *A* and *B*.

A
B
10 ft
128°
10 ft
C

8. A weight is supported by cables attached to both ends of a balance beam, as shown in the figure. What angles are formed between the beam and the cables?

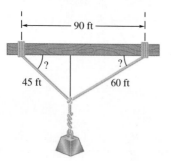

90 ft
?
?
45 ft
60 ft

9. A satellite travelling in a circular orbit 1600 km above earth is due to pass directly over a tracking station at noon.* (See the figure.) Assume that the satellite takes 2 hr to make an orbit and that the radius of the earth is 6400 km. Find the distance between the satellite and the tracking station at 12:03 P.M.

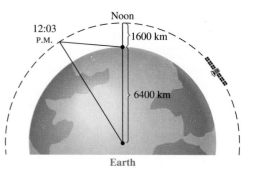

Noon
12:03 P.M.
1600 km
6400 km
Earth

10. Two factories blow their whistles at exactly 5:00. A man hears the two blasts at 3 seconds and 6 seconds after 5:00, respectively. The angle between his lines of sight to the two factories is 42.2°. If sound travels 344 m per sec, how far apart are the factories?

11. A parallelogram has sides of length 25.9 cm and 32.5 cm. The longer diagonal has a length of 57.8 cm. Find the angle opposite the diagonal.

12. A person in a plane flying a straight course observes a mountain at a bearing 24.1° to the right of its course. At that time the plane is 7.92 km from the mountain. A short time later, the bearing to the mountain becomes 32.7°. How far is the airplane from the mountain when the second bearing is taken?

---

*Bernice Kastner, Ph.D., *Spacemathematics*. National Aeronautics and Space Administration (NASA), 1985.

To help predict eruptions from the volcano *Mauna Loa* on the island of Hawaii, scientists keep track of the volcano's movement by using a "super triangle" with vertices on the three volcanoes shown on the map at the right. (For example, in a recent year, Mauna Loa moved 6 inches, a result of increasing internal pressure.) Refer to the map to work Exercises 13 and 14.

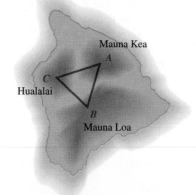

Mauna Kea
A
C
Hualalai
B
Mauna Loa

13. *AB* = 22.47928 mi, *AC* = 28.14276 mi, *A* = 58.56989°;   find *BC*

14. *AB* = 22.47928 mi, *BC* = 25.24983 mi, *A* = 58.56989°;   find *B*

15. Refer to Figure 7.8. If you attempt to find any angle of a triangle using the values $a = 3$, $b = 4$, and $c = 10$ with the law of cosines, what happens?

16. A familiar saying is "The shortest distance between two points is a straight line." Explain how this relates to the geometric property that states that the sum of the lengths of any two sides of a triangle must be greater than the remaining side.

Solve each of the following triangles. See Examples 2 and 3.

17. $C = 28.3°$, $b = 5.71$ in, $a = 4.21$ in

18. $A = 41.4°$, $b = 2.78$ yd, $c = 3.92$ yd

19. $C = 45.6°$, $b = 8.94$ m, $a = 7.23$ m

20. $A = 67.3°$, $b = 37.9$ km, $c = 40.8$ km

21. $A = 80° 40'$, $b = 143$ cm, $c = 89.6$ cm

22. $C = 72° 40'$, $a = 327$ ft, $b = 251$ ft

23. $B = 74.80°$, $a = 8.919$ in, $c = 6.427$ in

24. $C = 59.70°$, $a = 3.725$ mi, $b = 4.698$ mi

25. $A = 112.8°$, $b = 6.28$ m, $c = 12.2$ m

26. $B = 168.2°$, $a = 15.1$ cm, $c = 19.2$ cm

27. $C = 24° 49'$, $a = 251.3$ m, $b = 318.7$ m

28. $B = 52° 28'$, $a = 7598$ in, $c = 6973$ in

Find all the angles in each of the following triangles. See Example 3.

29. $a = 3.0$ ft, $b = 5.0$ ft, $c = 6.0$ ft

30. $a = 4.0$ ft, $b = 5.0$ ft, $c = 8.0$ ft

31. $a = 9.3$ cm, $b = 5.7$ cm, $c = 8.2$ cm

32. $a = 28$ ft, $b = 47$ ft, $c = 58$ ft

33. $a = 42.9$ m, $b = 37.6$ m, $c = 62.7$ m

34. $a = 189$ yd, $b = 214$ yd, $c = 325$ yd

35. $AB = 1240$ ft, $AC = 876$ ft, $BC = 918$ ft

36. $AB = 298$ m, $AC = 421$ m, $BC = 324$ m

Find the area of each of the following triangles. See Example 4.

37. $a = 12$ m, $b = 16$ m, $c = 25$ m

38. $a = 22$ in, $b = 45$ in, $c = 31$ in

39. $a = 154$ cm, $b = 179$ cm, $c = 183$ cm

40. $a = 25.4$ yd, $b = 38.2$ yd, $c = 19.8$ yd

41. $a = 76.3$ ft, $b = 109$ ft, $c = 98.8$ ft

42. $a = 15.89$ in, $b = 21.74$ in, $c = 10.92$ in

43. $a = 74.14$ ft, $b = 89.99$ ft, $c = 51.82$ ft

44. $a = 1.096$ km, $b = 1.142$ km, $c = 1.253$ km

Solve each of the following problems.

45. A painter needs to cover a triangular region 75 m by 68 m by 85 m. A can of paint covers 75 sq m of area. How many cans (to the next higher number of cans) will be needed?

46. How many cans of paint would be needed in Exercise 45 if the region were 8.2 m by 9.4 m by 3.8 m?

47. Find the area of the Bermuda Triangle, if the sides of the triangle have the approximate lengths 850 miles, 925 miles, and 1300 miles.

48. Find the area of a triangle in a rectangular coordinate plane whose vertices are $(0, 0)$, $(3, 4)$, and $(-8, 6)$ using Heron's area formula.

Use the fact that $\cos A = \dfrac{b^2 + c^2 - a^2}{2bc}$ to show that each of the following is true.

49. $1 + \cos A = \dfrac{(b + c + a)(b + c - a)}{2bc}$

50. $1 - \cos A = \dfrac{(a - b + c)(a + b - c)}{2bc}$

51. $\cos \dfrac{A}{2} = \sqrt{\dfrac{s(s - a)}{bc}}$ $\left(\text{Hint: } \cos \dfrac{A}{2} = \sqrt{\dfrac{1 + \cos A}{2}}\right)$

52. $\sin \dfrac{A}{2} = \sqrt{\dfrac{(s - b)(s - c)}{bc}}$ $\left(\text{Hint: } \sin \dfrac{A}{2} = \sqrt{\dfrac{1 - \cos A}{2}}\right)$

53. The area of a triangle having sides b and c and angle A is given by $(1/2)bc \sin A$. Show that this result can be written as

$$\sqrt{\frac{1}{2}bc(1 + \cos A) \cdot \frac{1}{2}bc(1 - \cos A)}.$$

54. Use the results of Exercises 49–53 to prove Heron's area formula.

55. Let a and b be the equal sides of an isosceles triangle. Prove that $c^2 = 2a^2(1 - \cos C)$.

56. Let point D on side AB of triangle ABC be such that CD bisects angle C. Show that $AD/DB = b/a$.

57. Use the law of cosines to prove that if one angle of a triangle is obtuse, then the side opposite the obtuse angle is longer than either of the other two sides.

58. In addition to the law of sines and the law of cosines, there is a **law of tangents.** In any triangle ABC,

$$\frac{\tan\frac{1}{2}(A - B)}{\tan\frac{1}{2}(A + B)} = \frac{a - b}{a + b}.$$

Verify this law for the triangle ABC with $a = 2$, $b = 2\sqrt{3}$, $A = 30°$, and $B = 60°$.

59. Prove the law of tangents by referring to Exercise 30 in Section 7.2 and applying identities found in Section 5.7.

60. Verify that, for triangle ABC,

$$\frac{\cos A}{a} + \frac{\cos B}{b} + \frac{\cos C}{c} = \frac{a^2 + b^2 + c^2}{2abc}.$$

7.4 ——— VECTORS

The measures of all six parts of a triangle can be found, given at least one side and any two other measures. This section and the next show applications of this work to *vectors*. In this section the basic ideas of vectors are presented, and in the next section the law of sines and the law of cosines are applied to vector problems.

Many quantities in mathematics involve magnitudes, such as 45 pounds or 60 miles per hour. These quantities are called **scalars.** Other quantities, called **vector quantities,** involve both magnitude and direction. Typical vector quantities are velocity, acceleration, and force.

A vector quantity is often represented with a directed line segment, which is called a **vector.** The length of the vector represents the magnitude of the vector quantity. The direction of the vector, indicated with an arrowhead, represents the direction of the quantity. For example, the vector in Figure 7.12 represents a force of 10 pounds applied at an angle of 30° from the horizontal.

The symbol for a vector is often printed in boldface type. When writing vectors by hand, it is customary to use an arrow over the letter or letters. Thus **OP** and $\overrightarrow{OP}$ both represent vector OP. Vectors may be named with either one lower-case or uppercase letter, or two uppercase letters. When two letters are used, the first indicates the *initial point* and the second indicates the *terminal point* of the

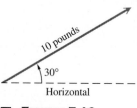

10 pounds

30°

Horizontal

■ **FIGURE 7.12**

vector. Knowing these points gives the direction of the vector. For example, vectors **OP** and **PO** in Figure 7.13 are not the same vector. They have the same magnitude, but they have opposite directions. The magnitude of vector **OP** is written |**OP**|.

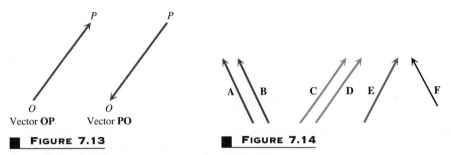

Vector **OP**   Vector **PO**

■ **FIGURE 7.13**   ■ **FIGURE 7.14**

Two vectors are *equal* if and only if they both have the same direction and the same magnitude. In Figure 7.14 vectors **A** and **B** are equal, as are vectors **C** and **D**.

As Figure 7.14 shows, equal vectors need not coincide, but they must be parallel. Vectors **A** and **E** are unequal because they do not have the same direction, while **A** ≠ **F** because they have different magnitudes, as indicated by their different lengths.

To find the *sum* of two vectors **A** and **B,** written **A** + **B**, place the initial point of vector **B** at the terminal point of vector **A,** as shown in Figure 7.15. The vector with the same initial point as **A** and the same terminal point as **B** is the sum **A** + **B.** The sum of two vectors is also a vector.

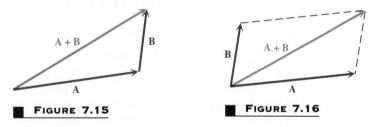

■ **FIGURE 7.15**   ■ **FIGURE 7.16**

Another way to find the sum of two vectors is to use the **parallelogram rule.** Place vectors **A** and **B** so that their initial points coincide. Then complete a parallelogram that has **A** and **B** as two adjacent sides. The diagonal of the parallelogram with the same initial point as **A** and **B** is the same vector sum **A** + **B** found by the definition. See Figure 7.16.

The vector sum **A** + **B** is the **resultant** of vectors **A** and **B.** Each of the vectors **A** and **B** is a **component** of vector **A** + **B.** In many practical applications, such as surveying, it is necessary to break a vector into its **vertical** and **horizontal components.** These components are two vectors, one vertical and one horizontal, whose resultant is the original vector. As shown in Figure 7.17, vector **OR** is the vertical component and vector **OS** is the horizontal component of **OP.**

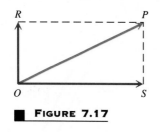

■ **FIGURE 7.17**

For every vector **v** there is a vector −**v** with the same magnitude as **v** but opposite direction. Vector −**v** is the **opposite** of **v.** See Figure 7.18. The sum of **v** and −**v** has magnitude 0 and is a **zero vector.** As with real numbers, to *subtract* vector **B** from vector **A,** find the vector sum **A** + (−**B**). See Figure 7.19.

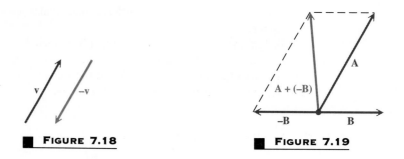

■ **FIGURE 7.18**   ■ **FIGURE 7.19**

The **scalar product** of a real number (or scalar) k and a vector **u** is the vector k**u**, which has magnitude $|k|$ times the magnitude of **u.** As shown in Figure 7.20, k**u** has the same direction as **u** if $k > 0$, and the opposite direction if $k < 0$.

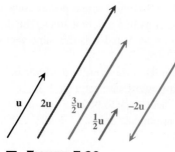

■ **FIGURE 7.20**

■ *Example 1*

FINDING
MAGNITUDES OF
VERTICAL AND
HORIZONTAL
COMPONENTS

Vector **w** has magnitude 25.0 and is inclined at an angle of 40° from the horizontal. Find the magnitudes of the horizontal and vertical components of the vector.

In Figure 7.21, the vertical component is labeled **v** and the horizontal component is labeled **u.** Vectors **u, v,** and **w** form a right triangle. In this right triangle,

$$\sin 40° = \frac{|\mathbf{v}|}{|\mathbf{w}|} = \frac{|\mathbf{v}|}{25.0},$$

and

$$|\mathbf{v}| = 25.0 \sin 40° = 16.1.$$

In the same way,

$$\cos 40° = \frac{|\mathbf{u}|}{25.0},$$

with

$$|\mathbf{u}| = 25.0 \cos 40° = 19.2. \quad ■$$

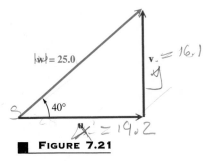

$|\mathbf{w}| = 25.0$

$\mathbf{v} = 16.1$

40°

■ FIGURE 7.21

It is helpful to review some of the properties of parallelograms when studying vectors. A parallelogram is a quadrilateral whose opposite sides are parallel. The opposite sides and opposite angles of a parallelogram are equal, and consecutive angles of a parallelogram are supplementary. The diagonals of a parallelogram bisect each other, but do not necessarily bisect the angles of the parallelogram.

Some of these properties are used in the following example.

■ *Example 2*

FINDING THE
MAGNITUDE OF THE
RESULTANT OF TWO
VECTORS IN AN
APPLICATION

Two forces of 15 newtons and 22 newtons (a newton is a unit of force used in physics) act at a point in the plane. If the angle between the forces is 100°, find the magnitude of the resultant force.

As shown in Figure 7.22, a parallelogram that has the forces as adjacent sides can be formed. The angles of the parallelogram adjacent to angle P each measure 80°, since adjacent angles of a parallelogram are supplementary. (Angle *SPQ* measures 100°.) Opposite sides of the parallelogram are equal in length. The resultant force divides the parallelogram into two triangles. Use the law of cosines to get

$$|\mathbf{v}|^2 = 15^2 + 22^2 - 2(15)(22)\cos 80°$$
$$|\mathbf{v}| = 24. \quad ■$$

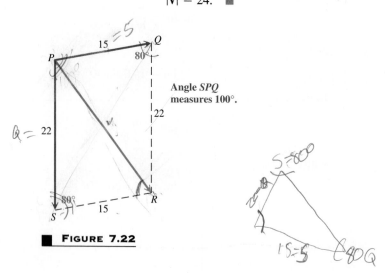

Angle *SPQ*
measures 100°.

■ **FIGURE 7.22**

7.4 EXERCISES ■

Exercises 1–4 refer to the following vectors.

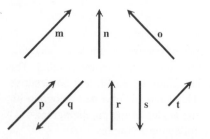

1. Name all pairs of vectors that appear to be equal.
2. Name all pairs of vectors that are opposites.
3. Name all pairs of vectors where the first is a scalar multiple of the other, with the scalar positive.
4. Name all pairs of vectors where the first is a scalar multiple of the other, with the scalar negative.

Exercises 5–22 refer to the vectors pictured here.

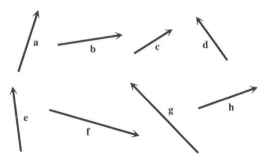

*Draw a sketch to represent each of the following vectors. For example, find **a** + **e** by placing **a** and **e** so that their initial points coincide. Then use the parallelogram rule to find the resultant, shown in the figure.*

5. −**b**	**6.** −**g**	**7.** 3**a**	**8.** 2**h**	**9.** **a** + **c**
10. **a** + **b**	**11.** **h** + **g**	**12.** **e** + **f**	**13.** **a** + **h**	**14.** **b** + **d**
15. **h** + **d**	**16.** **a** + **f**	**17.** **a** − **c**	**18.** **d** − **e**	**19.** **a** + (**b** + **c**)
20. (**a** + **b**) + **c**	**21.** **c** + **d**	**22.** **d** + **c**		

23. From the results of Exercises 19 and 20, do you think vector addition is associative?

24. From the results of Exercises 21 and 22, do you think vector addition is commutative?

*For each pair of vectors **u** and **w** with angle θ between them, sketch the resultant.*

25. |**u**| = 12, |**w**| = 20, θ = 27°   **26.** |**u**| = 8, |**w**| = 12, θ = 20°   **27.** |**u**| = 20, |**w**| = 30, θ = 30°

28. |**u**| = 27, |**w**| = 50, θ = 12°   **29.** |**u**| = 50, |**w**| = 70, θ = 40°

You will need a calculator to work many of the remaining exercises in this section.

*For each of the following, vector **v** has the given magnitude and direction. Find the magnitude of the horizontal and vertical components of **v**, if α is the inclination vector from the horizontal. See Example 1.*

30. α = 20°, |**v**| = 50   **31.** α = 38°, |**v**| = 12   **32.** α = 70°, |**v**| = 150

33. α = 50°, |**v**| = 26   **34.** α = 35° 50′, |**v**| = 47.8   **35.** α = 27° 30′, |**v**| = 15.4

36. α = 59° 40′, |**v**| = 78.9   **37.** α = 128.5°, |**v**| = 198   **38.** α = 146.3°, |**v**| = 238

In each of the following, two forces act at a point in the plane. The angle between the two forces is given. Find the magnitude of the resultant force. See Example 2.

39. Forces of 250 and 450 newtons, forming an angle of 85°

40. Forces of 19 and 32 newtons, forming an angle of 118°

41. Forces of 17.9 and 25.8 lb, forming an angle of 105° 30′

42. Forces of 75.6 and 98.2 lb, forming an angle of 82° 50′

43. Forces of 116 and 139 lb, forming an angle of 140° 50′

44. Forces of 37.8 and 53.7 lb, forming an angle of 68° 30′

274   *A vector **v** with horizontal component x and vertical component y may be written $\langle x, y \rangle$. We can define vector addition and scalar multiplication as follows.*

> *For real numbers a, b, c, d, and k,*
> $$\langle a, b \rangle + \langle c, d \rangle = \langle a + c, b + d \rangle$$
> $$k\langle a, b \rangle = \langle ka, kb \rangle.$$

Let $\mathbf{u} = \langle a_1, b_1 \rangle$, $\mathbf{v} = \langle a_2, b_2 \rangle$, $\mathbf{w} = \langle a_3, b_3 \rangle$, and $\mathbf{0} = \langle 0, 0 \rangle$. Let k be any real number. Prove each of the following statements.

45. $\mathbf{u} + \mathbf{v} = \mathbf{v} + \mathbf{u}$ **46.** $\mathbf{u} + (\mathbf{v} + \mathbf{w}) = (\mathbf{u} + \mathbf{v}) + \mathbf{w}$ **47.** $-1(\mathbf{u}) = -\mathbf{u}$

48. $k(\mathbf{u} + \mathbf{v}) = k\mathbf{u} + k\mathbf{v}$ **49.** $\mathbf{u} + \mathbf{0} = \mathbf{u}$ **50.** $\mathbf{u} + (-\mathbf{u}) = \mathbf{0}$

*If we define the unit vectors **i** and **j** as follows,*

$$\mathbf{i} = \langle 1, 0 \rangle, \qquad \mathbf{j} = \langle 0, 1 \rangle,$$

*then the vector $\mathbf{v} = \langle a, b \rangle$ may be written as a linear combination of **i** and **j**:*

$$\mathbf{v} = a\mathbf{i} + b\mathbf{j}.$$

*Let $\mathbf{u} = a_1\mathbf{i} + b_1\mathbf{j}$ and $\mathbf{v} = a_2\mathbf{i} + b_2\mathbf{j}$. The dot product, or inner product, of **u** and **v**, written $\mathbf{u} \cdot \mathbf{v}$, is defined as*

$$\mathbf{u} \cdot \mathbf{v} = a_1a_2 + b_1b_2.$$

Find $\mathbf{u} \cdot \mathbf{v}$ for each of the following pairs of vectors.

51. $\mathbf{u} = 6\mathbf{i} - 2\mathbf{j}$ and $\mathbf{v} = -3\mathbf{i} + 2\mathbf{j}$ **52.** $\mathbf{u} = 3\mathbf{i} + 2\mathbf{j}$ and $\mathbf{v} = -3\mathbf{i} + 7\mathbf{j}$

53. $\mathbf{u} = \langle -6, 8 \rangle$ and $\mathbf{v} = \langle 3, -4 \rangle$ **54.** $\mathbf{u} = \langle 0, -2 \rangle$ and $\mathbf{v} = \langle -2, 6 \rangle$

55. Let α be the angle between the vectors **u** and **v**, where $0° \leq \alpha \leq 180°$. Show that $\mathbf{u} \cdot \mathbf{v} = |\mathbf{u}| \cdot |\mathbf{v}| \cdot \cos \alpha$.

Use the result of Exercise 55 to find the angle between each of the following pairs of vectors.

56. $\mathbf{u} = \langle -2, 5 \rangle$ and $\mathbf{v} = \langle 3, -4 \rangle$ **57.** $\mathbf{u} = \langle 1, 8 \rangle$ and $\mathbf{v} = \langle 2, -5 \rangle$

58. $\mathbf{u} = \langle -6, -2 \rangle$ and $\mathbf{v} = \langle 3, -1 \rangle$

*Prove each of the following properties of the dot product. Assume that **u**, **v**, and **w** are vectors, and k is a nonzero real number.*

59. $\mathbf{u} \cdot \mathbf{v} = \mathbf{v} \cdot \mathbf{u}$ **60.** $\mathbf{u} \cdot \mathbf{u} = |\mathbf{u}|^2$

61. $\mathbf{u} \cdot (\mathbf{v} + \mathbf{w}) = \mathbf{u} \cdot \mathbf{v} + \mathbf{u} \cdot \mathbf{w}$ **62.** $(k \cdot \mathbf{u}) \cdot \mathbf{v} = k \cdot (\mathbf{u} \cdot \mathbf{v})$

63. If **u** and **v** are not **0**, and if $\mathbf{u} \cdot \mathbf{v} = 0$, then **u** and **v** are perpendicular.

64. If **u** and **v** are perpendicular, then $\mathbf{u} \cdot \mathbf{v} = 0$.

The previous section covered methods for finding the resultant of two vectors. In many applications it is necessary to find a vector that will counterbalance the resultant. This opposite vector is called the *equilibrant:* the **equilibrant** of vector **u** is the vector **−u.**

▪ *Example 1*

FINDING THE
MAGNITUDE AND
DIRECTION OF AN
EQUILIBRANT

Find the magnitude of the equilibrant of forces of 48 newtons and 60 newtons acting on a point *A,* if the angle between the forces is 50°. Then find the angle between the equilibrant and the 48-newton force.

In Figure 7.23, the equilibrant is **−v.** The magnitude of **v,** and hence of **−v,** is found by using triangle *ABC* and the law of cosines:

$$|\mathbf{v}|^2 = 48^2 + 60^2 - 2(48)(60) \cos 130°$$
$$= 9606.5,$$

and

$$|\mathbf{v}| = 98 \text{ newtons}$$

to two significant digits.

The required angle, labeled α in Figure 7.23, can be found by subtracting angle *CAB* from 180°. Use the law of sines to find angle *CAB.*

$$\frac{98}{\sin 130°} = \frac{60}{\sin CAB}$$

$$\sin CAB = .46900680$$
$$CAB = 28°$$

Finally,

$$\alpha = 180° - 28°$$
$$= 152°. \quad ▪$$

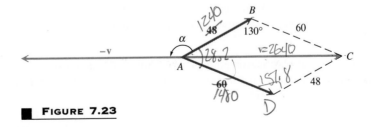

▪ FIGURE 7.23

276

■ *Example 2*

FINDING THE
MAGNITUDE OF A
FORCE

Find the force required to pull a 50-pound weight up a ramp inclined at 20° to the horizontal.

In Figure 7.24, the vertical 50-pound force represents the force of gravity. The component **BC** represents the force with which the body pushes against the ramp, while the component **BF** represents a force that would pull the body up the ramp. Since vectors **BF** and **AC** are equal, |**AC**| gives the required force.

Vectors **BF** and **AC** are parallel, so angle *EBD* equals angle *A*. Since angle *BDE* and angle *C* are right angles, triangles *CBA* and *DEB* have two corresponding angles equal and so are similar triangles. Therefore, angle *ABC* equals angle *E*, which is 20°. From right triangle *ABC*,

$$\sin 20° = \frac{|\mathbf{AC}|}{50}$$

$$|\mathbf{AC}| = 50 \sin 20°$$

$$|\mathbf{AC}| = 17.$$

To the nearest pound, a 17-pound force will be required to pull the weight up the ramp. ■

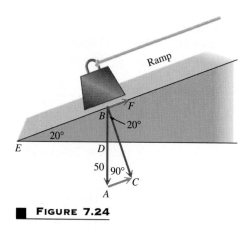

■ **FIGURE 7.24**

Problems involving bearing (defined in Section 2.5) can also be worked with vectors, as shown in the next example.

■ *Example 3*

APPLYING VECTORS
TO A NAVIGATION
PROBLEM

A ship leaves port on a bearing of 28° and travels 8.2 miles. The ship then turns due east and travels 4.3 miles. How far is the ship from port? What is its bearing from port?

In Figure 7.25, vectors **PA** and **AE** represent the ship's path. The magnitude and bearing of the resultant **PE** can be found as follows. Triangle *PNA* is a right triangle, so angle $NAP = 90° - 28° = 62°$. Then angle $PAE = 180° - 62° = 118°$. Use the law of cosines to find $|\mathbf{PE}|$, the magnitude of vector $\mathbf{PE}$.

$$|\mathbf{PE}|^2 = 8.2^2 + 4.3^2 - 2(8.2)(4.3) \cos 118°$$

$$|\mathbf{PE}|^2 = 118.84$$

Therefore, $\quad\quad |\mathbf{PE}| = 10.9,$

or 11 miles, rounded to two significant digits.

To find the bearing of the ship from port, first find angle APE. Use the law of sines, along with the value of $|\mathbf{PE}|$ before rounding.

$$\frac{\sin APE}{4.3} = \frac{\sin 118°}{10.9}$$

$$\sin APE = \frac{4.3 \sin 118°}{10.9}$$

$$\text{angle } APE = 20.4°$$

After rounding, angle APE is 20°, and the ship is 11 miles from port on a bearing of $28° + 20° = 48°$. ■

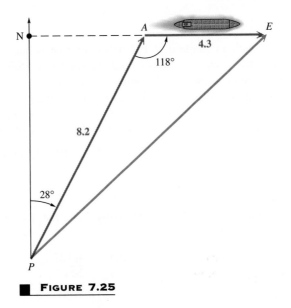

In air navigation, the **airspeed** of a plane is its speed relative to the air, while the **groundspeed** is its speed relative to the ground. Because of wind, these two speeds are usually different. The groundspeed of the plane is represented by the vector sum of the airspeed and windspeed vectors. See Figure 7.26.

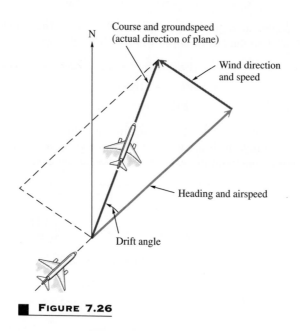

Course and groundspeed
(actual direction of plane)

Wind direction
and speed

Heading and airspeed

Drift angle

■ **FIGURE 7.26**

■ *Example 4*

**APPLYING VECTORS
TO A NAVIGATION
PROBLEM**

A plane with an airspeed of 192 miles per hour is headed on a bearing of 121°. A north wind is blowing (from north to south) at 15.9 miles per hour. Find the groundspeed and the actual bearing of the plane.

  In Figure 7.27 the groundspeed is represented by $|\mathbf{x}|$. We must find angle α to find the bearing, which will be $121° + \alpha$. From Figure 7.27, angle *BCO* equals angle *AOC*, which equals 121°. Find $|\mathbf{x}|$ by the law of cosines.

$$|\mathbf{x}|^2 = 192^2 + 15.9^2 - 2(192)(15.9) \cos 121°$$
$$|\mathbf{x}|^2 = 40{,}261$$

Therefore, $|\mathbf{x}| = 200.7,$

or 201 miles per hour. Now find α by using the law of sines. As before, use the value of $|\mathbf{x}|$ before rounding.

$$\frac{\sin \alpha}{15.9} = \frac{\sin 121°}{200.7}$$
$$\sin \alpha = .06792320$$
$$\alpha = 3.89°$$

After rounding, α is 3.9°. The groundspeed is about 201 miles per hour, on a bearing of 125°, to three significant digits. ■

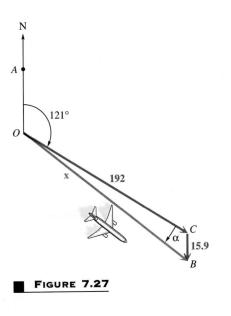

▪ **FIGURE 7.27**

7.5 EXERCISES ▪

A calculator will be necessary to solve the problems in this exercise set.

Solve each of the following problems. See Examples 1–4.

1. Two forces of 692 newtons and 423 newtons act at a point. The resultant force is 786 newtons. Find the angle between the forces.

2. Two forces of 128 lb and 253 lb act at a point. The equilibrant is 320 lb. Find the angle between the forces.

3. A force of 25 lb is required to push an 80-lb crate up a hill. What angle does the hill make with the horizontal?

4. Find the force required to keep a 3000-lb car parked on a hill that makes an angle of 15° with the horizontal.

5. To build the pyramids in Egypt, it is believed that giant causeways were built to transport the building materials to the site. One such causeway is said to have been 3000 ft long, with a slope of about 2.3°. How much force would be required to pull a 60-ton monolith along this causeway?

6. A force of 500 lb is required to pull a boat up a ramp inclined at 18° with the horizontal. How much does the boat weigh?

7. Two tugboats are pulling a disabled speedboat into port with forces of 1240 lb and 1480 lb. The angle between these forces is 28.2°. Find the direction and magnitude of the equilibrant.

8. Two people are carrying a box. One person exerts a force of 150 lb at an angle of 62.4° with the horizontal. The other person exerts a force of 114 lb at an angle of 54.9°. Find the weight of the box.

9. A crate is supported by two ropes. One rope makes an angle of 46° 20′ with the horizontal and has a tension of 89.6 lb on it. The other rope is horizontal. Find the weight of the crate and the tension in the horizontal rope.

10. Three forces acting at a point are in equilibrium. The forces are 980 lb, 760 lb, and 1220 lb. Find the angles between the directions of the forces. (*Hint:* Arrange the forces to form the sides of a triangle.)

280

11. A force of 176 lb makes an angle of 78° 50′ with a second force. The resultant of the two forces makes an angle of 41° 10′ with the first force. Find the magnitude of the second force and of the resultant.

12. A force of 28.7 lb makes an angle of 42° 10′ with a second force. The resultant of the two forces makes an angle of 32° 40′ with the first force. Find the magnitudes of the second force and of the resultant.

13. A plane flies 650 mph on a bearing of 175.3°. A 25 mph wind, from a direction of 266.6°, blows against the plane. Find the resulting bearing of the plane.

14. A pilot wants to fly on a bearing of 74.9°. By flying due east, he finds that a 42 mph wind, blowing from the south, puts him on course. Find the airspeed and the groundspeed.

15. Starting at point *A*, a ship sails 18.5 km on a bearing of 189°, then turns and sails 47.8 km on a bearing of 317°. Find the distance of the ship from point *A*.

16. Two towns 21 mi apart are separated by a dense forest. (See the figure.) To travel from town *A* to town *B*, a person must go 17 mi on a bearing of 325°, then turn and continue for 9 mi to reach town *B*. Find the bearing of *B* from *A*.

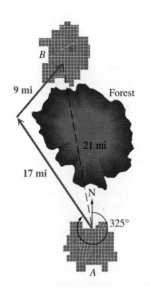

17. An airline route from San Francisco to Honolulu is on a bearing of 233°. A jet flying at 450 mph on that bearing flies into a wind blowing at 39 mph from a direction of 114°. Find the resulting bearing and groundspeed of the plane.

18. A pilot is flying at 168 mph. She wants her flight path to be on a bearing of 57° 40′. A wind is blowing from the south at 27.1 mph. Find the bearing the pilot should fly, and find the plane's groundspeed.

19. What bearing and airspeed are required for a plane to fly 400 mi due north in 2.5 hr if the wind is blowing from a direction of 328° at 11 mph?

20. A plane is headed due south with an airspeed of 192 mph. A wind from a direction of 78° is blowing at 23 mph. Find the groundspeed and resulting bearing of the plane.

21. An airplane is headed on a bearing of 174° at an airspeed of 240 km per hr. A 30 km per hr wind is blowing from a direction of 245°. Find the groundspeed and resulting bearing of the plane.

22. A ship sailing due east in the North Atlantic has been warned to change course to avoid a group of icebergs. The captain turns and sails on a bearing of 62° for a while, then changes course again to a bearing of 115° until the ship reaches its original course. (See the figure.) How much farther did the ship have to travel to avoid the icebergs?

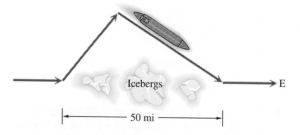

23. The aircraft carrier *Tallahassee* is travelling at sea on a steady course with a bearing of 30° at 32 mph. Patrol planes on the carrier have enough fuel for 2.6 hr of flight when travelling at a speed of 520 mph. One of the pilots takes off on a bearing of 338° and then turns and heads in a straight line, so as to be able to catch the carrier and land on the deck at the exact instant that his fuel runs out. If the pilot left at 2 P.M., at what time did he turn to head for the carrier?

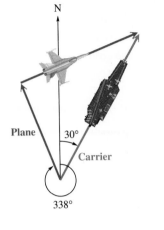

24. A car going around a banked curve is subject to the forces shown in the figure. If the radius of the curve is 100 ft, what value of θ to the nearest degree would allow an automobile to travel around the curve at a speed of 40 ft per sec without depending on friction?

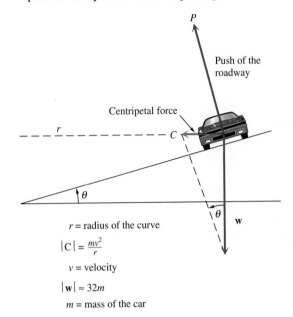

r = radius of the curve

$$|\mathbf{C}| = \frac{mv^2}{r}$$

v = velocity

$|\mathbf{w}| \approx 32m$

m = mass of the car

CHAPTER 7   SUMMARY ▪

SECTION	KEY IDEAS
7.1 Oblique Triangles and the Law of Sines	**Law of Sines** In any triangle *ABC*, with sides *a*, *b*, and *c*, $$\frac{a}{\sin A} = \frac{b}{\sin B}, \quad \frac{a}{\sin A} = \frac{c}{\sin C}, \quad \text{and} \quad \frac{b}{\sin B} = \frac{c}{\sin C}.$$ **Area of a Triangle** The area of a triangle is given by half the product of the lengths of two sides and the sine of the angle between the two sides. $$K = \frac{1}{2} bc \sin A, \quad K = \frac{1}{2} ab \sin C, \quad K = \frac{1}{2} ac \sin B$$

SECTION	KEY IDEAS
7.3 The Law of Cosines	**Law of Cosines** In any triangle ABC, with sides a, b, and c, $$a^2 = b^2 + c^2 - 2bc \cos A$$ $$b^2 = a^2 + c^2 - 2ac \cos B$$ $$c^2 = a^2 + b^2 - 2ab \cos C.$$ **Heron's Area Formula** If a triangle has sides of lengths a, b, and c, and if the semiperimeter is $$s = \frac{1}{2}(a + b + c),$$ then the area of the triangle is $$K = \sqrt{s(s - a)(s - b)(s - c)}.$$
7.4 Vectors	**Sum of Two Vectors** **Opposite Vectors**

A calculator will be necessary to work most of the exercises in this set.

Use the law of sines to find the indicated part of each triangle ABC.

1. $C = 74° 10'$, $c = 96.3$ m, $B = 39° 30'$; find b

2. $A = 129° 40'$, $a = 127$ ft, $b = 69.8$ ft; find B

3. $C = 51° 20'$, $c = 68.3$ m, $b = 58.2$ m; find B

4. $a = 165$ m, $A = 100.2°$, $B = 25.0°$; find b

5. $B = 39° 50'$, $b = 268$ m, $a = 340$ m; find A

6. $C = 79° 20'$, $c = 97.4$ mm, $a = 75.3$ mm; find A

7. If we are given a, A, and C in a triangle ABC, does the possibility of the ambiguous case exist? If not, explain why.

8. Can triangle ABC exist if $a = 4.7$, $b = 2.3$, and $c = 7.0$? If not, explain why. Answer this question without using trigonometry.

Use the law of cosines to find the indicated part of each triangle ABC.

9. $a = 86.14$ in, $b = 253.2$ in, $c = 241.9$ in; find A

10. $B = 120.7°$, $a = 127$ ft, $c = 69.8$ ft; find b

11. $A = 51° 20'$, $c = 68.3$ m, $b = 58.2$ m; find a

12. $a = 14.8$ m, $b = 19.7$ m, $c = 31.8$ m; find B

13. $A = 46° 10'$, $b = 184$ cm, $c = 192$ cm; find a

14. $a = 7.5$ ft, $b = 12.0$ ft, $c = 6.9$ ft; find C

Solve each triangle ABC having the given information.

15. $A = 25° 10'$, $a = 6.92$ yd, $b = 4.82$ yd

16. $A = 61.7°$, $a = 78.9$ m, $b = 86.4$ m

17. $a = 27.6$ cm, $b = 19.8$ cm, $C = 42° 30'$

18. $a = 94.6$ yd, $b = 123$ yd, $c = 109$ yd

Find the area of each triangle ABC with the given information.

19. $b = 840.6$ m, $c = 715.9$ m, $A = 149° 18'$

20. $a = 6.90$ ft, $b = 10.2$ ft, $C = 35° 10'$

21. $a = .913$ km, $b = .816$ km, $c = .582$ km

22. $a = 43$ m, $b = 32$ m, $c = 51$ m

The following identities involve all six parts of a triangle, ABC, and are thus useful for checking answers.

Newton's formula $\dfrac{a+b}{c} = \dfrac{\cos\frac{1}{2}(A-B)}{\sin\frac{1}{2}C}$ Mollweide's formula $\dfrac{a-b}{c} = \dfrac{\sin\frac{1}{2}(A-B)}{\cos\frac{1}{2}C}$

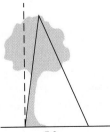

A
$30°$
$b = 7\sqrt{3}$ $c = 14$
$60°$
C $a = 7$ B

23. Apply Newton's formula to the triangle shown in the figure to verify the accuracy of the information.

24. Apply Mollweide's formula to the triangle shown in the figure to verify the accuracy of the information.

Solve each of the following problems.

25. Raoul plans to paint a triangular wall in his A-frame cabin. Two sides measure 7 m each, and the third side measures 6 m. How much paint will he need to buy if a can of paint covers 7.5 sq m?

26. A lot has the shape of a quadrilateral. (See the figure.) What is its area?

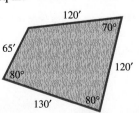

120′
70°
65′
120′
80°
130′
80°

27. A tree leans at an angle of 8.0° from the vertical. (See the figure.) From a point 7.0 m from the bottom of the tree, the angle of elevation to the top of the tree is 68°. How tall is the tree?

7.0 m

28. A hill makes an angle of 14.3° with the horizontal. From the base of the hill, the angle of elevation to the top of a tree on top of the hill is 27.2°. The distance along the hill from the base to the tree is 212 ft. Find the height of the tree.

29. A ship is sailing east. At one point, the bearing of a submerged rock is 45° 20′. After sailing 15.2 mi, the bearing of the rock has become 308° 40′. Find the distance of the ship from the rock at the latter point.

30. From an airplane flying over the ocean, the angle of depression to a submarine lying just under the surface is 24° 10′. At the same moment the angle of depression from the airplane to a battleship is 17° 30′. (See the figure.) The distance from the airplane to the battleship is 5120 ft. Find the distance between the battleship and the submarine. (Assume the airplane, submarine, and battleship are in a vertical plane.)

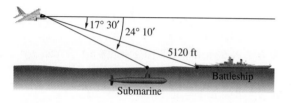

31. Two boats leave a dock together. Each travels in a straight line. The angle between their courses measures 54° 10′. One boat travels 36.2 km per hr, and the other travels 45.6 km per hr. How far apart will they be after 3 hr?

32. Find the lengths of both diagonals of a parallelogram with adjacent sides of 12 cm and 15 cm if the angle between these sides is 33°.

33. To measure the distance through a mountain for a proposed tunnel, a point C is chosen that can be reached from each end of the tunnel. (See the figure.) If $AC = 3800$ m, $BC = 2900$ m, and angle $C = 110°$, find the length of the tunnel.

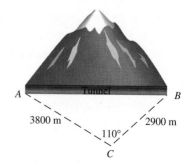

34. A baseball diamond is a square, 90 ft on a side, with home plate and the three bases as vertices. The pitcher's rubber is located 60.5 ft from home plate. Find the distance from the pitcher's rubber to each of the bases.

35. The Vietnam Veterans' Memorial in Washington, D.C., is in the shape of an unenclosed isosceles triangle (that is, V-shaped) with equal sides of length 246.75 feet and the angle between these sides measuring 125° 12′. Find the distance between the ends of the two equal sides.

36. If angle C of a triangle ABC measures 90°, what does the law of cosines $c^2 = a^2 + b^2 - 2ab \cos C$ become?

In Exercises 37–39, use the vectors pictured here. Find each of the following.

37. a + b

38. a − b

39. a + 3c

40. True or false: Opposite angles of a parallelogram are congruent.

41. True or false: A diagonal of a parallelogram must bisect two angles of the parallelogram.

Find the horizontal and vertical components of each of the following vectors, where α is the inclination of the vector from the horizontal.

42. $\alpha = 45°$, magnitude 50

43. $\alpha = 75°$, magnitude 69.2

44. $\alpha = 154° 20′$, magnitude 964

Given two forces and the angle between them, find the magnitude of the resultant force.

45. Forces of 15 and 23 lb, forming an angle of 87°

46. Forces of 142 and 215 newtons, forming an angle of 112°

47. Forces of 85.2 and 69.4 newtons, forming an angle of 58° 20′

48. Forces of 475 and 586 lb, forming an angle of 78° 20′

Solve each of the following problems.

49. One rope pulls a barge directly east with a force of 100 newtons. Another rope pulls the barge to the northeast with a force of 200 newtons. Find the resultant force acting on the barge and its direction relative to the first rope.

50. Paula and Steve are pulling their daughter Jessie on a sled. Steve pulls with a force of 18 lb at an angle of 10°. Paula pulls with a force of 12 lb at an angle of 15°. What is the weight of Jessie and the sled? See the figure. (*Hint:* Find the resultant force.)

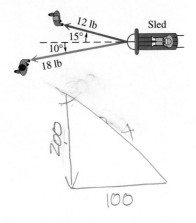

51. A 186-lb force just keeps a 2800-lb car from rolling down a hill. What angle does the hill make with the horizontal?

52. A plane has an airspeed of 520 mph. The pilot wishes to fly on a bearing of 310°. A wind of 37 mph is blowing from a bearing of 212°. What direction should the pilot fly, and what will be her actual speed?

53. A boat travels 15 km per hr in still water. The boat is travelling across a large river, on a bearing of 130°. The current in the river, coming from the west, has a speed of 7 km per hr. Find the resulting speed of the boat and its resulting direction of travel.

54. A long-distance swimmer starts out swimming a steady 3.2 mph due north. A 5.1 mph current is flowing on a bearing of 12°. What is the swimmer's resulting bearing and speed?

COMPLEX NUMBERS AND POLAR EQUATIONS

So far this text has dealt only with real numbers. The set of real numbers, however, does not include enough numbers for our needs. For example, there is no real number solution of the equation $x^2 + 1 = 0$. This chapter discusses a set of numbers having the real numbers as a subset, the set of *complex numbers.*

8.1 ─────── **OPERATIONS ON COMPLEX NUMBERS**

Methods of solving quadratic equations have been known since early Babylonian days. However, when faced with solving equations such as $x^2 = -9$, mathematicians were faced with the dilemma that no real numbers have squares that are negative. In order to resolve this, a new number system, the *complex numbers,* was developed. The imaginary unit i is defined as follows.

DEFINITION OF i	$i = \sqrt{-1}$ or $i^2 = -1$

A **complex number** is a number that has the form $a + bi$, where a and b are real numbers. The form $a + bi$ is called the **standard,** or **rectangular form** of the complex number. The real number a is called the **real part,** and the real number b is called the **imaginary part.** Each real number is a complex number, since a real number a may be thought of as the complex number $a + 0i$. The set of real numbers is a subset of the set of complex numbers. See Figure 8.1.

Complex numbers	Rational numbers $\frac{4}{9}, -\frac{5}{8}, \frac{11}{7}$	Irrational numbers
$8 - i$		$-\sqrt{8}$
$3 - i\sqrt{2}$	Integers $-11, -6, -4$	$\sqrt{15}$
$4i$		$\sqrt{23}$
$-11i$	Whole numbers 0	π
$i\sqrt{7}$		$\frac{\pi}{4}$
$1 + \pi i$	Natural numbers $1, 2, 3, 4,$ $5, 37, 50$	e

Real numbers are shaded.

■ **FIGURE 8.1**

NOTE	The form $a + ib$ is often used for symbols such as $i\sqrt{5}$, since $\sqrt{5}i$ could be too easily mistaken for $\sqrt{5i}$.

The square root of a negative number can be written as the product of a real number and i, using the definition of $\sqrt{-a}$ which follows.

| DEFINITION OF $\sqrt{-a}$ | For positive real numbers a, $$\sqrt{-a} = i\sqrt{a}.$$ |

For example, $\sqrt{-16} = i\sqrt{16} = 4i$, and $\sqrt{-75} = i\sqrt{75} = 5i\sqrt{3}$.

The first example shows how complex numbers may occur as solutions of quadratic equations.

■ *Example 1*

SOLVING QUADRATIC EQUATIONS WITH COMPLEX SOLUTIONS

Solve each equation for its complex solutions.

(a) $x^2 = -9$

Take the square root on both sides, remembering that we must find both roots, indicated by the $\pm$ sign.

$$x^2 = -9$$
$$x = \pm\sqrt{-9}$$
$$x = \pm i\sqrt{9} \qquad \sqrt{-a} = i\sqrt{a}$$
$$x = \pm 3i \qquad \sqrt{9} = 3$$

(b) $9x^2 + 5 = 6x$

Write the equation in standard form, $9x^2 - 6x + 5 = 0$. We will use the quadratic formula, with $a = 9$, $b = -6$, and $c = 5$.

$$x = \frac{-b \pm \sqrt{b^2 - 4ac}}{2a} \qquad \text{Quadratic formula}$$

$$= \frac{-(-6) \pm \sqrt{(-6)^2 - 4(9)(5)}}{2(9)} \qquad a = 9, b = -6, c = 5$$

$$= \frac{6 \pm \sqrt{-144}}{18}$$

$$= \frac{6 \pm 12i}{18} \qquad \sqrt{-144} = 12i$$

$$= \frac{6(1 \pm 2i)}{18} \qquad \text{Factor.}$$

$$= \frac{1 \pm 2i}{3} \qquad \text{Lowest terms}$$

The solutions may be written in standard form as

$$\frac{1}{3} \pm \frac{2}{3}i. \quad ■$$

NOTE	In the quadratic formula, the expression $b^2 - 4ac$ is called the *discriminant*. If a, b, and c are real numbers, where $a \neq 0$, the quadratic equation $ax^2 + bx + c = 0$ has two real solutions if the discriminant is positive, one real solution if it is 0, and two nonreal, complex solutions if the discriminant is negative. This last case was seen in Example 1(b).

When working with negative radicands, it is very important to use the definition $\sqrt{-a} = i\sqrt{a}$ before using any of the other rules for radicals. In particular, the rule $\sqrt{c} \cdot \sqrt{d} = \sqrt{cd}$ is valid only when c and d are not both negative. For example, multiplying $\sqrt{-2} \cdot \sqrt{-32}$ to get $\sqrt{64}$ (which is incorrect) gives 8, but

$$\sqrt{-2} \cdot \sqrt{-32} = i\sqrt{2} \cdot i\sqrt{32}$$
$$= i^2\sqrt{64}$$
$$= (-1)8$$
$$= -8,$$

which is the correct result.

■ *Example 2*

SIMPLIFYING PRODUCTS AND QUOTIENTS WITH NEGATIVE RADICANDS

Express each product or quotient as a real number, or a product of a real number and i.

(a) $\sqrt{-7} \cdot \sqrt{-7} = i\sqrt{7} \cdot i\sqrt{7}$
$$= i^2 \cdot (\sqrt{7})^2$$
$$= (-1) \cdot 7 = -7$$

(b) $\sqrt{-6} \cdot \sqrt{-10} = i\sqrt{6} \cdot i\sqrt{10} = i^2 \cdot \sqrt{6 \cdot 10} = -1 \cdot 2\sqrt{15} = -2\sqrt{15}$

(c) $\dfrac{\sqrt{-20}}{\sqrt{-2}} = \dfrac{i\sqrt{20}}{i\sqrt{2}} = \sqrt{\dfrac{20}{2}} = \sqrt{10}$

(d) $\dfrac{\sqrt{-48}}{\sqrt{24}} = \dfrac{i\sqrt{48}}{\sqrt{24}} = i\sqrt{2}$ ■

Addition and subtraction of complex numbers is defined in a manner similar to these operations on binomials.

SUM AND DIFFERENCE OF COMPLEX NUMBERS	For complex numbers $a + bi$ and $c + di$, $$(a + bi) + (c + di) = (a + c) + (b + d)i$$ and $$(a + bi) - (c + di) = (a - c) + (b - d)i.$$

To add complex numbers, add their real parts and add their imaginary parts. Subtraction is accomplished in a similar manner.

■ *Example 3*

ADDING AND
SUBTRACTING
COMPLEX NUMBERS

Add or subtract as indicated.

(a) $(3 - 4i) + (-2 + 6i) = [3 + (-2)] + [-4 + 6]i = 1 + 2i$

(b) $(-9 + 7i) + (3 - 15i) = -6 - 8i$

(c) $(-4 + 3i) - (6 - 7i) = (-4 - 6) + [3 - (-7)]i = -10 + 10i$

(d) $(12 - 5i) - (8 - 3i) = 4 - 2i$ ■

The *product* of two complex numbers can be found by multiplying as if the numbers were binomials and using the fact that $i^2 = -1$, as follows.

$$(a + bi)(c + di) = ac + adi + bic + bidi$$
$$= ac + adi + bci + bdi^2$$
$$= ac + (ad + bc)i + bd(-1)$$
$$= (ac - bd) + (ad + bc)i$$

Thus, the product of the complex numbers $a + bi$ and $c + di$ is defined as follows.

PRODUCT OF
COMPLEX
NUMBERS

For complex numbers $a + bi$ and $c + di$,

$$(a + bi)(c + di) = (ac - bd) + (ad + bc)i.$$

The formal definition is usually not used when multiplying complex numbers. It is usually easier just to multiply as with binomials.

■ *Example 4*

MULTIPLYING
COMPLEX NUMBERS

Find each product.

(a) $(5 - 4i)(7 - 2i) = 5(7) + 5(-2i) - 4i(7) - 4i(-2i)$
$$= 35 - 10i - 28i + 8i^2$$
$$= 35 - 38i + 8(-1) \qquad \text{Replace } i^2 \text{ with } -1.$$
$$= 27 - 38i \qquad \text{Combine terms.}$$

(b) $(3 - i)(3 + i) = 9 + 3i - 3i - i^2$
$$= 9 - (-1) = 10$$ ■

The factors in Example 4(b) are called *conjugates*. The **conjugate** of the complex number $a + bi$ is the complex number $a - bi$. Notice that the product of a pair of conjugates is the difference of squares, so Example 4(b) could have been written as $(3 - i)(3 + i) = 3^2 - i^2 = 9 - (-1) = 10$. The product of conjugates is always a real number. Specifically, the product is the sum of the squares of the real and imaginary parts.

$$(a + bi)(a - bi) = a^2 + b^2$$

Conjugates are used in the division of complex numbers.

RULE FOR THE QUOTIENT OF COMPLEX NUMBERS	To divide complex numbers, multiply both the numerator and the denominator (divisor) by the conjugate of the denominator.

■ *Example 5*

DIVIDING COMPLEX NUMBERS

Find each quotient.

(a) $\dfrac{3 + 2i}{5 - i}$

Multiply the numerator and denominator by $5 + i$, the conjugate of $5 - i$.

$$\frac{3 + 2i}{5 - i} = \frac{(3 + 2i)(5 + i)}{(5 - i)(5 + i)}$$

$$= \frac{15 + 3i + 10i + 2i^2}{26} \qquad (5 - i)(5 + i) = 5^2 + 1^2 = 26$$

$$= \frac{13 + 13i}{26} = \frac{1}{2} + \frac{1}{2}i$$

To check this answer, show that

$$(5 - i)\left(\frac{1}{2} + \frac{1}{2}i\right) = 3 + 2i.$$

(b) $\dfrac{3}{i} = \dfrac{3(-i)}{i(-i)}$ $\quad -i$ is the conjugate of i.

$$= \frac{-3i}{-i^2}$$

$$= \frac{-3i}{1} \qquad -i^2 = -(-1) = 1$$

$$= -3i \quad ■$$

The fact that i^2 is equal to -1 can be used to find higher powers of i.

$$i^0 = 1 \qquad\qquad\qquad i^4 = i^2 \cdot i^2 = (-1)(-1) = 1$$

$$i^1 = i \qquad\qquad\qquad i^5 = i \cdot i^4 = i \cdot 1 = i$$

$$i^2 = -1 \qquad\qquad\qquad i^6 = i^2 \cdot i^4 = (-1) \cdot 1 = -1$$

$$i^3 = i \cdot i^2 = i(-1) = -i \qquad i^7 = i^3 \cdot i^4 = (-i) \cdot 1 = -i$$

As these examples show, the powers of i rotate through the four numbers $1, i, -1,$ and $-i$. Larger powers of i can be simplified by using the fact that $i^4 = 1$. For example $i^{75} = (i^4)^{18} \cdot i^3 = 1^{18} \cdot i^3 = 1 \cdot i^3 = i^3 = -i$.

■ *Example 6*
SIMPLIFYING
POWERS OF *i*

Find each power of i.

(a) $i^{12} = (i^4)^3 = 1^3 = 1$

(b) $i^{39} = i^{36} \cdot i^3$

$= (i^4)^9 \cdot i^3$

$= 1^9 \cdot (-i)$

$= -i$

(c) $i^{-2} = \dfrac{1}{i^2} = \dfrac{1}{-1} = -1$ ■

8.1 EXERCISES ■

Simplify each of the following. See Example 2.

1. $\sqrt{-4}$ **2.** $\sqrt{-49}$ **3.** $\sqrt{-\dfrac{25}{9}}$ **4.** $\sqrt{-\dfrac{1}{81}}$

5. $\sqrt{-150}$ **6.** $\sqrt{-180}$ **7.** $\sqrt{-80}$ **8.** $\sqrt{-72}$

9. $\sqrt{-3} \cdot \sqrt{-3}$ **10.** $\sqrt{-2} \cdot \sqrt{-2}$ **11.** $\sqrt{-5} \cdot \sqrt{-6}$ **12.** $\sqrt{-27} \cdot \sqrt{-3}$

13. $\dfrac{\sqrt{-12}}{\sqrt{-8}}$ **14.** $\dfrac{\sqrt{-15}}{\sqrt{-3}}$ **15.** $\dfrac{\sqrt{-24}}{\sqrt{72}}$ **16.** $\dfrac{\sqrt{-27}}{\sqrt{9}}$

Solve each of the following quadratic equations. See Example 1.

17. $x^2 = -16$ **18.** $y^2 = -36$ **19.** $z^2 + 12 = 0$ **20.** $w^2 + 48 = 0$

21. $3x^2 + 4x + 2 = 0$ **22.** $2k^2 + 3k = -2$ **23.** $m^2 - 6m + 14 = 0$ **24.** $p^2 + 4p + 11 = 0$

25. $4z^2 = 4z - 7$ **26.** $9a^2 + 7 = 6a$ **27.** $m^2 + 1 = -m$ **28.** $y^2 = 2y - 2$

29. Explain how to determine whether a quadratic equation of the form $ax^2 + bx + c = 0$ has nonreal, complex solutions without actually solving the equation.

30. What is wrong with the following?

$$\sqrt{-5} \cdot \sqrt{-5} = \sqrt{-5(-5)}$$
$$= \sqrt{25}$$
$$= 5$$

After explaining what is wrong, evaluate the expression correctly.

31. Suppose that a quadratic equation $ax^2 + bx + c = 0$ has all real coefficients. Also, suppose that a solution is $4 + (1/2)i$. What must the other solution be? Why?

32. Explain why any real number is also a complex number.

Perform the following operations and express all results in standard form. See Examples 3–5.

33. $(2 - 5i) + (3 + 2i)$ **34.** $(5 - i) + (3 + 4i) = 8 + 3i$ **35.** $(-2 + 3i) - (3 + i)$

36. $(4 + 6i) - (-2 - i)$ **37.** $(1 - i) - (5 - 2i)$ **38.** $(-2 + 6i) - (-3 - 8i) = 1 + 14i$

39. $(2 + i)(3 - 2i)$ **40.** $(-2 + 3i)(4 - 2i) = -2 + 16i$ **41.** $(2 + 4i)(-1 + 3i)$

42. $(1 + 3i)(2 - 5i)$ **43.** $(2 - i)(2 + i)$ **44.** $(5 + 4i)(5 - 4i)$

45. $\dfrac{5}{i}$ **46.** $\dfrac{-4}{i}$ **47.** $\dfrac{4 - 3i}{4 + 3i}$ **48.** $\dfrac{5 + 6i}{5 - 6i}$ **49.** $\dfrac{4 + i}{6 + 2i}$ **50.** $\dfrac{3 - 2i}{5 + 3i}$

$$\dfrac{3 - 2i}{5 + 3i} \cdot \dfrac{5 - 3i}{5 - 3i} = \dfrac{15 - 19i + 6i^2}{25 - 9i^2} =$$

Simplify each power of i. See Example 6.

51. i^{12} = / **52.** i^9 **53.** i^{18} **54.** i^{99} **55.** i^{-3} **56.** i^{-5} **57.** i^{-10} **58.** i^{-40}

59. A student makes the following statement. "I can simplify a large positive power of i by dividing the exponent by 4, and looking at the remainder. If the remainder is 0, it simplifies to 1, if the remainder is 1, it simplifies to i, if the remainder is 2, it simplifies to -1, and if the remainder is 3, it simplifies to $-i$." Explain why this statement is true.

60. Using the procedure of Exercise 59, why don't we have to consider getting a remainder of 4?

Two complex numbers $a + bi$ and $c + di$ are equal if and only if $a = c$ and $b = d$. Use this definition of equality of complex numbers to solve each of the following equations for x and y.

61. $2x + yi = 4 - 3i$ **62.** $x + 3yi = 5 + 2i$ **63.** $7 - 2yi = 14x - 30i$

64. $-5 + yi = x + 6i$ **65.** $x + yi = (2 + 3i)(4 - 2i)$ **66.** $x + yi = (5 - 7i)(1 + i)$

In work with alternating current, complex numbers are used to describe current, I, voltage, E, and impedance, Z (the opposition to current). These three quantities are related by the equation $E = IZ$. Thus, if any two of these quantities are known, the third can be found. In each of the following problems, solve the equation $E = IZ$ for the missing variable.

67. $I = 8 + 6i, Z = 6 + 3i$ **68.** $I = 10 + 6i, Z = 8 + 5i$

69. $I = 7 + 5i, E = 28 + 54i$ **70.** $E = 35 + 55i, Z = 6 + 4i$

71. Show that $\dfrac{\sqrt{2}}{2} + \dfrac{\sqrt{2}}{2} i$ is a square root of i.   **72.** Show that $-\dfrac{\sqrt{3}}{2} + \dfrac{1}{2} i$ is a cube root of i.

8.2 ——————— # TRIGONOMETRIC FORM OF COMPLEX NUMBERS

To graph a complex number such as $2 - 3i$, the familiar coordinate system must be modified. One way to do this is by calling the horizontal axis the **real axis** and the vertical axis the **imaginary axis.** Then complex numbers can be graphed in this **complex plane,** as shown in Figure 8.2 for the complex number $2 - 3i$.

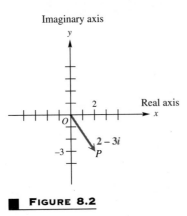

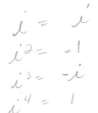

■ **FIGURE 8.2**

$\dfrac{15 - 19i - 6}{34} = \dfrac{9 - 19i}{34}$

Each nonzero complex number graphed in this way determines a unique directed line segment, the segment from the origin to the point representing the complex number. Recall from Chapter 7 that such directed line segments (like **OP** of Figure 8.2) are called vectors.

The previous section showed how to find the sum of two complex numbers, such as $4 + i$ and $1 + 3i$.

$$(4 + i) + (1 + 3i) = 5 + 4i$$

Graphically, the sum of two complex numbers is represented by the vector that is the resultant of the vectors corresponding to the two numbers. The vectors representing the complex numbers $4 + i$ and $1 + 3i$ and the resultant vector that represents their sum, $5 + 4i$, are shown in Figure 8.3.

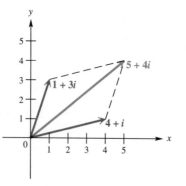

■ **FIGURE 8.3**

■ *Example 1*
**EXPRESSING THE
SUM OF COMPLEX
NUMBERS
GRAPHICALLY**

Find the sum of $6 - 2i$ and $-4 - 3i$. Graph both complex numbers and their resultant.

The sum is found by adding the two numbers.

$$(6 - 2i) + (-4 - 3i) = 2 - 5i$$

The graphs are shown in Figure 8.4. ■

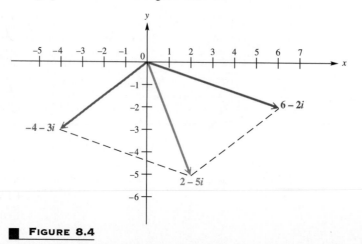

■ **FIGURE 8.4**

Figure 8.5 shows the complex number $x + yi$ that corresponds to a vector **OP** with direction θ and magnitude r. The following relationships among r, θ, x, and y can be verified from Figure 8.5.

RELATIONSHIPS AMONG x, y, r, AND θ	$x = r \cos \theta$ $\qquad r = \sqrt{x^2 + y^2}$ $y = r \sin \theta$ $\qquad \tan \theta = \dfrac{y}{x}$, if $x \neq 0$

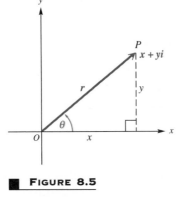

■ **FIGURE 8.5**

Substituting $x = r \cos \theta$ and $y = r \sin \theta$ from the results above into $x + yi$ gives

$$x + yi = r \cos \theta + (r \sin \theta)i$$
$$= r(\cos \theta + i \sin \theta).$$

TRIGONOMETRIC OR POLAR FORM OF A COMPLEX NUMBER	The expression $$r(\cos \theta + i \sin \theta)$$ is called the **trigonometric form** or **polar form** of the complex number $x + yi$. The expression $\cos \theta + i \sin \theta$ is sometimes abbreviated cis θ. Using this notation, $$\underline{r(\cos \theta + i \sin \theta)} \text{ is written as } r \text{ cis } \theta.$$

The number r is called the **modulus** or **absolute value** of $x + yi$, while θ is the **argument** of $x + yi$. In this section we will choose the value of θ in the interval $[0°, 360°)$. However, angles coterminal with such angles are also possible; that is, the argument for a particular complex number is not unique.

■ *Example 2*

CONVERTING FROM
TRIGONOMETRIC
FORM TO STANDARD
FORM

Express $2(\cos 300° + i \sin 300°)$ in standard form.
Since $\cos 300° = 1/2$ and $\sin 300° = -\sqrt{3}/2$,

$$2(\cos 300° + i \sin 300°) = 2\left(\frac{1}{2} - i\frac{\sqrt{3}}{2}\right) = 1 - i\sqrt{3}. \quad ■$$

NOTE

In the examples of this section, we will write arguments using degree measure. Arguments may also be written with radian measure.

In order to convert from standard form to trigonometric form, the following procedure is used.

STEPS FOR
CONVERTING FROM
STANDARD TO
TRIGONOMETRIC
FORM

1. Sketch a graph of the number in the complex plane.
2. Find r by using the equation $r = \sqrt{x^2 + y^2}$.
3. Find θ by using the equation $\tan \theta = y/x$, $x \neq 0$.

CAUTION

Errors often occur in Step 3 described above. Be sure that the correct quadrant for θ is chosen by referring to the graph sketched in Step 1.

■ *Example 3*

CONVERTING FROM
STANDARD FORM TO
TRIGONOMETRIC
FORM

Write the following complex numbers in trigonometric form.

(a) $-\sqrt{3} + i$

Start by sketching the graph of $-\sqrt{3} + i$ in the complex plane, as shown in Figure 8.6.

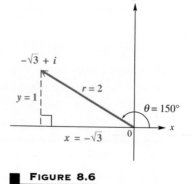

■ **FIGURE 8.6**

Next, find r. Since $x = -\sqrt{3}$ and $y = 1$.

$$r = \sqrt{x^2 + y^2} = \sqrt{(-\sqrt{3})^2 + 1^2} = \sqrt{3 + 1} = 2.$$

Then find θ.

$$\tan \theta = \frac{y}{x} = \frac{1}{-\sqrt{3}} = -\frac{\sqrt{3}}{3}$$

Since $\tan \theta = -\sqrt{3}/3$, the reference angle for θ is 30°. From the sketch we see that θ is in quadrant II, so $\theta = 180° - 30° = 150°$. Therefore, in trigonometric form,

$$-\sqrt{3} + i = 2(\cos 150° + i \sin 150°)$$
$$= 2 \operatorname{cis} 150°.$$

(b) $-3i$

The sketch of $-3i$ is shown in Figure 8.7.

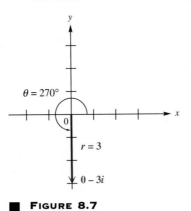

■ **FIGURE 8.7**

Since $-3i = 0 - 3i$, we have $x = 0$ and $y = -3$. Find r as follows.

$$r = \sqrt{0^2 + (-3)^2} = \sqrt{0 + 9} = \sqrt{9} = 3$$

We cannot find θ by using $\tan \theta = y/x$, since $x = 0$. In a case like this, refer to the graph and determine the argument directly from the sketch. A value for θ here is 270°. In trigonometric form,

$$-3i = 3(\cos 270° + i \sin 270°)$$
$$= 3 \operatorname{cis} 270°. \quad ■$$

NOTE

In Examples 2 and 3 we gave answers in both forms: $r(\cos \theta + i \sin \theta)$ and $r \operatorname{cis} \theta$. These forms will be used interchangeably throughout the rest of this chapter.

In the final example we see how a calculator can be used to convert between trigonometric and standard form.

298

■ *Example 4*

CONVERTING
BETWEEN
TRIGONOMETRIC AND
STANDARD FORMS
USING A
CALCULATOR

(a) Write the complex number 6(cos 115° + i sin 115°) in standard form.

Since 115° does not have a special angle as a reference angle, we cannot find exact values for cos 115° and sin 115°. Use a calculator set in the degree mode to find cos 115° = −.42261826 and sin 115° = .90630779. Therefore, in standard form,

$$6(\cos 115° + i \sin 115°) = 6(-.42261826 + .90630779i)$$
$$= -2.5357096 + 5.4378467i$$

(b) Write 5 − 4i in trigonometric form.

A sketch of 5 − 4i shows that θ must be in quadrant IV. See Figure 8.8.

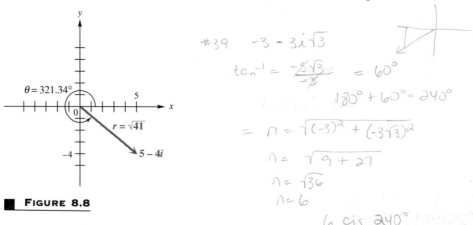

#39   −3 − 3$i\sqrt{3}$

$\tan^{-1} = \dfrac{-3\sqrt{3}}{-3} = 60°$

$180° + 60° = 240°$

$= n = \sqrt{(-3)^2 + (-3\sqrt{3})^2}$

$n = \sqrt{9 + 27}$

$n = \sqrt{36}$

$n = 6$

$6 \text{ cis } 240°$

■ FIGURE 8.8

Here $r = \sqrt{5^2 + (-4)^2} = \sqrt{41}$ and tan θ = −4/5. Use a calculator to find that one measure of θ is −38.66°. In order to express θ in the interval [0, 360°), we find that θ = 360° − 38.66° = 321.34°. Use these results to get

$$5 - 4i = \sqrt{41}(\cos 321.34° + i \sin 321.34°). \quad ■$$

8.2 EXERCISES ■ ————————————————————————

Graph each of the following complex numbers. See Example 1.

1. −2 + 3i **2.** −4 + 5i **3.** 8 − 5i **4.** 6 − 5i **5.** 2 − 2$i\sqrt{3}$

6. 4$\sqrt{2}$ + 4$i\sqrt{2}$ **7.** −4i **8.** 3i **9.** −8   **10.** 2

11. What must be true in order for a complex number to also be a real number?

12. If a real number is graphed in the complex plane, on what axis does the vector lie?

13. A complex number of the form $a + bi$ will have its corresponding vector lying on the y-axis provided $a =$ _____ .

14. The modulus of a complex number represents the _____ of the vector representing it in the complex plane.

Find the resultant of each of the following pairs of complex numbers. See Example 1.

15. $4 - 3i, -1 + 2i$ **16.** $2 + 3i, -4 - i$ **17.** $5 - 6i, -2 + 3i$ **18.** $7 - 3i, -4 + 3i$

19. $-3, 3i$ **20.** $6, -2i$ **21.** $2 + 6i, -2i$ **22.** $4 - 2i, 5$

23. $7 + 6i, 3i$ **24.** $-5 - 8i, -1$

Write the following complex numbers in standard form. See Example 2.

25. $2(\cos 45° + i \sin 45°)$ **26.** $4(\cos 60° + i \sin 60°)$ **27.** $10(\cos 90° + i \sin 90°)$

28. $8(\cos 270° + i \sin 270°)$ **29.** $4(\cos 240° + i \sin 240°)$ **30.** $2(\cos 330° + i \sin 330°)$

31. $(\cos 30° + i \sin 30°)$ **32.** $3(\cos 150° + i \sin 150°)$ **33.** $5 \text{ cis } 300°$

34. $6 \text{ cis } 135°$ **35.** $\sqrt{2} \text{ cis } 180°$ **36.** $\sqrt{3} \text{ cis } 315°$

Write each of the following complex numbers in trigonometric form $r(\cos \theta + i \sin \theta)$, with θ in the interval $[0°, 360°)$. See Example 3.

37. $3 - 3i$ **38.** $-2 + 2i\sqrt{3}$ **39.** $-3 - 3i\sqrt{3}$ **40.** $1 + i\sqrt{3}$ **41.** $\sqrt{3} - i$

42. $4\sqrt{3} + 4i$ **43.** $-5 - 5i$ **44.** $-\sqrt{2} + i\sqrt{2}$ **45.** $2 + 2i$ **46.** $-\sqrt{3} + i$

47. -4 **48.** $5i$ **49.** $-2i$ **50.** 7

Perform the following conversions, using a calculator as necessary. See Example 4.

	Standard Form	Trigonometric Form
51.	$2 + 3i$	_____
52.	_____	$(\cos 35° + i \sin 35°)$
53.	_____	$3(\cos 250° + i \sin 250°)$
54.	$-4 + i$	_____
55.	$12i$	_____
56.	_____	$3(\cos 180° + i \sin 180°)$
57.	$3 + 5i$	_____
58.	_____	$\text{cis } 110.5°$

(handwritten:)

#26 $4 \text{ cis } 60°$
$= 4\left(\frac{1}{2} + i\frac{\sqrt{3}}{2}\right)$
$= 2 + 2\sqrt{3}i$

#34 $6 \text{ cis } 135°$
$= 6\left(-\frac{\sqrt{2}}{2} + i\frac{\sqrt{2}}{2}\right)$
$= -3\sqrt{2} + 3\sqrt{2}i$

8.3 — **PRODUCT AND QUOTIENT THEOREMS**

The product of the two complex numbers $1 + i\sqrt{3}$ and $-2\sqrt{3} + 2i$ can be found by the method shown in Section 8.1.

$$(1 + i\sqrt{3})(-2\sqrt{3} + 2i) = -2\sqrt{3} + 2i - 2i(3) + 2i^2\sqrt{3}$$
$$= -2\sqrt{3} + 2i - 6i - 2\sqrt{3}$$
$$= -4\sqrt{3} - 4i$$

This same product also can be found by first converting the complex numbers $1 + i\sqrt{3}$ and $-2\sqrt{3} + 2i$ to trigonometric form. Using the method explained in the previous section,

$$1 + i\sqrt{3} = 2(\cos 60° + i \sin 60°)$$

and

$$-2\sqrt{3} + 2i = 4(\cos 150° + i \sin 150°).$$

(handwritten:)

#38 $\tan^{-1} = \frac{2\sqrt{3}}{-2} = 60°$

$180° - 60° = 120°$

$r = \sqrt{(-2)^2 + (2\sqrt{3})^2}$

$r = \sqrt{4 + 12}$

$r = \sqrt{16}$

$r = 4$

$4(\cos 120° + i \sin 120°)$

If the trigonometric forms are now multiplied together and if the trigonometric identities for the cosine and the sine of the sum of two angles are used, the result is

$$[2(\cos 60° + i \sin 60°)][4(\cos 150° + i \sin 150°)]$$

$$= 2 \cdot 4(\cos 60° \cdot \cos 150° + i \sin 60° \cdot \cos 150°$$
$$+ i \cos 60° \cdot \sin 150° + i^2 \sin 60° \cdot \sin 150°)$$

$$= 8[(\cos 60° \cdot \cos 150° - \sin 60° \cdot \sin 150°)$$
$$+ i(\sin 60° \cdot \cos 150° + \cos 60° \cdot \sin 150°)]$$

$$= 8[\cos (60° + 150°) + i \sin (60° + 150°)]$$

$$= 8(\cos 210° + i \sin 210°).$$

The modulus of the product, 8, is equal to the product of the moduli of the factors, $2 \cdot 4$, while the argument of the product, 210°, is the sum of the arguments of the factors, $60° + 150°$.

As we would expect, the product obtained upon multiplying by the first method is the standard form of the product obtained upon multiplying by the second method.

$$8(\cos 210° + i \sin 210°) = 8\left(-\frac{\sqrt{3}}{2} - \frac{1}{2}i\right)$$

$$= -4\sqrt{3} - 4i$$

Generalizing, the product of the two complex numbers, $r_1(\cos \theta_1 + i \sin \theta_1)$ and $r_2(\cos \theta_2 + i \sin \theta_2)$, is

$$[r_1(\cos \theta_1 + i \sin \theta_1)] \cdot [r_2(\cos \theta_2 + i \sin \theta_2)]$$

$$= r_1 r_2(\cos \theta_1 \cos \theta_2 + i \sin \theta_1 \cos \theta_2 + i \cos \theta_1 \sin \theta_2 + i^2 \sin \theta_1 \sin \theta_2)$$

$$= r_1 r_2[(\cos \theta_1 \cos \theta_2 - \sin \theta_1 \sin \theta_2) + i(\sin \theta_1 \cos \theta_2 + \cos \theta_1 \sin \theta_2)]$$

$$= r_1 r_2[\cos (\theta_1 + \theta_2) + i \sin (\theta_1 + \theta_2)].$$

This work is summarized in the following *product theorem*.

PRODUCT THEOREM If $r_1(\cos \theta_1 + i \sin \theta_1)$ and $r_2(\cos \theta_2 + i \sin \theta_2)$ are any two complex numbers, then

$$[r_1(\cos \theta_1 + i \sin \theta_1)] \cdot [r_2(\cos \theta_2 + i \sin \theta_2)]$$
$$= r_1 r_2[\cos (\theta_1 + \theta_2) + i \sin (\theta_1 + \theta_2)].$$

In compact form, this is written

$$(r_1 \operatorname{cis} \theta_1)(r_2 \operatorname{cis} \theta_2) = r_1 r_2 \operatorname{cis} (\theta_1 + \theta_2).$$

■ *Example 1*

USING THE
PRODUCT THEOREM

Find the product of 3(cos 45° + i sin 45°) and 2(cos 135° + i sin 135°).
Using the product theorem,

$$[3(\cos 45° + i \sin 45°)][2(\cos 135° + i \sin 135°)]$$
$$= 3 \cdot 2[\cos (45° + 135°) + i \sin (45° + 135°)]$$
$$= 6(\cos 180° + i \sin 180°),$$

which can be expressed as $6(-1 + i \cdot 0) = 6(-1) = -6$. The two complex numbers in this example are complex factors of -6. ■

Using the method shown in Section 8.1, in standard form the quotient of the complex numbers $1 + i\sqrt{3}$ and $-2\sqrt{3} + 2i$ is

$$\frac{1 + i\sqrt{3}}{-2\sqrt{3} + 2i} = \frac{(1 + i\sqrt{3})(-2\sqrt{3} - 2i)}{(-2\sqrt{3} + 2i)(-2\sqrt{3} - 2i)}$$
$$= \frac{-2\sqrt{3} - 2i - 6i - 2i^2\sqrt{3}}{12 - 4i^2}$$
$$= \frac{-8i}{16} = -\frac{1}{2}i.$$

Writing $1 + i\sqrt{3}$, $-2\sqrt{3} + 2i$, and $-\frac{1}{2}i$ in trigonometric form gives

$$1 + i\sqrt{3} = 2(\cos 60° + i \sin 60°)$$
$$-2\sqrt{3} + 2i = 4(\cos 150° + i \sin 150°)$$
$$-\frac{1}{2}i = \frac{1}{2}[(\cos (-90°) + i \sin (-90°))].$$

The modulus of the quotient, 1/2, is the quotient of the two moduli, 2 and 4. The argument of the quotient, $-90°$, is the difference of the two arguments, $60° - 150° = -90°$. It would be easier to find the quotient of these two complex numbers in trigonometric form than in standard form. Generalizing from this example leads to another theorem. The proof is similar to the proof of the product theorem, after the numerator and denominator are multiplied by the conjugate of the denominator.

QUOTIENT
THEOREM

If $r_1(\cos \theta_1 + i \sin \theta_1)$ and $r_2(\cos \theta_2 + i \sin \theta_2)$ are complex numbers, where $r_2(\cos \theta_2 + i \sin \theta_2) \neq 0$, then

$$\frac{r_1(\cos \theta_1 + i \sin \theta_1)}{r_2(\cos \theta_2 + i \sin \theta_2)} = \frac{r_1}{r_2}[\cos (\theta_1 - \theta_2) + i \sin (\theta_1 - \theta_2)].$$

In compact form, this is written

$$\frac{r_1 \text{ cis } \theta_1}{r_2 \text{ cis } \theta_2} = \frac{r_1}{r_2} \text{ cis } (\theta_1 - \theta_2).$$

302

■ *Example 2*
USING THE
QUOTIENT THEOREM

Find the quotient

$$\frac{10 \text{ cis } (-60°)}{5 \text{ cis } 150°}.$$

Write the result in standard form.
By the quotient theorem,

$$\frac{10 \text{ cis } (-60°)}{5 \text{ cis } 150°} = \frac{10}{5} \text{ cis } (-60° - 150°) \qquad \text{Quotient theorem}$$

$$= 2 \text{ cis } (-210°) \qquad \text{Subtract.}$$

$$= 2[\cos (-210°) + i \sin (-210°)]$$

$$= 2\left[-\frac{\sqrt{3}}{2} + i\left(\frac{1}{2}\right) \right] \qquad \cos (-210°) = -\frac{\sqrt{3}}{2}; \\ \sin (-210°) = \frac{1}{2}$$

$$= -\sqrt{3} + i \qquad \text{Standard form} \quad ■$$

■ *Example 3*
USING THE
PRODUCT AND
QUOTIENT
THEOREMS WITH A
CALCULATOR

Use a calculator to find the following. Write the results in standard form.

(a) $(9.3 \text{ cis } 125.2°)(2.7 \text{ cis } 49.8°)$
By the product theorem,

$$(9.3 \text{ cis } 125.2°)(2.7 \text{ cis } 49.8°) = (9.3)(2.7) \text{ cis } (125.2° + 49.8°)$$

$$= 25.11 \text{ cis } 175°$$

$$= 25.11(\cos 175° + i \sin 175°)$$

$$= 25.11(-.99619470 + i(.08715574))$$

$$= -25.014449 + 2.1884807i.$$

(b) $\dfrac{10.42\left(\cos \dfrac{3\pi}{4} + i \sin \dfrac{3\pi}{4} \right)}{5.21 \left(\cos \dfrac{\pi}{5} + i \sin \dfrac{\pi}{5} \right)}$

Use the quotient theorem.

$$\frac{10.42\left(\cos \frac{3\pi}{4} + i \sin \frac{3\pi}{4} \right)}{5.21 \left(\cos \frac{\pi}{5} + i \sin \frac{\pi}{5} \right)} = \frac{10.42}{5.21}\left[\cos \left(\frac{3\pi}{4} - \frac{\pi}{5}\right) + i \sin \left(\frac{3\pi}{4} - \frac{\pi}{5}\right) \right]$$

$$= 2\left(\cos \frac{11\pi}{20} + i \sin \frac{11\pi}{20} \right)$$

$$= -.31286893 + 1.9753767i \quad ■$$

8.3 EXERCISES ■

Find each of the following products. Write each product in standard form. See Example 1.

1. $[3(\cos 60° + i \sin 60°)][2(\cos 90° + i \sin 90°)]$

2. $[4(\cos 30° + i \sin 30°)][5(\cos 120° + i \sin 120°)]$

3. $[2(\cos 45° + i \sin 45°)][2(\cos 225° + i \sin 225°)]$

4. $[8(\cos 300° + i \sin 300°)][5(\cos 120° + i \sin 120°)]$

5. $[4(\cos 60° + i \sin 60°)][6(\cos 330° + i \sin 330°)]$

6. $[8(\cos 210° + i \sin 210°)][2(\cos 330° + i \sin 330°)]$

7. $[5 \operatorname{cis} 90°][3 \operatorname{cis} 45°]$

8. $[6 \operatorname{cis} 120°][5 \operatorname{cis} (-30°)]$

9. $[\sqrt{3} \operatorname{cis} 45°][\sqrt{3} \operatorname{cis} 225°]$

10. $[\sqrt{2} \operatorname{cis} 300°][\sqrt{2} \operatorname{cis} 270°]$

Find each of the following quotients. Write each quotient in standard form. See Example 2.

11. $\dfrac{4(\cos 120° + i \sin 120°)}{2(\cos 150° + i \sin 150°)}$

12. $\dfrac{10(\cos 225° + i \sin 225°)}{5(\cos 45° + i \sin 45°)}$

13. $\dfrac{16(\cos 300° + i \sin 300°)}{8(\cos 60° + i \sin 60°)}$

14. $\dfrac{24(\cos 150° + i \sin 150°)}{2(\cos 30° + i \sin 30°)}$

15. $\dfrac{3 \operatorname{cis} 305°}{9 \operatorname{cis} 65°}$

16. $\dfrac{12 \operatorname{cis} 293°}{6 \operatorname{cis} 23°}$

17. $\dfrac{8}{\sqrt{3} + i}$

18. $\dfrac{2i}{-1 - i\sqrt{3}}$

19. $\dfrac{-i}{1 + i}$

20. $\dfrac{1}{2 - 2i}$

21. $\dfrac{2\sqrt{6} - 2i\sqrt{2}}{\sqrt{2} - i\sqrt{6}}$

22. $\dfrac{4 + 4i}{2 - 2i}$

Use a calculator to perform the indicated operations. Give answers in standard form. See Example 3.

23. $[2.5(\cos 35° + i \sin 35°)] [3.0(\cos 50° + i \sin 50°)]$

24. $[4.6(\cos 12° + i \sin 12°)][2.0(\cos 13° + i \sin 13°)]$

25. $(12 \operatorname{cis} 18.5°)(3 \operatorname{cis} 12.5°)$

26. $(4 \operatorname{cis} 19.25°)(7 \operatorname{cis} 41.75°)$

27. $\dfrac{45(\cos 127° + i \sin 127°)}{22.5(\cos 43° + i \sin 43°)}$

28. $\dfrac{30(\cos 130° + i \sin 130°)}{10(\cos 21° + i \sin 21°)}$

29. $\left[2 \operatorname{cis} \dfrac{5\pi}{9}\right]^2$

30. $\left[24.3\left(\cos \dfrac{7\pi}{12} + i \sin \dfrac{7\pi}{12}\right)\right]^2$

31. Without actually performing the operations, state why the products $[2(\cos 45° + i \sin 45°)] \cdot [5(\cos 90° + i \sin 90°)]$ and $[2(\cos (-315°) + i \sin (-315°))][5(\cos (-270°) + i \sin (-270°))]$ are the same.

32. Notice that $(r \operatorname{cis} \theta)^2 = (r \operatorname{cis} \theta)(r \operatorname{cis} \theta) = r^2 \operatorname{cis} (\theta + \theta) = r^2 \operatorname{cis} 2\theta$. State in your own words how we can square a complex number in trigonometric form. (In the next section, we will develop this idea more fully.)

33. The alternating current in an electric inductor is

$$I = \dfrac{E}{Z}$$

amperes, where E is the voltage and $Z = R + X_L i$ is the impedance. If $E = 8(\cos 20° + i \sin 20°)$, $R = 6$, and $X_L = 3$, find the current. Give the answer in standard form.

34. The current I in a circuit with voltage E, resistance R, capacitive reactance X_c, and inductive reactance X_L is

$$I = \dfrac{E}{R + (X_L - X_c)i}.$$

Find I if $E = 12(\cos 25° + i \sin 25°)$, $R = 3$, $X_L = 4$, and $X_c = 6$. Give the answer in standard form.

Graphs of polar equations can be plotted on graphing calculators such as the TI-81 by using *parametric equations.* If the coordinates (x, y) of the points on a curve are each defined in terms of a number t, we have parametric equations, with t as the *parameter.*

To graph the equation $r = 1 + \cos \theta$, the cardioid found in Example 3 of Section 8.5, we must use T rather than θ, since the calculator is designed this way. Press the [MODE] key of the calculator and choose the parametric (Param) and radian (Rad) modes. Now press the [Y=] key and notice that the display differs from that of the function mode in that we can enter up to three pairs of values of X and Y parametrically. A polar equation $r = f(T)$ can be graphed by using the conversion formulas $X = f(T) \cos T$ and $Y = f(T) \sin T$, so enter the following equations:

$$X_{1T} = (1 + \cos T) \cos T$$
$$Y_{1T} = (1 + \cos T) \sin T.$$

(The parameter T is entered by using the [X|T] key.) Press [GRAPH] to obtain the cardioid seen in Figure 8.17. By zooming in we can change the viewing screen to get a better picture of the graph.

You may wish to experiment with your graphing calculator by graphing the following polar equations using the method described above. They are taken from Examples and Exercises in Section 8.5. Check your results by comparing them to the given figure numbers or the answer section in the back of the text.

1. $r = 3 \cos 2T$ (Figure 8.19)

2. $r = \dfrac{4}{1 + \sin T}$ (Figure 8.22)

3. $r = 3 + \cos T$ (Exercise 15)

4. $r = 4 \cos 2T$ (Exercise 17)

5. $r^2 = 4 \cos 2T$ (Exercise 19)

6. $r = 4(1 - \cos T)$ (Exercise 21)

7. $r = 2 \sin T \tan T$ (Exercise 23)

8. $r = 2 \sec T$ (Exercise 33)

LOGARITHMS

Before small calculators became easily available, computations involving multiplication, division, powers, and roots were performed by using the properties of logarithms. Now those problems can be worked simply on a calculator. We still study them, however, since properties of logarithms are useful in applications, as we shall see in this chapter.

Logarithmic functions are the inverses of exponential functions. This chapter briefly reviews both logarithmic and exponential functions; the properties of logarithms, exponential and logarithmic equations; and applications involving exponential and logarithmic equations.

9.1 —— EXPONENTIAL AND LOGARITHMIC FUNCTIONS; PROPERTIES OF LOGARITHMS

Properties of exponents are studied in elementary algebra courses. These properties of exponents are listed below.

PROPERTIES OF EXPONENTS

For real numbers a, b, m, and n, where no denominators are zero,

$$a^m \cdot a^n = a^{m+n} \qquad (ab)^m = a^m b^m$$

$$\frac{a^m}{a^n} = a^{m-n} \qquad (a^m)^n = a^{mn}$$

$$a^0 = 1 \quad (\text{if } a \neq 0).$$

The next two definitions are also helpful in work with exponents.

DEFINITIONS OF a^{-n} AND $a^{m/n}$

If a is a nonzero real number, then

$$a^{-n} = \frac{1}{a^n}.$$

For all integers m and positive integers n, and for all numbers a for which the roots exist,

$$a^{m/n} = \sqrt[n]{a^m} = (\sqrt[n]{a})^m.$$

■ *Example 1*

USING PROPERTIES AND DEFINITIONS INVOLVING EXPONENTS

Use the properties and definitions listed above to simplify each of the following.

(a) $2^5 \cdot 2^9 \cdot 2^7 = 2^{5+9+7} = 2^{21}$

(b) $3^{-2} = \frac{1}{3^2} = \frac{1}{9}$

(c) $5^{-4} = \frac{1}{5^4} = \frac{1}{625}$

(d) $\frac{4^9}{4^{-2}} = 4^{9-(-2)} = 4^{11}$

(e) $(7^3 \cdot 3^4)^2 = (7^3)^2 (3^4)^2 = 7^6 \cdot 3^8$

(f) $2^0 + (-5)^0 - 3^0 = 1 + 1 - 1 = 1$

(g) $4^{3/2} = (4^{1/2})^3 = (\sqrt{4})^3 = 2^3 = 8$ ■

An **exponential function** is a function in which the independent variable appears in the exponent, as in $y = 2^x$, where x is the independent variable.

EXPONENTIAL FUNCTION	An exponential function is a function of the form $$y = a^x,$$ where $a > 0$, and $a \neq 1$.

The equation $y = 2^x$ defines a typical exponential function. To graph this function, first make a table of values of x and y, as follows.

x	-3	-2	-1	0	1	2	3	4
2^x	$\frac{1}{8}$	$\frac{1}{4}$	$\frac{1}{2}$	1	2	4	8	16

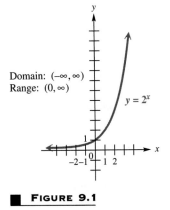

Domain: $(-\infty, \infty)$
Range: $(0, \infty)$

$y = 2^x$

Plotting these points and drawing a smooth curve through them produces the graph shown in Figure 9.1. This graph is typical of the graphs of exponential functions of the form $y = a^x$, where $a > 1$. The larger the value of a, the faster the graph rises. As the graph suggests, the domain is the set of all real numbers but the range includes only positive real numbers.

■ **FIGURE 9.1**

■ *Example 2*
GRAPHING AN EXPONENTIAL FUNCTION

Graph $y = \left(\dfrac{1}{2}\right)^x$.

Again, find selected ordered pairs and plot them to obtain the graph shown in Figure 9.2. Some ordered pairs are given in the table below. This graph is typical of the graphs of exponential functions of the form $y = a^x$, where $0 < a < 1$. Again, the domain is the set of real numbers and the range is the set of positive real numbers. ■

x	y
-3	8
-2	4
-1	2
0	1
1	1/2
2	1/4
3	1/8

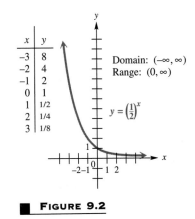

Domain: $(-\infty, \infty)$
Range: $(0, \infty)$

$y = \left(\dfrac{1}{2}\right)^x$

■ **FIGURE 9.2**

Based on these examples, the following generalizations can be made about the graphs of exponential functions of the form $y = a^x$.

GRAPH OF $y = a^x$	1. The graph will always contain the point $(0, 1)$. 2. When $a > 1$, the graph will *rise* from left to right. When $0 < a < 1$, the graph will *fall* from left to right. In both cases, the graph goes from the second quadrant to the first. 3. The graph will approach the x-axis, but never touch it. (It is an asymptote.) 4. The domain is $(-\infty, \infty)$ and the range is $(0, \infty)$.

Since neither horizontal nor vertical lines intersect the graph of $y = 2^x$ in more than one point, $y = 2^x$ is a one-to-one function and has an inverse. The inverse of $y = 2^x$ is $x = 2^y$. To get the equation of the inverse, x and y were exchanged as first discussed in Section 6.1.

The equation $x = 2^y$ cannot be solved for y by the methods studied so far. The following definition is used to solve such equations for y.

DEFINITION OF LOGARITHM	For all positive numbers a, with $a \neq 1$, $$y = \log_a x \quad \text{means} \quad x = a^y.$$

Log is an abbreviation for **logarithm.** For example, the logarithmic statement

$$5 = \log_2 32 \quad \text{has the same meaning as} \quad 32 = 2^5$$

(read $5 = \log_2 32$ as "5 is the logarithm of 32 to the base 2"). The number 5 is the exponent on 2 that gives the result 32.

By the definition given here, exponential statements can be turned into logarithmic statements, and logarithmic statements into exponential statements.

■ *Example 3*

COMPARING EXPONENTIAL AND LOGARITHMIC FORMS

The chart below lists several pairs of equivalent statements. The same statement is written in both exponential and logarithmic form.

Exponential Form	Logarithmic Form
$3^4 = 81$	$4 = \log_3 81$
$5^3 = 125$	$3 = \log_5 125$
$2^{-4} = \dfrac{1}{16}$	$-4 = \log_2 \dfrac{1}{16}$
$5^{-1} = \dfrac{1}{5}$	$-1 = \log_5 \dfrac{1}{5}$
$\left(\dfrac{3}{4}\right)^{-2} = \dfrac{16}{9}$	$-2 = \log_{3/4} \dfrac{16}{9}$

■

By using the definition of logarithm, the logarithmic function with base a is defined as follows.

LOGARITHMIC FUNCTION	If $a > 0$, $a \neq 1$, and $x > 0$, then $$y = \log_a x$$ is the logarithmic function with base a.

As mentioned above, exponential and logarithmic functions are inverses of each other. Since the domain of an exponential function is the set of all real numbers, $(-\infty, \infty)$, the range of a logarithmic function is also the set of all real numbers. In the same way, both the range of an exponential function and the domain of a logarithmic function are the set of all positive real numbers, $(0, \infty)$. Because of this domain, only positive numbers have logarithms.

The graph of $y = 2^x$ is shown in Figure 9.3. To get the graph of its inverse, $y = \log_2 x$, reflect $y = 2^x$ about the line $y = x$, as shown in Figure 9.3. The graph is typical of graphs of functions of the form $y = \log_a x$, where $a > 1$.

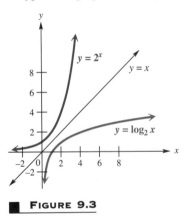

■ **FIGURE 9.3**

■ *Example 4*
GRAPHING A
LOGARITHMIC
FUNCTION

Graph $y = \log_{1/2} x$.

Some ordered pairs for this function are shown in the table at the right. Plotting these points and connecting them with a curve leads to the graph in Figure 9.4. This graph is typical of graphs of functions of the form $y = \log_a x$, where $0 < a < 1$. ■

x	y
8	-3
4	-2
2	-1
1	0
1/2	1
1/4	2

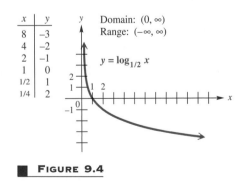

Domain: $(0, \infty)$
Range: $(-\infty, \infty)$

■ **FIGURE 9.4**

Based on the graphs of the functions $y = \log_2 x$ in Figure 9.3 and $y = \log_{1/2} x$ in Figure 9.4, the following generalizations can be made about the graphs of logarithmic functions of the form $y = \log_a x$.

GRAPH OF
$y = \log_a x$

1. The graph will always contain the point $(1, 0)$.
2. When $a > 1$, the graph will *rise* from left to right, from the fourth quadrant to the first. When $0 < a < 1$, the graph will *fall* from left to right, from the first quadrant to the fourth.
3. The graph will approach the y-axis, but never touch it. (It is an asymptote.)
4. The domain is $(0, \infty)$ and the range is $(-\infty, \infty)$.

Compare these generalizations to the similar ones for exponential functions given earlier.

PROPERTIES OF LOGARITHMS The basic properties of logarithms that make them useful in mathematics are summarized below.

PROPERTIES OF
LOGARITHMS

If x and y are positive real numbers, r is any real number, and a is any positive real number, with $a \neq 1$, then

(a) $\log_a xy = \log_a x + \log_a y$ (b) $\log_a \dfrac{x}{y} = \log_a x - \log_a y$

(c) $\log_a x^r = r \log_a x$ (d) $\log_a a = 1$

(e) $\log_a 1 = 0$.

To prove part (a), let $\log_a x = M$ and $\log_a y = N$. Then by the definition of logarithm,

$$a^M = x \quad \text{and} \quad a^N = y,$$

and
$$(a^M)(a^N) = xy,$$

or
$$a^{M+N} = xy$$

and
$$\log_a xy = M + N.$$

Substituting for M and N gives the desired result,

$$\log_a xy = \log_a x + \log_a y.$$

The proofs of parts (b) and (c) are left for the exercises. Parts (d) and (e) come directly from the definition of logarithm.

■ *Example 5*

USING THE
PROPERTIES OF
LOGARITHMS

The properties given above are used to rewrite each of the following logarithmic expressions.

(a) $\log_6 9 \cdot 4 = \log_6 9 + \log_6 4$

(b) $\log_{12} \dfrac{17}{18} = \log_{12} 17 - \log_{12} 18$

(c) $\log_5 9^8 = 8 \cdot \log_5 9$

(d) $\log_7 \sqrt{19} = \log_7 19^{1/2} = \dfrac{1}{2} \cdot \log_7 19$ ■

■ *Example 6*

USING THE
PROPERTIES OF
LOGARITHMS

Assume that

$$\log_{10} 2 = .3010 \quad \text{and} \quad \log_{10} 3 = .4771.$$

Find the base 10 logarithms of 4, 5, 6, and 12.
Use the properties of logarithms.

(a) $\log_{10} 4 = \log_{10} 2^2 = 2 \cdot \log_{10} 2$
$$= 2(.3010) = .6020$$

(b) $\log_{10} 5 = \log_{10} \dfrac{10}{2} = \log_{10} 10 - \log_{10} 2$
$$= 1 - .3010 = .6990$$

(c) $\log_{10} 6 = \log_{10} 2 \cdot 3 = \log_{10} 2 + \log_{10} 3$
$$= .3010 + .4771 = .7781$$

(d) $\log_{10} 12 = \log_{10} 4 \cdot 3 = \log_{10} 4 + \log_{10} 3$
$$= .6020 + .4771 = 1.0791$$ ■

9.1 EXERCISES ■

Use the properties of exponents to simplify each of the following. Express answers using only positive exponents. Leave answers in exponential form. See Example 1.

1. $3^4 \cdot 3^7$

2. $4^5 \cdot 4^8$

3. $\dfrac{7^4}{7^3}$

4. $\dfrac{9^5}{9^2}$

5. $\dfrac{6^4 \cdot 6^5}{6^6}$

6. $\dfrac{9^{12} \cdot 9^4}{9^{11}}$

7. $(8^9)^2$

8. $(3^4)^3$

9. $\dfrac{(2^5)^2 \cdot 2^4}{2^6}$

10. $\dfrac{(3^6)^3 \cdot 3^7}{3^8}$

11. 2^{-3}

12. 4^{-2}

13. $3^{-7} \cdot 3^8$

14. $5^{-9} \cdot 5^6$

15. $\dfrac{9^{-8} \cdot 9^2}{(9^2)^{-2}}$

16. $\dfrac{(4^{-3})^2 \cdot 4^5}{(4^3)^{-3}}$

17. $25^{1/2}$

18. $8^{2/3}$

19. $16^{3/4}$

20. $9^{3/2}$

Graph each of the following exponential functions. See Example 2.

21. $y = \left(\dfrac{1}{3}\right)^x$ **22.** $y = \left(\dfrac{1}{4}\right)^x$ **23.** $y = 2^{x+1}$ **24.** $y = 3^{x+1}$

25. $y = 2^{-x}$ **26.** $y = 3^{-x}$ **27.** $y = 3^x - 3$ **28.** $y = 2^x - 2$

Write each of the following in logarithmic form. See Example 3.

29. $3^5 = 243$ **30.** $2^7 = 128$ **31.** $10^4 = 10{,}000$

32. $8^2 = 64$ **33.** $6^{-2} = \dfrac{1}{36}$ **34.** $2^{-3} = \dfrac{1}{8}$

Write each of the following in exponential form. See Example 3.

35. $\log_4 16 = 2$ **36.** $\log_5 25 = 2$ **37.** $\log_{10} 1000 = 3$

38. $\log_7 2401 = 4$ **39.** $\log_{3/4} \dfrac{9}{16} = 2$ **40.** $\log_{5/8} \dfrac{125}{512} = 3$

41. Explain why logarithms of negative numbers are not defined.

42. Explain why 1 is not allowed as a base for a logarithm.

43. Why is it that for any valid base a, the logarithm to the base a of 1 must be 0?

44. Compare the summary of facts about the graphs of $y = a^x$ and $y = \log_a x$ given in this section. Make a list of the facts that reinforce the idea that they are inverse functions.

Graph each of the following. See Example 4.

45. $y = \log_4 x$ **46.** $y = \log_{1/4} x$ **47.** $y = \log_{1/3} x$ **48.** $y = \log_3 x$

49. $y = \log_2 (x - 1)$ **50.** $y = \log_2 x^2$ **51.** $y = 1 + \log_3 x$ **52.** $y = 1 - \log_3 x$

Simplify each of the following using the properties of logarithms. See Example 5.

53. $\log_6 9 + \log_6 5$ **54.** $\log_2 7 + \log_2 3$ **55.** $\log_5 12 - \log_5 7$

56. $\log_9 17 - \log_9 23$ **57.** $\log_2 7 + \log_2 4 - \log_2 5$ **58.** $\log_7 8 + \log_7 16 - \log_7 3$

59. $4 \cdot \log_3 2$ **60.** $3 \cdot \log_5 2$ **61.** $\dfrac{1}{2} \cdot \log_8 7$

62. $\dfrac{1}{2} \cdot \log_2 14$ **63.** $4 \cdot \log_2 3 - 2 \cdot \log_2 5$ **64.** $3 \cdot \log_4 5 - 4 \cdot \log_4 2$

Use the properties of logarithms to find each of the following. See Example 6. Assume that

$$\log_{10} 2 = .3010 \qquad \log_{10} 7 = .8451$$
$$\log_{10} 3 = .4771 \qquad \log_{10} 11 = 1.0414.$$

65. $\log_{10} 14$ **66.** $\log_{10} 22$ **67.** $\log_{10} 28$ **68.** $\log_{10} 63$ **69.** $\log_{10} 2^8$

70. $\log_{10} 3^7$ **71.** $\log_{10} 11^3$ **72.** $\log_{10} \sqrt{3}$ **73.** $\log_{10} \sqrt[3]{11}$ **74.** $\log_{10} \sqrt[4]{3}$

75. Prove property (b) of logarithms. (*Hint:* The proof is very similar to that of property (a).)

76. Prove property (c) of logarithms.

Logarithms are important in many applications of mathematics to everyday problems, particularly in biology, engineering, economics, and social science. In this section we show how to find numerical approximations for logarithms. Traditionally, base 10 logarithms have been used most extensively, since our number system is base 10. Logarithms to base 10 are called **common logarithms** and $\log_{10} x$ is abbreviated as simply $\log x$, where the base is understood to be 10.

While extensive tables of common logarithms have been available for many years, today common logarithms are nearly always evaluated using calculators. Nevertheless, a table of common logarithms is included in this text (Table 2), and instructions for using the table are given in Appendix B.

The following example gives the results of evaluating some common logarithms using a calculator with a log key. (This may be a second function key on some calculators.) Just enter the number, then touch the log key. (For some calculators, this order may be different.) We will give all logarithms to four decimal places.

▪ *Example 1*

EVALUATING
COMMON
LOGARITHMS

Evaluate each logarithm.

(a) $\log 327.1 = 2.5147$

(b) $\log 437{,}000 = 5.6405$

(c) $\log .0615 = -1.2111$

In Example 1(c), $\log .0615 = -1.2111$, a negative result. The common logarithm of a number between 0 and 1 is always negative because the logarithm is the exponent on 10 that produces the number. For example,

$$10^{-1.2111} = .0615.$$

If the exponent (the logarithm) were positive, the result would be greater than 1, since $10^0 = 1$. ▪

In Example 1(a), the number 327.1, whose common logarithm is 2.5147, is called the common **antilogarithm** (abbreviated *antilog*) of 2.5147. In the same way, 437,000, in Example 1(b), is the common antilogarithm of 5.6405. In general, in the expression $y = \log_{10} x$, x is the antilog.

NOTE From Example 1,

antilogarithm   logarithm

$$\log \overbrace{327.1} = \overbrace{2.5147}.$$

In exponential form, this is written as

logarithm   antilogarithm

$$10^{\overbrace{2.5147}} = \overbrace{327.1},$$

which shows that the logarithm 2.5147 is just an exponent. A common logarithm is the exponent on 10 that produces the antilogarithm.

■ *Example 2*
FINDING COMMON ANTILOGARITHMS

Find the common antilogarithm of each of the following.

(a) 2.6454

Use the fact that for an antilogarithm N, $10^{2.6454} = N$. Some calculators have a 10^x key. With such a calculator, enter 2.6454, then touch the 10^x key to get the antilogarithm:

$$N \approx 442.$$

On other calculators, antilogarithms are found using the INV and log keys: enter 2.6454, touch the INV key, then touch the log key to get 442. INV indicates "inverse," so the INV key followed by the log key gives the inverse of the logarithm, which is the exponential expression 10^x.

(b) −2.3686

The antilogarithm is N, where

$$10^{-2.3686} = N.$$

Using the 10^x key or the INV and log keys gives

$$N \approx .00428.$$

(c) 1.5203

The antilogarithm is 33.14 to four significant figures. ■

Most calculators will work by one of the methods we have described. There are a few that may not. If yours is one of those, see your instruction booklet. Your teacher also may be able to help you.

In chemistry, the **pH** of a solution is defined as follows.

$$pH = -\log [H_3O^+],$$

where $[H_3O^+]$ is the hydronium ion concentration in moles per liter.

The pH is a measure of the acidity or alkalinity of a solution, with water, for example, having a pH of 7. In general, acids have pH numbers less than 7, and alkaline solutions have pH values greater than 7.

■ *Example 3*
| FINDING pH

Find the pH of grapefruit with a hydronium ion concentration of 6.3×10^{-4}.
Use the definition of pH.

$$\begin{aligned}
\text{pH} &= -\log{(6.3 \times 10^{-4})} \\
&= -(\log{6.3} + \log{10^{-4}}) \qquad \text{Logarithm of a product} \\
&= -[.7993 - 4] \\
&= -.7993 + 4 \approx 3.2
\end{aligned}$$

It is customary to round pH values to the nearest tenth. ■

■ *Example 4*
| FINDING
| HYDRONIUM ION
| CONCENTRATION

Find the hydronium ion concentration of drinking water with a pH of 6.5.

$$\text{pH} = 6.5 = -\log{[H_3O^+]}$$
$$\log{[H_3O^+]} = -6.5$$

Since the antilogarithm of -6.5 is 3.2×10^{-7},

$$[H_3O^+] = 3.2 \times 10^{-7}. \quad ■$$

THE NUMBER *e* AND NATURAL LOGARITHMS Consider the expression

$$\left(1 + \frac{1}{x}\right)^x.$$

If we let x take on larger and larger values, starting with 1 and continuing with powers of 10, we obtain the results shown below.

Value of x	Value of $\left(1 + \dfrac{1}{x}\right)^x$
1	2
10	2.59374246
100	2.70481383
1000	2.71692393
10,000	2.71814593
100,000	2.71826824
1,000,000	2.71828047

It seems that, as the value of x gets larger and larger, the value of $(1 + 1/x)^x$ approaches a number just larger than 2.7. This is indeed the case. It can be shown that this limiting number is approximately 2.718281828, and is called *e*, in honor of the Swiss mathematician Leonard Euler (1707–83).

VALUE OF *e* To nine decimal places,

$$e \approx 2.718281828.$$

The number *e* is positive and therefore can be used as the base of a logarithmic function. Logarithms to base *e* are called natural logarithms because they occur in biology and the social sciences in natural situations that involve growth or decay. The base *e* logarithm of *x* is written ln *x* (read "el en *x*"). A graph of $y = \ln x$, the natural logarithm function, is given in Figure 9.5.

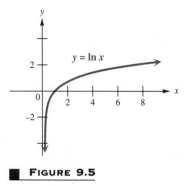

■ **FIGURE 9.5**

Natural logarithms can be found with a calculator that has an ln key or with a table of natural logarithms (Table 3 of Appendix B). Reading directly from this table,

$$\ln 55 = 4.0073,$$
$$\ln 1.9 = .6419,$$
and
$$\ln .4 = -.9163.$$

The next example shows how to find natural logarithms with a calculator that has a key labeled ln *x*.

■ *Example 5*
FINDING NATURAL LOGARITHMS

Find each of the following logarithms to four significant digits.

(a) ln .5841

Enter .5841 and touch the ln key to get

$$\ln .5841 \approx -.5377.$$

(Again, with some calculators, the steps are reversed.) As with common logarithms, a number between 0 and 1 has a negative natural logarithm.

(b) ln 192.7 = 5.261

(c) ln 10.84 = 2.383 ■

If your calculator has an e^x key, but not a key labeled ln x, find natural logarithms by entering the number, touching the INV key, and then touching the e^x key. This works because $y = e^x$ is the inverse function of $y = \ln x$ (or $y = \log_e x$). Antilogarithms of natural logarithms are found as those for common logarithms.

$$\underset{\downarrow}{\text{logarithm}} \qquad \underset{\downarrow}{\text{antilogarithm}}$$

$$\text{If} \quad \ln x = y, \quad \text{then} \quad e^y = x,$$

since the base of natural logarithms is e. Some calculators have a key labeled e^x. With others, use the INV key and the ln x key to evaluate a natural antilogarithm.

■ *Example 6*
FINDING NATURAL ANTILOGARITHMS

Find the natural antilogarithm of each of the following.

(a) 2.5017

Let N be the antilogarithm. Then

$$\ln N = 2.5017.$$

Find N by entering 2.5017, then touching the INV and ln x keys, getting

$$N = 12.20.$$

Alternatively, to use the e^x key, enter 2.5017 and touch the e^x key to get 12.20.

(b) -1.429

The antilogarithm is .2395.

(c) .0053

The antilogarithm is 1.0053.   ■

The next example shows a social science application of natural logarithms.

■ *Example 7*
APPLYING NATURAL LOGARITHMS

The number of years, $N(r)$, since two independently evolving languages split off from a common ancestral language is approximated by

$$N(r) = -5000 \ln r$$

where r is the percent of words from the ancestral language common to both languages now. Find $N(r)$ if $r = 70\%$.

Write 70% as .7 and find $N(.7)$.

$$N(.7) = -5000 \ln .7$$
$$\approx -5000(-.3567) \approx 1783$$

Approximately 1800 years have passed since the two languages separated.   ■

LOGARITHMS TO OTHER BASES A calculator (or a table) can be used to approximate the values of common logarithms (base 10) or natural logarithms (base e). However, in some applications it is convenient to use logarithms to other bases. The following rule is used to convert logarithms from one base to another.

CHANGE-OF-BASE RULE	If $a > 0$, $a \neq 1$, $b > 0$, $b \neq 1$, and $x > 0$, then $$\log_a x = \frac{\log_b x}{\log_b a}.$$

NOTE As an aid in remembering the change-of-base rule, notice that x is "above" a on both sides of the equation.

In order to prove the change-of-base rule, let $\log_a x = m$.

$$\log_a x = m$$
$$a^m = x \qquad \text{Change to exponential form.}$$
$$\log_b (a^m) = \log_b x \qquad \text{Take logarithms on both sides.}$$
$$m \log_b a = \log_b x \qquad \text{Use the power property.}$$
$$(\log_a x)(\log_b a) = \log_b x \qquad \text{Substitute for } m.$$
$$\log_a x = \frac{\log_b x}{\log_b a} \qquad \text{Divide both sides by } \log_b a.$$

Any positive number other than 1 can be used for base b in the change of base rule, but usually the only practical bases are e and 10, since calculators (or tables) give logarithms only for these two bases.

The next example shows how the change-of-base rule is used to find logarithms to bases other than 10 or e with a calculator.

■ *Example 8*

USING THE CHANGE-OF-BASE RULE

Find each logarithm using a calculator.

(a) $\log_5 12$

Use common logarithms and the rule for change-of-base.

$$\log_5 12 = \frac{\log 12}{\log 5} \approx \frac{1.0792}{.6990}$$

Use a calculator to evaluate this quotient.

$$\log_5 12 \approx \frac{1.0792}{.6990} \approx 1.5440$$

(b) $\log_2 134$

Use natural logarithms and the change-of-base rule.

$$\log_2 134 = \frac{\ln 134}{\ln 2}$$

$$\approx \frac{4.8978}{.6931}$$

$$\approx 7.0661 \quad ■$$

NOTE In Example 8, the logarithms that were evaluated in intermediate steps, such as log 12 and ln 2, were shown to four decimal places. However, the final answers were obtained *without* rounding off these intermediate values, using all the digits obtained with the calculator. In general, it is best to wait until the final step to round off the answer; otherwise, a build-up of round-off error may cause the final answer to have an incorrect final decimal place digit.

An actual application of base 2 logarithms concludes this section.

■ *Example 9*

FINDING THE DIVERSITY OF A SPECIES

One measure of the diversity of the species in an ecological community is given by the formula

$$H = -[P_1 \log_2 P_1 + P_2 \log_2 P_2 + \cdots + P_n \log_2 P_n],$$

where $P_1, P_2, \ldots, P_n$ are the proportions of a sample belonging to each of n species found in the sample. For example, in a community with two species, where there are 90 of one species and 10 of the other, $P_1 = 90/100 = .9$ and $P_2 = 10/100 = .1$. Thus,

$$H = -[.9 \log_2 .9 + .1 \log_2 .1].$$

Verify that $\log_2 .1 \approx -3.32$ and $\log_2 .9 \approx -.152$, so that

$$H \approx -[(.9)(-.152) + (.1)(-3.32)] \approx .469. \quad ■$$

9.2 EXERCISES ■

You will need to use a calculator for most of the problems in this exercise set.

Find each logarithm to four decimal places. See Examples 1 and 5. (Hint: Use the properties of logarithms first in Exercises 20–24.)

1. log 278

2. log 3460

3. log 3.28

4. log 57.3

5. log 9.83

6. log 589

7. log .327

8. log .0763

9. log .000672

10. log .00382

11. log 675,000

12. log 371,000,000

13. ln 5

14. ln 6

15. ln 697

16. ln 4.83

17. $\ln 1.72$ **18.** $\ln 38.5$ **19.** $\ln 97.6$ **20.** $\ln e^2$

21. $\ln e^{3.1}$ **22.** $\ln 5e^3$ **23.** $\ln 4e^2$ **24.** $\ln 3e^{-1}$

25. Enter -1 into a scientific calculator and touch the log key. What happens? Explain why this happens.

26. Repeat Exercise 25, but use the ln key instead.

27. Which one of the following is not equal to 1? Do not use a calculator.
     (a) $\log 1$ **(b)** $\log 10$ **(c)** $\ln e$ **(d)** $(\ln 10)^0$

28. Which one of the following is a negative number? Do not use a calculator.
     (a) $\log 125$ **(b)** $\ln 5$ **(c)** $\log 1.348$ **(d)** $\ln .322$

Use a calculator to find the common antilogarithm of each of the following numbers. Give answers to four decimal places. See Example 2.

29. $.5340$ **30.** $.9309$ **31.** 2.6571 **32.** $-.0691$ **33.** -1.118 **34.** -2.3893

Use a calculator to find the natural antilogarithm of each of the following numbers. Give answers to four decimal places. See Example 6.

35. 3.8494 **36.** 1.7938 **37.** 1.3962 **38.** $-.0259$ **39.** $-.4168$ **40.** -1.2190

Use common logarithms or natural logarithms and a calculator to find each of the following to four decimal places. See Example 8.

41. $\log_5 11$ **42.** $\log_6 15$ **43.** $\log_3 1.89$ **44.** $\log_4 7.21$ **45.** $\log_8 9.63$

46. $\log_2 3.42$ **47.** $\log_{11} 47.3$ **48.** $\log_{18} 51.2$ **49.** $\log_{50} 31.3$

50. Refer to Example 8 in this section. Work part (a) using natural logarithms, and work part (b) using common logarithms. Verify that the same final answers are obtained using these other bases.

Use the formula $\mathrm{pH} = -\log [H_3O^+]$ *to find the pH of substances with the given hydronium ion concentrations. See Example 3.*

51. Milk, 4×10^{-7} **52.** Sodium hydroxide (lye), 3.2×10^{-14}

53. Limes, 1.6×10^{-2} **54.** Crackers, 3.9×10^{-9}

Use the formula for pH to find the hydronium ion concentrations of substances with the given pH values. See Example 4.

55. Shampoo, 5.5 **56.** Beer, 4.8 **57.** Soda pop, 2.7 **58.** Wine, 3.4

Refer to Example 7 for Exercises 59–62. How many years have elapsed since the split if the following percents of the words of the ancestral language are common to both languages today?

59. 90% **60.** 50%

Find r if two languages split from a common ancestral language the following number of years ago.

61. About 1000   **62.** About 500

Work the following problems. For Exercises 63 and 64, see Example 9.

63. Suppose a sample of a small community shows two species with 50 individuals each. Find the index of diversity H.

64. A virgin forest in northwestern Pennsylvania has 4 species of large trees with the following proportions of each: hemlock, .521; beech, .324; birch, .081; maple, .074. Find the index of diversity H.

65. The number of species in a sample is given by

$$S(n) = .36 \ln \left(1 + \frac{n}{.36}\right).$$

Here n is the number of individuals in the sample and the constant .36 indicates the diversity of species in the community. Find $S(n)$ for the following values of n. Round to the nearest unit.
(a) 100   (b) 500   (c) 2000   (d) 10

66. In Exercise 65, find $S(n)$ if .36 is changed to .88. Use the following values of n. Round to the nearest unit.
(a) 50   (b) 100   (c) 250

In the central Sierra Nevada mountains of California, the percent of moisture that falls as snow rather than rain is approximated reasonably well by

$$p = 86.3 \ln h - 680,$$

where p is the percent of snow moisture at an altitude h in feet. (Assume h ≥ 3000.)

67. Find the percent of moisture that falls as snow at the following altitudes.
(a) 3000 feet   (b) 4000 feet   (c) 7000 feet

68. What altitude corresponds to 55 percent snow moisture?

9.3   EXPONENTIAL AND LOGARITHMIC EQUATIONS; FURTHER APPLICATIONS

Equations with variables as exponents are called **exponential equations.** Equations that contain logarithms are called **logarithmic equations.** This section reviews solving these types of equations. Solving these equations depends on the following two properties.

PROPERTIES OF EXPONENTIAL AND LOGARITHMIC EQUATIONS	**1.** If a is a positive number, with $a \neq 1$, and x and y are real numbers, then $$a^x = a^y \quad \text{if and only if} \quad x = y.$$ **2.** If x, y, and a are positive with $a \neq 1$, then $$x = y \quad \text{if and only if} \quad \log_a x = \log_a y.$$

Property 1 is used to solve the equation in the first example.

■ *Example 1*

SOLVING AN EQUATION WITH A VARIABLE EXPONENT

Solve $\left(\dfrac{1}{2}\right)^{2x-3} = 32^x.$

Notice that both bases can easily be written as powers of the same base: $1/2 = 2^{-1}$ and $32 = 2^5$. Begin by writing both sides of the equation as a power of 2.

$$\left(\frac{1}{2}\right)^{2x-3} = 32^x$$
$$(2^{-1})^{2x-3} = (2^5)^x$$
$$2^{-2x+3} = 2^{5x} \qquad (a^m)^n = a^{mn}$$

Use property 1 to set exponents equal, and solve.

$$-2x + 3 = 5x$$
$$3 = 7x$$
$$x = \frac{3}{7} \quad ■$$

If both sides of an equation such as the one in Example 1 cannot easily be written as powers of the same base, the equation can be solved by using property 2, as shown in the next example.

■ *Example 2*

SOLVING AN
EQUATION WITH A
VARIABLE
EXPONENT USING
LOGARITHMS

Solve $5^x = 15$.

Here it is not possible to easily write both sides as powers of the same base. Use property 2 above, and take common logarithms of both sides of the equation.

$$\log 5^x = \log 15$$
$$x \cdot \log 5 = \log 15 \qquad \text{Property (c) of logarithms}$$
$$x = \frac{\log 15}{\log 5}$$

The exact solution is $x = \log 15/\log 5$. A decimal approximation can be found by evaluating log 15 and log 5, and then dividing to get

$$x \approx \frac{1.1761}{.6990} \approx 1.683,$$

to the nearest thousandth. ■

■ *Example 3*

SOLVING AN
EQUATION
INVOLVING
LOGARITHMS

Solve $\log_2 (x - 3) + \log_2 x = \log_2 4$.

By property (a) of logarithms,

$$\log_2 (x - 3) + \log_2 x = \log_2 x(x - 3),$$

so the equation becomes

$$\log_2 x(x - 3) = \log_2 4.$$

Then, by property 2 above,

$$x(x - 3) = 4$$
$$x^2 - 3x = 4$$
$$x^2 - 3x - 4 = 0$$
$$(x - 4)(x + 1) = 0$$
$$x = 4 \quad \text{or} \quad x = -1.$$

Since -1 causes the expression $x - 3$ to be nonpositive, it is not a solution. The only solution is $x = 4$. In general, reject any proposed solution that leads to the logarithm of a nonpositive number in the original equation. ■

■ *Example 4*

SOLVING AN
EQUATION
INVOLVING
LOGARITHMS

Solve $\log_y 7 = 2$.

Use the definition of logarithms to convert this equation from logarithmic form to exponential form.

$$y^2 = 7$$

Take the square root of both sides.

$$y = \pm\sqrt{7}$$

Since the base of a logarithm must be positive,

$$y = \sqrt{7}. \quad ■$$

■ *Example 5*

SOLVING AN
EQUATION
INVOLVING
LOGARITHMS

Solve $\log_3 (1 - 2x) = \log_3 x - 1$.

Subtract $\log_3 x$ from both sides to get

$$\log_3 (1 - 2x) - \log_3 x = -1.$$

$$\log_3 \frac{1 - 2x}{x} = -1 \qquad \text{Property (b) of logarithms}$$

$$3^{-1} = \frac{1 - 2x}{x} \qquad \text{Definition of logarithm}$$

$$\frac{1}{3} = \frac{1 - 2x}{x}$$

$$x = 3 - 6x$$

$$7x = 3$$

$$x = \frac{3}{7}$$

Since 3/7 makes both $1 - 2x$ and x positive numbers, the solution is 3/7. ■

Exponential and logarithmic equations arise in many important applications, particularly in applications of population growth or decay. The next example is a problem of this type.

■ *Example 6*

SOLVING AN
APPLICATION OF
EXPONENTIAL
GROWTH

Suppose that the population of a city is given by

$$P = 1,000,000(2^{.02t}),$$

where t is time measured in months.

(a) Find the population at time $t = 0$.

Let $t = 0$ in the equation and solve for P.

$$P = 1,000,000(2^{.02t})$$

$$= 1,000,000 \cdot 2^{(.02)(0)} = 1,000,000 \cdot 2^0$$

$$= 1,000,000(1) = 1,000,000$$

The population is 1,000,000 when t is 0.

(b) How long will it take for the population to reach 1,300,000?
Let $P = 1,300,000$ and solve for t.

$$P = 1,000,000(2^{.02t})$$
$$1,300,000 = 1,000,000(2^{.02t})$$

Divide both sides by 1,000,000.

$$1.3 = 2^{.02t}$$

Take logarithms on both sides. Then use property (c) of logarithms on the right side.

$$\log 1.3 = \log 2^{.02t}$$
$$\log 1.3 = .02t \log 2$$

To solve for t, divide both sides of the equation by $.02 \log 2$. Then evaluate the quotient.

$$\frac{\log 1.3}{.02 \log 2} = t$$

$$t = \frac{.1139}{(.02)(.3010)}$$

$$t \approx 18.9$$

The population will reach 1,300,000 in about 18.9 or 19 months. ■

Another important application of exponential functions is in computing compound interest. When a bank pays compound interest, it pays interest on both the principal and the interest. If there are a finite number of compounding periods per year, the following formula is used.

COMPOUND AMOUNT FORMULA	If P dollars is deposited in an account paying an annual rate of interest r compounded (paid) n times per year, the account will contain $$A = P\left(1 + \frac{r}{n}\right)^{nt}$$ dollars after t years.

In the formula above, r is usually expressed as a decimal.

■ *Example 7*
SOLVING A
COMPOUND
INTEREST PROBLEM

How much money will there be in an account at the end of 5 years if $1000 is deposited at 6% compounded quarterly? (Assume no withdrawals are made.)
    Since interest is compounded quarterly, $n = 4$. The other values given in the problem are $P = 1000$, $r = .06$ (since $6\% = .06$), and $t = 5$. Substitute into the compound amount formula to get the value of A.

$$A = 1000\left(1 + \frac{.06}{4}\right)^{4 \cdot 5}$$

$$A = 1000(1.015)^{20}$$

Now use the y^x key on a calculator, and round the answer to the nearest cent.

$$A = 1346.86$$

The account will contain $1346.86. (The actual amount of interest earned is $1346.86 − $1000 = $346.86. Do you see why?) ■

Interest can be compounded annually, semiannually, quarterly, daily, and so on. The number of compounding periods can get larger and larger. If the value of n is allowed to approach infinity, we have an example of *continuous compounding*. However, the compound interest formula above cannot be used for continuous compounding, since there is no finite value for n. The formula for continuous compounding is an example of exponential growth, and is derived in advanced courses. It involves the number e, and is given below.

CONTINUOUS COMPOUNDING FORMULA
If P dollars is deposited at an annual rate of interest r compounded continuously for t years, the final amount on deposit is

$$A = Pe^{rt}$$

dollars.

■ *Example 8*
SOLVING A CONTINUOUS COMPOUND INTEREST PROBLEM

If $1000 is invested at 6% annual interest compounded continuously, find the following.

(a) the amount in the account after 5 years (assuming no withdrawals)
In the continuous compounding formula, let $P = 1000$, $r = .06$, and $t = 5$.

$$A = 1000e^{.06(5)}$$
$$= 1000e^{.3}$$
$$= 1000(1.34986) \quad \text{Use a calculator to find } e^{.3}.$$
$$= 1349.86$$

There will be $1349.86 in the account after 5 years.

(b) the time it would take for the amount to double
When the initial deposit of $1000 is doubled, there will be $2000 in the account. Let $A = 2000$, $P = 1000$, and $r = .06$ in the formula, and solve for t.

$$2000 = 1000e^{.06t}$$
$$2 = e^{.06t} \quad \text{Divide by 1000.}$$

This equation has the variable in the exponent. To solve for the variable, take the natural logarithm of both sides. (Use natural logarithms since the base is e.)

$$\ln 2 = \ln e^{.06t}$$

$$\ln 2 = .06t \qquad\qquad \text{In } e^k = k$$

$$t = \frac{\ln 2}{.06} \qquad\qquad \text{Divide by .06.}$$

$$t \approx \frac{.6931}{.06} \approx 11.6$$

It will take about 11.6 years for the amount to double. ■

NOTE In Example 8(b), the amount of the initial deposit does not matter. Under the conditions given, *any* amount of money would double in about 11.6 years.

9.3 EXERCISES ■

Solve each of the following equations. Do not use a calculator. See Example 1.

1. $5^x = 125$ **2.** $3^p = 81$ **3.** $32^k = 2$ **4.** $64^z = 4$

5. $100^r = 1000$ **6.** $32^m = 16$ **7.** $27^z = 81$ **8.** $49^r = 343$

9. $2^{5x} = 16^{x+2}$ **10.** $3^{6x} = 9^{2x+1}$ **11.** $25^{2k} = 125^{k+1}$ **12.** $36^{2p+1} = 6^{3p-2}$

Solve each of the following equations. Express answers to the nearest thousandth when an approximation is applicable. See Examples 2–5.

13. $10^p = 3$ **14.** $10^m = 7$

15. $10^{r+2} = 15$ **16.** $10^{m-3} = 28$

17. $6^k = 10$ **18.** $12^x = 10$

19. $14^z = 28$ **20.** $21^p = 63$

21. $7^a = 32$ **22.** $12^b = 70$

23. $15^x = 8^{2x+1}$ **24.** $21^{y+1} = 32^{2y}$

25. $16^{z+3} = 28^{2z-1}$ **26.** $40^{r-3} = 32^{3r+1}$

27. $\log x - \log (x - 14) = \log 8$ **28.** $\log (k - 3) = 1 + \log (k - 21)$

29. $\log z = 1 - \log (3z - 13)$ **30.** $\log x + \log (x + 2) = 0$

31. $\log_x 10 = 3$ **32.** $\log_k 25 = 5$

33. $2 + \log x = 0$ **34.** $-6 + \log 2k = 0$

35. For $a > 0$ and $a \neq 1$, if $a^x = a^y$, then $x = y$. Give an example to show why the restriction $a \neq 1$ is necessary for this statement to be true.

36. Explain why the exponential equation $5^x = 10$ cannot be solved using the method of Example 1 in this section.

37. Suppose that you overhear the following statement: "I must reject any negative answer that I obtain when I solve an equation involving logarithms." Is this correct? Explain.

38. What values of x could not possibly be solutions of the following equation? (Do not solve.)

$$\log (4x - 7) + \log (x^2 + 4) = 0$$

39. Under what conditions is $\log x^2 = 2 \log x$ true?

Solve each of the following problems. See Examples 6–8.

40. Suppose that the number of rabbits in a colony increases according to the relationship

$$y = y_0\, e^{.4t},$$

where t represents time in months and y_0 is the initial population of rabbits.
   (a) Find the number of rabbits present at time $t = 4$ if $y_0 = 100$.
   (b) How long will it take for the number of rabbits to triple?

41. A city in Ohio finds its residents moving into the countryside. Its population is declining according to the relationship

$$P = P_0\, e^{-.04t},$$

where t is time measured in years and P_0 is the population at time $t = 0$.
   (a) If $P_0 = 1{,}000{,}000$, find the population at time $t = 1$.
   (b) If $P_0 = 1{,}000{,}000$, estimate the time it will take for the population to be reduced to 750,000.
   (c) How long will it take for the population to be cut in half?

42. Suppose that a large cloud of radioactive debris from a nuclear explosion has floated over the Pacific Northwest, contaminating much of the hay supply. Consequently, farmers in the area are concerned that the cows eating this hay will give contaminated milk. (The tolerance level for radioactive iodine in milk is 0.) The percent of the initial amount of radioactive iodine still present in the hay after t days is approximated by

$$y = 100(2.7)^{-.1t},$$

where t is time measured in days.

(a) Some scientists believe that the hay is safe after the percent of radioactive iodine has declined to 10% of the original amount. Find the number of days before the hay can be used.
(b) Other scientists believe that the hay is not safe until the level of radioactive iodine has declined to only 1% of the original level. Find the number of days this would take.

43. Find the amount of money in an account after 12 years if $5000 is deposited at 7% annual interest compounded as follows.
   (a) annually
   (b) semiannually
   (c) quarterly
   (d) daily (Use $n = 365$.)
   (e) continuously

44. How much money will be in an account at the end of 8 years if $4500 is deposited at 6% annual interest compounded as follows?
   (a) annually
   (b) semiannually
   (c) quarterly
   (d) daily (Use $n = 365$.)
   (e) continuously

45. How much money must be deposited today to amount to $1000 in 10 years at 5% compounded continuously?

46. How much money must be deposited today to become $1850 in 40 years at 6.5% compounded continuously?

SECTION	KEY IDEAS
9.1 Exponential and Logarithmic Functions; Properties of Logarithms	**Exponential Function** An exponential function is a function of the form $$y = a^x,$$ where $a > 0$ and $a \neq 1$. **Definition of Logarithm** For all positive numbers a, with $a \neq 1$, $$y = \log_a x \quad \text{means} \quad x = a^y.$$ **Logarithmic Function** If $a > 0$, $a \neq 1$, and $x > 0$, then $$y = \log_a x$$ is the logarithmic function with base a. **Properties of Logarithms** If x and y are positive real numbers, r is any real number, and a is any positive real number, with $a \neq 1$, then (a) $\log_a xy = \log_a x + \log_a y$ (b) $\log_a \dfrac{x}{y} = \log_a x - \log_a y$ (c) $\log_a x^r = r \log_a x$ (d) $\log_a a = 1$ (e) $\log_a 1 = 0$.
9.2 Common and Natural Logarithms and Their Applications	**Common Logarithm** A logarithm to base 10 is called a common logarithm. $$\log_{10} x = \log x$$ e The irrational number e is approximately equal to 2.718281828. **Natural Logarithm** A logarithm to base e is called a natural logarithm. $$\log_e x = \ln x$$ **Change-of-Base Rule** If $a > 0$, $a \neq 1$, $b > 0$, $b \neq 1$, and $x > 0$, then $$\log_a x = \frac{\log_b x}{\log_b a}.$$

SECTION	KEY IDEAS
9.3 Exponential and Logarithmic Equations; Further Applications	**Properties of Exponential and Logarithmic Equations** **1.** If a is a positive number, with $a \neq 1$, and x and y are real numbers, then $$a^x = a^y \text{ if and only if } x = y.$$ **2.** If x, y, and a are positive, with $a \neq 1$, then $$x = y \text{ if and only if } \log_a x = \log_a y.$$ **Compound Amount Formula** If P dollars is deposited in an account paying an annual rate of interest r compounded (paid) n times per year, the account will contain $$A = P\left(1 + \frac{r}{n}\right)^{nt}$$ dollars after t years. **Continuous Compounding Formula** If P dollars is deposited at an annual rate of interest r compounded continuously for t years, the final amount on deposit is $$A = Pe^{rt}$$ dollars.

CHAPTER 9 REVIEW EXERCISES ■

Simplify each expression.

1. $\dfrac{3^4 \cdot 3^5}{3^6}$ **2.** $(4^3)^5$ **3.** $(2^3)^2 \cdot 2^5$ **4.** $3^{-12} \cdot 3^5$ **5.** $16^{3/4}$ **6.** $\dfrac{2^{1/3}}{2^{5/3}}$

Graph each function.

7. $y = 5^x$ **8.** $y = \left(\dfrac{1}{3}\right)^x$ **9.** $y = 2^{-x}$ **10.** $y = 3^{x+2}$

11. $y = \log_3 x$ **12.** $y = 1 + \log_2 x$ **13.** $y = \log_2 (2 - x)$ **14.** $y = \log_3 2x$

Convert to logarithmic form.

15. $2^9 = 512$ **16.** $10^{-2} = .01$ **17.** $\left(\dfrac{2}{3}\right)^{-1} = \dfrac{3}{2}$ **18.** $4^{3/2} = 8$

Write in exponential form.

19. $\log_3 81 = 4$

20. $\log_4 2 = \dfrac{1}{2}$

21. $\log_{1/3} 3 = -1$

22. $\log_{2/5} \dfrac{8}{125} = 3$

Use the properties of logarithms to simplify each expression.

23. $\log_5 4 + \log_5 3$

24. $\log_6 8 + \log_6 4$

25. $5 \log_2 x$

26. $\dfrac{1}{2} \log_{10} p$

27. $2 \log_3 x + 4 \log_3 y$

28. $\dfrac{2}{3} \ln x^3 - \dfrac{1}{4} \ln y$

Use a calculator to find the common logarithm of each number. Give answers to four decimal places.

29. 2.48

30. 63.5

31. 1240

32. .00421

33. 5,270,000

34. .643

Use a calculator to find the common antilogarithm of each number. Give answers to four decimal places.

35. 3.4314

36. .7497

37. -1.7721

38. $-.3507$

Use a calculator to find each of the following natural logarithms to four decimal places.

39. $\ln 15$

40. $\ln 120$

41. $\ln .12$

42. $\ln .03$

Solve each equation without a calculator.

43. $3^x = 27$

44. $25^p = 625$

45. $2^{m-4} = 8^{2m}$

46. $49^{-k+3} = 7^{5k}$

Solve each equation. Use a calculator and give answers as decimals rounded to the nearest thousandth when applicable.

47. $5^n = 33$

48. $4^{y+2} = 17$

49. $\log x = 2 - \log (x - 3)$

50. $\ln p - \ln (p + 3) = 4$

51. $\log_t 16 = 2$

52. $\ln k = 1$

Solve each problem.

53. The amount of a radioactive specimen present at time t (measured in days) is
$$A(t) = 5000(3)^{-.02t},$$
where $A(t)$ is measured in grams. Find the *half-life* of the specimen, that is, the time when only half of the specimen remains.

54. The number of fish in a pond is
$$N(x) = 10 \log_3 (x + 1),$$
where x is time in months. How long will it take for the number of fish to reach 30?

55. If $20,000 is deposited at 7% annual interest compounded quarterly, how much will be in the account after 5 years, assuming no withdrawals are made?

56. How much will $10,000 compounded continuously at 6% annual interest amount to in three years?

57. In your own words, explain the meaning of $\log_a x$.

58. Explain why the value of $\ln e$ is 1.

APPENDICES

APPENDIX A   USING THE TABLE OF
TRIGONOMETRIC FUNCTION VALUES

Because of the widespread availability of scientific calculators, using tables to find trigonometric function values is very seldom required. However, we do include a table in this text. The following examples show how Table 1 is used to find values of trigonometric functions for θ in the intervals [0°, 90°] or [0, π/2]. For θ in [0°, 45°] or [0, π/4], find θ by reading down one of the first two columns. Values for θ in [45°, 90°] or [π/4, π/2] are given by reading *up* the last two columns and referring to the names of the trigonometric functions at the *bottom* of the table. The values of θ in degrees are given to the nearest ten minutes. To use the table for decimal degrees, it is necessary to convert to degrees and minutes, to the nearest ten minutes, as in part (c) of the first example.

Keep in mind that this table does not produce the accuracy afforded by even the least powerful scientific calculators.

■ *Example 1*

USING THE TABLE
TO FIND FUNCTION
VALUES

Find the following function values.

(a) sin 49° 10′

Look in the last column for 49° 10′ reading up Table 1 and looking for sin at the bottom of the table. You should find

$$\sin 49° \, 10' = .7566.$$

(b) sec .2414

Look in the second column for .2414 radians and across the top of the table for the column headed sec.

$$\sec .2414 = 1.030$$

(c) tan 17.3°

Convert 17.3° to degrees and minutes as follows.

$$17.3° = 17° + \frac{3°}{10}$$

$$= 17° + \frac{18°}{60}$$

$$= 17° \, 18' \approx 17° \, 20'$$

to the nearest ten minutes. Now find the entry for 17° 20′ in the tan column.

$$\tan 17.3° \approx \tan 17° \, 20' = .3121 \quad ■$$

347

Table 1 can also be used backwards to find θ given a trigonometric function value of θ.

■ *Example 2*

USING THE TABLE
TO FIND ANGLE
MEASURES

Find a value of θ in degrees that satisfies each of the following.

(a) $\cos \theta = .9063$

Look in the body of the table in a column headed cos at the top or with cos at the bottom. Since the entry is found in a column with cos at the top, read over to the first column to find $\theta = 25° \ 00'$.

(b) $\csc \theta = 1.117$

The entry 1.117 is found in a column with csc at the bottom. Read across to the last column to see that

$$\theta = 63° \ 30'.$$

Notice that the row for 63° 30′ is *above* the row for 63° 00′, since we read *up* the table in the last column. ■

APPENDIX B USING THE TABLE OF COMMON LOGARITHMS

This appendix gives a brief explanation of Table 2, the table of common logarithms. To find logarithms using the table, first write the number in scientific notation; that is, as a product of a number between 1 and 10, and a power of 10. For example, to find log 423, first write 423 in scientific notation.

$$\log 423 = \log (4.23 \times 10^2)$$

By the product property for logarithms,

$$\log 423 = \log 4.23 + \log 10^2.$$

The logarithm of 10^2 is 2 (the exponent). To find log 4.23, use Table 2, the table of common logarithms. (A portion of that table is reproduced here.) Read down on the left to the row headed 4.2. Then read across to the column headed by 3 to get .6263. From the table

$$\log 423 \approx .6263 + 2 = 2.6263.$$

x	0	1	2	3	4	5	6	7	8	9
4.0	.6021	.6031	.6042	.6053	.6064	.6075	.6085	.6096	.6107	.6117
4.1	.6128	.6138	.6149	.6160	.6170	.6180	.6191	.6201	.6212	.6222
4.2	.6232	.6243	.6253	.6263	.6274	.6284	.6294	.6304	.6314	.6325
4.3	.6335	.6345	.6355	.6365	.6375	.6385	.6395	.6405	.6415	.6425
4.4	.6435	.6444	.6454	.6464	.6474	.6484	.6493	.6503	.6513	.6522

The decimal part of the logarithm, .6263, is called the **mantissa.** The integer part, 2, is called the **characteristic.**

■ *Example 1*

USING THE TABLE
TO FIND COMMON
LOGARITHMS

Find the following common logarithms.

(a) log 437,000

Since $437,000 = 4.37 \times 10^5$, the characteristic is 5. From the portion of the logarithm table given above, the mantissa of log 437,000 is .6405, and log 437,000 = 5 + .6405 = 5.6405.

(b) log .0415

Express .0415 as 4.15×10^{-2}. The characteristic is -2. The mantissa, from the table, is .6180, and

$$\log .0415 = -2 + .6180 = -1.3820.$$

To retain the mantissa, the answer may be given as $-2 + .6180$, or $.6180 - 2$. A calculator with a logarithm key gives the result as -1.3820. ■

In Example 1(a), the number 437,000, whose logarithm is 5.6405, is called the **antilogarithm** (abbreviated *antilog*) of 5.6405. In the same way, in Example 1(b), .0415 is the antilogarithm of the logarithm $.6180 - 2$.

From Example 1,

$$\log 437,000 = 5.6405.$$

In exponential form, this is equivalent to

$$437,000 = 10^{5.6405}.$$

This shows that the antilogarithm, 437,000, is just an exponential.

■ *Example 2*

USING THE TABLE
TO FIND
ANTILOGARITHMS

Find the antilogarithm of each of the following logarithms.

(a) 2.6454

To find the number whose logarithm is 2.6454, first look in the table for a mantissa of .6454. The number whose mantissa is .6454 is 4.42. Since the characteristic is 2, the antilogarithm is $4.42 \times 10^2 = 442$.

(b) $.6314 - 3$

The mantissa is .6314 and the characteristic is -3. In the table, the number with a mantissa of .6314 is 4.28, so the antilogarithm is $4.28 \times 10^{-3} = .00428$. ■

The table of logarithms contains logarithms to four decimal places. This table can be used to find the logarithm of any positive number containing three significant digits. **Linear interpolation** is used to find logarithms of numbers containing four significant digits. In Example 3 we show how this is done.

■ *Example 3*

USING LINEAR
INTERPOLATION TO
FIND A LOGARITHM

Find log 4238.

Since

$$4230 < 4238 < 4240,$$

$$\log 4230 < \log 4238 < \log 4240.$$

Find log 4230 and log 4240 in the table.

$$10\left\{ 8\left\{ \begin{array}{l} \log 4230 = 3.6263 \\ \log 4238 = \\ \log 4240 = 3.6274 \end{array} \right\}.0011 \right.$$

Since 4238 is 8/10 of the way from 4230 to 4240, take 8/10 of .0011, the difference between the two logarithms.

$$\frac{8}{10}(.0011) \approx .0009$$

Now add .0009 to log 4230 to get

$$\log 4238 = 3.6263 + .0009$$
$$= 3.6272.$$

Check the answer by consulting a more accurate logarithm table or use a calculator to find that log 4238 = 3.62716 (to five places). ■

■ *Example 4*

**USING LINEAR
INTERPOLATION TO
FIND A LOGARITHM**

Find log .02386.

Arrange the work as follows.

$$10\left\{ 6\left\{ \begin{array}{l} \log .0238 = .3766 - 2 \\ \log .02386 = \\ \log .0239 = .3784 - 2 \end{array} \right\}.0018 \right.$$

$$\frac{6}{10}(.0018) \approx .0011$$

Now add .3766 − 2 and .0011.

$$
\begin{array}{r}
.3766 - 2 \\
.0011 \\
\hline
\log .02386 = .3777 - 2
\end{array}
$$ ■

Interpolation can also be used to find antilogarithms with four significant digits.

■ *Example 5*

**USING LINEAR
INTERPOLATION TO
FIND AN
ANTILOGARITHM**

Find the antilogarithm of 3.5894.

From the table, the mantissas closest to .5894 are .5888 and .5899. Find the antilogarithms of 3.5888 and 3.5899.

$$.0011\left\{ .0006\left\{ \begin{array}{l} 3.5888 = \log 3880 \\ 3.5894 = \\ 3.5899 = \log 3890 \end{array} \right\}10 \right.$$

$$\left(\frac{.0006}{.0011}\right)10 \approx .5(10) = 5$$

Add the 5 to 3880, giving

$$\log 3885 = 3.5894.$$

The antilogarithm of 3.5894 is 3885. ■

TABLE 1 TRIGONOMETRIC FUNCTION VALUES ∎

351

TABLE 1 Trigonometric Function Values

θ (degrees)	θ (radians)	sin θ	cos θ	tan θ	cot θ	sec θ	csc θ		
0° 00′	.0000	.0000	1.0000	.0000	—	1.000	—	1.5708	90° 00′
10	.0029	.0029	1.0000	.0029	343.8	1.000	343.8	1.5679	50
20	.0058	.0058	1.0000	.0058	171.9	1.000	171.9	1.5650	40
30	.0087	.0087	1.0000	.0087	114.6	1.000	114.6	1.5621	30
40	.0116	.0116	.9999	.0116	85.94	1.000	85.95	1.5592	20
50	.0145	.0145	.9999	.0145	68.75	1.000	68.76	1.5563	10
1° 00′	.0175	.0175	.9998	.0175	57.29	1.000	57.30	1.5533	89° 00′
10	.0204	.0204	.9998	.0204	49.10	1.000	49.11	1.5504	50
20	.0233	.0233	.9997	.0233	42.96	1.000	42.98	1.5475	40
30	.0262	.0262	.9997	.0262	38.19	1.000	38.20	1.5446	30
40	.0291	.0291	.9996	.0291	34.37	1.000	34.38	1.5417	20
50	.0320	.0320	.9995	.0320	31.24	1.001	31.26	1.5388	10
2° 00′	.0349	.0349	.9994	.0349	28.64	1.001	28.65	1.5359	88° 00′
10	.0378	.0378	.9993	.0378	26.43	1.001	26.45	1.5330	50
20	.0407	.0407	.9992	.0407	24.54	1.001	24.56	1.5301	40
30	.0436	.0436	.9990	.0437	22.90	1.001	22.93	1.5272	30
40	.0465	.0465	.9989	.0466	21.47	1.001	21.49	1.5243	20
50	.0495	.0494	.9988	.0495	20.21	1.001	20.23	1.5213	10
3° 00′	.0524	.0523	.9986	.0524	19.08	1.001	19.11	1.5184	87° 00′
10	.0553	.0552	.9985	.0553	18.07	1.002	18.10	1.5155	50
20	.0582	.0581	.9983	.0582	17.17	1.002	17.20	1.5126	40
30	.0611	.0610	.9981	.0612	16.35	1.002	16.38	1.5097	30
40	.0640	.0640	.9980	.0641	15.60	1.002	15.64	1.5068	20
50	.0669	.0669	.9978	.0670	14.92	1.002	14.96	1.5039	10
4° 00′	.0698	.0698	.9976	.0699	14.30	1.002	14.34	1.5010	86° 00′
10	.0727	.0727	.9974	.0729	13.73	1.003	13.76	1.4981	50
20	.0756	.0756	.9971	.0758	13.20	1.003	13.23	1.4952	40
30	.0785	.0785	.9969	.0787	12.71	1.003	12.75	1.4923	30
40	.0814	.0814	.9967	.0816	12.25	1.003	12.29	1.4893	20
50	.0844	.0843	.9964	.0846	11.83	1.004	11.87	1.4864	10
5° 00′	.0873	.0872	.9962	.0875	11.43	1.004	11.47	1.4835	85° 00′
10	.0902	.0901	.9959	.0904	11.06	1.004	11.10	1.4806	50
20	.0931	.0929	.9957	.0934	10.71	1.004	10.76	1.4777	40
30	.0960	.0958	.9954	.0963	10.39	1.005	10.43	1.4748	30
40	.0989	.0987	.9951	.0992	10.08	1.005	10.13	1.4719	20
50	.1018	.1016	.9948	.1022	9.788	1.005	9.839	1.4690	10
		cos θ	sin θ	cot θ	tan θ	csc θ	sec θ	θ (radians)	θ (degrees)

TABLE 1   Trigonometric Function Values (continued)

θ (degrees)	θ (radians)	sin θ	cos θ	tan θ	cot θ	sec θ	csc θ		
6° 00′	.1047	.1045	.9945	.1051	9.514	1.006	9.567	1.4661	**84° 00′**
10	.1076	.1074	.9942	.1080	9.255	1.006	9.309	1.4632	50
20	.1105	.1103	.9939	.1110	9.010	1.006	9.065	1.4603	40
30	.1134	.1132	.9936	.1139	8.777	1.006	8.834	1.4573	30
40	.1164	.1161	.9932	.1169	8.556	1.007	8.614	1.4544	20
50	.1193	.1190	.9929	.1198	8.345	1.007	8.405	1.4515	10
7° 00′	.1222	.1219	.9925	.1228	8.144	1.008	8.206	1.4486	**83° 00′**
10	.1251	.1248	.9922	.1257	7.953	1.008	8.016	1.4457	50
20	.1280	.1276	.9918	.1287	7.770	1.008	7.834	1.4428	40
30	.1309	.1305	.9914	.1317	7.596	1.009	7.661	1.4399	30
40	.1338	.1334	.9911	.1346	7.429	1.009	7.496	1.4370	20
50	.1376	.1363	.9907	.1376	7.269	1.009	7.337	1.4341	10
8° 00′	.1396	.1392	.9903	.1405	7.115	1.010	7.185	1.4312	**82° 00′**
10	.1425	.1421	.9899	.1435	6.968	1.010	7.040	1.4283	50
20	.1454	.1449	.9894	.1465	6.827	1.011	6.900	1.4254	40
30	.1484	.1478	.9890	.1495	6.691	1.011	6.765	1.4224	30
40	.1513	.1507	.9886	.1524	6.561	1.012	6.636	1.4195	20
50	.1542	.1536	.9881	.1554	6.435	1.012	6.512	1.4166	10
9° 00′	.1571	.1564	.9877	.1584	6.314	1.012	6.392	1.4137	**81° 00′**
10	.1600	.1593	.9872	.1614	6.197	1.013	6.277	1.4108	50
20	.1629	.1622	.9868	.1644	6.084	1.013	6.166	1.4079	40
30	.1658	.1650	.9863	.1673	5.976	1.014	6.059	1.4050	30
40	.1687	.1679	.9858	.1703	5.871	1.014	5.955	1.4021	20
50	.1716	.1708	.9853	.1733	5.769	1.015	5.855	1.3992	10
10° 00′	.1745	.1736	.9848	.1763	5.671	1.015	5.759	1.3963	**80° 00′**
10	.1774	.1765	.9843	.1793	5.576	1.016	5.665	1.3934	50
20	.1804	.1794	.9838	.1823	5.485	1.016	5.575	1.3904	40
30	.1833	.1822	.9833	.1853	5.396	1.017	5.487	1.3875	30
40	.1862	.1851	.9827	.1883	5.309	1.018	5.403	1.3846	20
50	.1891	.1880	.9822	.1914	5.226	1.018	5.320	1.3817	10
11° 00′	.1920	.1908	.9816	.1944	5.145	1.019	5.241	1.3788	**79° 00′**
10	.1949	.1937	.9811	.1974	5.066	1.019	5.164	1.3759	50
20	.1978	.1965	.9805	.2004	4.989	1.020	5.089	1.3730	40
30	.2007	.1994	.9799	.2035	4.915	1.020	5.016	1.3701	30
40	.2036	.2022	.9793	.2065	4.843	1.021	4.945	1.3672	20
50	.2065	.2051	.9787	.2095	4.773	1.022	4.876	1.3643	10
		cos θ	sin θ	cot θ	tan θ	csc θ	sec θ	θ (radians)	θ (degrees)

TABLE 1 TRIGONOMETRIC FUNCTION VALUES ∎

TABLE 1 Trigonometric Function Values (continued)

θ (degrees)	θ (radians)	sin θ	cos θ	tan θ	cot θ	sec θ	csc θ		
12° 00′	.2094	.2079	.9781	.2126	4.705	1.022	4.810	1.3614	**78° 00′**
10	.2123	.2108	.9775	.2156	4.638	1.023	4.745	1.3584	50
20	.2153	.2136	.9769	.2186	4.574	1.024	4.682	1.3555	40
30	.2182	.2164	.9763	.2217	4.511	1.024	4.620	1.3526	30
40	.2211	.2193	.9757	.2247	4.449	1.025	4.560	1.3497	20
50	.2240	.2221	.9750	.2278	4.390	1.026	4.502	1.3468	10
13° 00′	.2269	.2250	.9744	.2309	4.331	1.026	4.445	1.3439	**77° 00′**
10	.2298	.2278	.9737	.2339	4.275	1.027	4.390	1.3410	50
20	.2327	.2306	.9730	.2370	4.219	1.028	4.336	1.3381	40
30	.2356	.2334	.9724	.2401	4.165	1.028	4.284	1.3352	30
40	.2385	.2363	.9717	.2432	4.113	1.029	4.232	1.3323	20
50	.2414	.2391	.9710	.2462	4.061	1.030	4.182	1.3294	10
14° 00′	.2443	.2419	.9703	.2493	4.011	1.031	4.134	1.3265	**76° 00′**
10	.2473	.2447	.9696	.2524	3.962	1.031	4.086	1.3235	50
20	.2502	.2476	.9689	.2555	3.914	1.032	4.039	1.3206	40
30	.2531	.2504	.9681	.2586	3.867	1.033	3.994	1.3177	30
40	.2560	.2532	.9674	.2617	3.821	1.034	3.950	1.3148	20
50	.2589	.2560	.9667	.2648	3.776	1.034	3.906	1.3119	10
15° 00′	.2618	.2588	.9659	.2679	3.732	1.035	3.864	1.3090	**75° 00′**
10	.2647	.2616	.9652	.2711	3.689	1.036	3.822	1.3061	50
20	.2676	.2644	.9644	.2742	3.647	1.037	3.782	1.3032	40
30	.2705	.2672	.9636	.2773	3.606	1.038	3.742	1.3003	30
40	.2734	.2700	.9628	.2805	3.566	1.039	3.703	1.2974	20
50	.2763	.2728	.9621	.2836	3.526	1.039	3.665	1.2945	10
16° 00′	.2793	.2756	.9613	.2867	3.487	1.040	3.628	1.2915	**74° 00′**
10	.2822	.2784	.9605	.2899	3.450	1.041	3.592	1.2886	50
20	.2851	.2812	.9596	.2931	3.412	1.042	3.556	1.2857	40
30	.2880	.2840	.9588	.2962	3.376	1.043	3.521	1.2828	30
40	.2909	.2868	.9580	.2994	3.340	1.044	3.487	1.2799	20
50	.2938	.2896	.9572	.3026	3.305	1.045	3.453	1.2770	10
17° 00′	.2967	.2924	.9563	.3057	3.271	1.046	3.420	1.2741	**73° 00′**
10	.2996	.2952	.9555	.3089	3.237	1.047	3.388	1.2712	50
20	.3025	.2979	.9546	.3121	3.204	1.048	3.356	1.2683	40
30	.3054	.3007	.9537	.3153	3.172	1.049	3.326	1.2654	30
40	.3083	.3035	.9528	.3185	3.140	1.049	3.295	1.2625	20
50	.3113	.3062	.9520	.3217	3.108	1.050	3.265	1.2595	10
		cos θ	sin θ	cot θ	tan θ	csc θ	sec θ	θ (radians)	θ (degrees)

TABLE 1   Trigonometric Function Values (continued)

θ (degrees)	θ (radians)	sin θ	cos θ	tan θ	cot θ	sec θ	csc θ		
18° 00′	.3142	.3090	.9511	.3249	3.078	1.051	3.236	1.2566	**72° 00′**
10	.3171	.3118	.9502	.3281	3.047	1.052	3.207	1.2537	50
20	.3200	.3145	.9492	.3314	3.018	1.053	3.179	1.2508	40
30	.3229	.3173	.9483	.3346	2.989	1.054	3.152	1.2479	30
40	.3258	.3201	.9474	.3378	2.960	1.056	3.124	1.2450	20
50	.3287	.3228	.9465	.3411	2.932	1.057	3.098	1.2421	10
19° 00′	.3316	.3256	.9455	.3443	2.904	1.058	3.072	1.2392	**71° 00′**
10	.3345	.3283	.9446	.3476	2.877	1.059	3.046	1.2363	50
20	.3374	.3311	.9436	.3508	2.850	1.060	3.021	1.2334	40
30	.3403	.3338	.9426	.3541	2.824	1.061	2.996	1.2305	30
40	.3432	.3365	.9417	.3574	2.798	1.062	2.971	1.2275	20
50	.3462	.3393	.9407	.3607	2.773	1.063	2.947	1.2246	10
20° 00′	.3491	.3420	.9397	.3640	2.747	1.064	2.924	1.2217	**70° 00′**
10	.3520	.3448	.9387	.3673	2.723	1.065	2.901	1.2188	50
20	.3549	.3475	.9377	.3706	2.699	1.066	2.878	1.2159	40
30	.3578	.3502	.9367	.3739	2.675	1.068	2.855	1.2130	30
40	.3607	.3529	.9356	.3772	2.651	1.069	2.833	1.2101	20
50	.3636	.3557	.9346	.3805	2.628	1.070	2.812	1.2072	10
21° 00′	.3665	.3584	.9336	.3839	2.605	1.071	2.790	1.2043	**69° 00′**
10	.3694	.3611	.9325	.3872	2.583	1.072	2.769	1.2014	50
20	.3723	.3638	.9315	.3906	2.560	1.074	2.749	1.1985	40
30	.3752	.3665	.9304	.3939	2.539	1.075	2.729	1.1956	30
40	.3782	.3692	.9293	.3973	2.517	1.076	2.709	1.1926	20
50	.3811	.3719	.9283	.4006	2.496	1.077	2.689	1.1897	10
22° 00′	.3840	.3746	.9272	.4040	2.475	1.079	2.669	1.1868	**68° 00′**
10	.3869	.3773	.9261	.4074	2.455	1.080	2.650	1.1839	50
20	.3898	.3800	.9250	.4108	2.434	1.081	2.632	1.1810	40
30	.3927	.3827	.9239	.4142	2.414	1.082	2.613	1.1781	30
40	.3956	.3854	.9228	.4176	2.394	1.084	2.595	1.1752	20
50	.3985	.3881	.9216	.4210	2.375	1.085	2.577	1.1723	10
23° 00′	.4014	.3907	.9205	.4245	2.356	1.086	2.559	1.1694	**67° 00′**
10	.4043	.3934	.9194	.4279	2.337	1.088	2.542	1.1665	50
20	.4072	.3961	.9182	.4314	2.318	1.089	2.525	1.1636	40
30	.4102	.3987	.9171	.4348	2.300	1.090	2.508	1.1606	30
40	.4131	.4014	.9159	.4383	2.282	1.092	2.491	1.1577	20
50	.4160	.4041	.9147	.4417	2.264	1.093	2.475	1.1548	10
		cos θ	sin θ	cot θ	tan θ	csc θ	sec θ	θ (radians)	θ (degrees)

TABLE 1 TRIGONOMETRIC FUNCTION VALUES ■

355

TABLE 1 Trigonometric Function Values (continued)

θ (degrees)	θ (radians)	sin θ	cos θ	tan θ	cot θ	sec θ	csc θ		
24° 00′	.4189	.4067	.9135	.4452	2.246	1.095	2.459	1.1519	**66° 00′**
10	.4218	.4094	.9124	.4487	2.229	1.096	2.443	1.1490	50
20	.4247	.4120	.9112	.4522	2.211	1.097	2.427	1.1461	40
30	.4276	.4147	.9100	.4557	2.194	1.099	2.411	1.1432	30
40	.4305	.4173	.9088	.4592	2.177	1.100	2.396	1.1403	20
50	.4334	.4200	.9075	.4628	2.161	1.102	2.381	1.1374	10
25° 00′	.4363	.4226	.9063	.4663	2.145	1.103	2.366	1.1345	**65° 00′**
10	.4392	.4253	.9051	.4699	2.128	1.105	2.352	1.1316	50
20	.4422	.4279	.9038	.4734	2.112	1.106	2.237	1.1286	40
30	.4451	.4305	.9026	.4770	2.097	1.108	2.323	1.1257	30
40	.4480	.4331	.9013	.4806	2.081	1.109	2.309	1.1228	20
50	.4509	.4358	.9001	.4841	2.066	1.111	2.295	1.1199	10
26° 00′	.4538	.4384	.8988	.4877	2.050	1.113	2.281	1.1170	**64° 00′**
10	.4567	.4410	.8975	.4913	2.035	1.114	2.268	1.1141	50
20	.4596	.4436	.8962	.4950	2.020	1.116	2.254	1.1112	40
30	.4625	.4462	.8949	.4986	2.006	1.117	2.241	1.1083	30
40	.4654	.4488	.8936	.5022	1.991	1.119	2.228	1.1054	20
50	.4683	.4514	.8923	.5059	1.977	1.121	2.215	1.1025	10
27° 00′	.4712	.4540	.8910	.5095	1.963	1.122	2.203	1.0996	**63° 00′**
10	.4741	.4566	.8897	.5132	1.949	1.124	2.190	1.0966	50
20	.4771	.4592	.8884	.5169	1.935	1.126	2.178	1.0937	40
30	.4800	.4617	.8870	.5206	1.921	1.127	2.166	1.0908	30
40	.4829	.4643	.8857	.5243	1.907	1.129	2.154	1.0879	20
50	.4858	.4669	.8843	.5280	1.894	1.131	2.142	1.0850	10
28° 00′	.4887	.4695	.8829	.5317	1.881	1.133	2.130	1.0821	**62° 00′**
10	.4916	.4720	.8816	.5354	1.868	1.134	2.118	1.0792	50
20	.4945	.4746	.8802	.5392	1.855	1.136	2.107	1.0763	40
30	.4974	.4772	.8788	.5430	1.842	1.138	2.096	1.0734	30
40	.5003	.4797	.8774	.5467	1.829	1.140	2.085	1.0705	20
50	.5032	.4823	.8760	.5505	1.816	1.142	2.074	1.0676	10
29° 00′	.5061	.4848	.8746	.5543	1.804	1.143	2.063	1.0647	**61° 00′**
10	.5091	.4874	.8732	.5581	1.792	1.145	2.052	1.0617	50
20	.5120	.4899	.8718	.5619	1.780	1.147	2.041	1.0588	40
30	.5149	.4924	.8704	.5658	1.767	1.149	2.031	1.0559	30
40	.5178	.4950	.8689	.5696	1.756	1.151	2.020	1.0530	20
50	.5207	.4975	.8675	.5735	1.744	1.153	2.010	1.0501	10
		cos θ	sin θ	cot θ	tan θ	csc θ	sec θ	θ (radians)	θ (degrees)

TABLE 1 Trigonometric Function Values (continued)

θ (degrees)	θ (radians)	sin θ	cos θ	tan θ	cot θ	sec θ	csc θ		
30° 00′	.5236	.5000	.8660	.5774	1.732	1.155	2.000	1.0472	**60° 00′**
10	.5265	.5025	.8646	.5812	1.720	1.157	1.990	1.0443	50
20	.5294	.5050	.8631	.5851	1.709	1.159	1.980	1.0414	40
30	.5323	.5075	.8616	.5890	1.698	1.161	1.970	1.0385	30
40	.5352	.5100	.8601	.5930	1.686	1.163	1.961	1.0356	20
50	.5381	.5125	.8587	.5969	1.675	1.165	1.951	1.0327	10
31° 00′	.5411	.5150	.8572	.6009	1.664	1.167	1.942	1.0297	**59° 00′**
10	.5440	.5175	.8557	.6048	1.653	1.169	1.932	1.0268	50
20	.5469	.5200	.8542	.6088	1.643	1.171	1.923	1.0239	40
30	.5498	.5225	.8526	.6128	1.632	1.173	1.914	1.0210	30
40	.5527	.5250	.8511	.6168	1.621	1.175	1.905	1.0181	20
50	.5556	.5275	.8496	.6208	1.611	1.177	1.896	1.0152	10
32° 00′	.5585	.5299	.8480	.6249	1.600	1.179	1.887	1.0123	**58° 00′**
10	.5614	.5324	.8465	.6289	1.590	1.181	1.878	1.0094	50
20	.5643	.5348	.8450	.6330	1.580	1.184	1.870	1.0065	40
30	.5672	.5373	.8434	.6371	1.570	1.186	1.861	1.0036	30
40	.5701	.5398	.8418	.6412	1.560	1.188	1.853	1.0007	20
50	.5730	.5422	.8403	.6453	1.550	1.190	1.844	.9977	10
33° 00′	.5760	.5446	.8387	.6494	1.540	1.192	1.836	.9948	**57° 00′**
10	.5789	.5471	.8371	.6536	1.530	1.195	1.828	.9919	50
20	.5818	.5495	.8355	.6577	1.520	1.197	1.820	.9890	40
30	.5847	.5519	.8339	.6619	1.511	1.199	1.812	.9861	30
40	.5876	.5544	.8323	.6661	1.501	1.202	1.804	.9832	20
50	.5905	.5568	.8307	.6703	1.492	1.204	1.796	.9803	10
34° 00′	.5934	.5592	.8290	.6745	1.483	1.206	1.788	.9774	**56° 00′**
10	.5963	.5616	.8274	.6787	1.473	1.209	1.781	.9745	50
20	.5992	.5640	.8258	.6830	1.464	1.211	1.773	.9716	40
30	.6021	.5664	.8241	.6873	1.455	1.213	1.766	.9687	30
40	.6050	.5688	.8225	.6916	1.446	1.216	1.758	.9657	20
50	.6080	.5712	.8208	.6959	1.437	1.218	1.751	.9628	10
35° 00′	.6109	.5736	.8192	.7002	1.428	1.221	1.743	.9599	**55° 00′**
10	.6138	.5760	.8175	.7046	1.419	1.223	1.736	.9570	50
20	.6167	.5783	.8158	.7089	1.411	1.226	1.729	.9541	40
30	.6196	.5807	.8141	.7133	1.402	1.228	1.722	.9512	30
40	.6225	.5831	.8124	.7177	1.393	1.231	1.715	.9483	20
50	.6254	.5854	.8107	.7221	1.385	1.233	1.708	.9454	10
		cos θ	sin θ	cot θ	tan θ	csc θ	sec θ	θ (radians)	θ (degrees)

TABLE 1 TRIGONOMETRIC FUNCTION VALUES ■

TABLE 1 Trigonometric Function Values (continued)

θ (degrees)	θ (radians)	sin θ	cos θ	tan θ	cot θ	sec θ	csc θ		
36° 00′	.6283	.5878	.8090	.7265	1.376	1.236	1.701	.9425	**54° 00′**
10	.6312	.5901	.8073	.7310	1.368	1.239	1.695	.9396	50
20	.6341	.5925	.8056	.7355	1.360	1.241	1.688	.9367	40
30	.6370	.5948	.8039	.7400	1.351	1.244	1.681	.9338	30
40	.6400	.5972	.8021	.7445	1.343	1.247	1.675	.9308	20
50	.6429	.5995	.8004	.7490	1.335	1.249	1.668	.9279	10
37° 00′	.6458	.6018	.7986	.7536	1.327	1.252	1.662	.9250	**53° 00′**
10	.6487	.6041	.7969	.7581	1.319	1.255	1.655	.9221	50
20	.6516	.6065	.7951	.7627	1.311	1.258	1.649	.9192	40
30	.6545	.6088	.7934	.7673	1.303	1.260	1.643	.9163	30
40	.6574	.6111	.7916	.7720	1.295	1.263	1.636	.9134	20
50	.6603	.6134	.7898	.7766	1.288	1.266	1.630	.9105	10
38° 00′	.6632	.6157	.7880	.7813	1.280	1.269	1.624	.9076	**52° 00′**
10	.6661	.6180	.7862	.7860	1.272	1.272	1.618	.9047	50
20	.6690	.6202	.7844	.7907	1.265	1.275	1.612	.9018	40
30	.6720	.6225	.7826	.7954	1.257	1.278	1.606	.8988	30
40	.6749	.6248	.7808	.8002	1.250	1.281	1.601	.8959	20
50	.6778	.6271	.7790	.8050	1.242	1.284	1.595	.8930	10
39° 00′	.6807	.6293	.7771	.8098	1.235	1.287	1.589	.8901	**51° 00′**
10	.6836	.6316	.7753	.8146	1.228	1.290	1.583	.8872	50
20	.6865	.6338	.7735	.8195	1.220	1.293	1.578	.8843	40
30	.6894	.6361	.7716	.8243	1.213	1.296	1.572	.8814	30
40	.6923	.6383	.7698	.8292	1.206	1.299	1.567	.8785	20
50	.6952	.6406	.7679	.8342	1.199	1.302	1.561	.8756	10
40° 00′	.6981	.6428	.7660	.8391	1.192	1.305	1.556	.8727	**50° 00′**
10	.7010	.6450	.7642	.8441	1.185	1.309	1.550	.8698	50
20	.7039	.6472	.7623	.8491	1.178	1.312	1.545	.8668	40
30	.7069	.6494	.7604	.8541	1.171	1.315	1.540	.8639	30
40	.7098	.6517	.7585	.8591	1.164	1.318	1.535	.8610	20
50	.7127	.6539	.7566	.8642	1.157	1.322	1.529	.8581	10
41° 00′	.7156	.6561	.7547	.8693	1.150	1.325	1.524	.8552	**49° 00′**
10	.7185	.6583	.7528	.8744	1.144	1.328	1.519	.8523	50
20	.7214	.6604	.7509	.8796	1.137	1.332	1.514	.8494	40
30	.7243	.6626	.7490	.8847	1.130	1.335	1.509	.8465	30
40	.7272	.6648	.7470	.8899	1.124	1.339	1.504	.8436	20
50	.7301	.6670	.7451	.8952	1.117	1.342	1.499	.8407	10
		cos θ	sin θ	cot θ	tan θ	csc θ	sec θ	θ (radians)	θ (degrees)

TABLE 1 Trigonometric Function Values (continued)

θ (degrees)	θ (radians)	$\sin \theta$	$\cos \theta$	$\tan \theta$	$\cot \theta$	$\sec \theta$	$\csc \theta$		
42° 00′	.7330	.6691	.7431	.9004	1.111	1.346	1.494	.8378	**48° 00′**
10	.7359	.6713	.7412	.9057	1.104	1.349	1.490	.8348	50
20	.7389	.6734	.7392	.9110	1.098	1.353	1.485	.8319	40
30	.7418	.6756	.7373	.9163	1.091	1.356	1.480	.8290	30
40	.7447	.6777	.7353	.9217	1.085	1.360	1.476	.8261	20
50	.7476	.6799	.7333	.9271	1.079	1.364	1.471	.8232	10
43° 00′	.7505	.6820	.7314	.9325	1.072	1.367	1.466	.8203	**47° 00′**
10	.7534	.6841	.7294	.9380	1.066	1.371	1.462	.8174	50
20	.7563	.6862	.7274	.9435	1.060	1.375	1.457	.8145	40
30	.7592	.6884	.7254	.9490	1.054	1.379	1.453	.8116	30
40	.7621	.6905	.7234	.9545	1.048	1.382	1.448	.8087	20
50	.7560	.6926	.7214	.9601	1.042	1.386	1.444	.8058	10
44° 00′	.7679	.6947	.7193	.9657	1.036	1.390	1.440	.8029	**46° 00′**
10	.7709	.6967	.7173	.9713	1.030	1.394	1.435	.7999	50
20	.7738	.6988	.7153	.9770	1.024	1.398	1.431	.7970	40
30	.7767	.7009	.7133	.9827	1.018	1.402	1.427	.7941	30
40	.7796	.7030	.7112	.9884	1.012	1.406	1.423	.7912	20
50	.7825	.7050	.7092	.9942	1.006	1.410	1.418	.7883	10
45° 00′	.7854	.7071	.7071	1.000	1.000	1.414	1.414		
		$\cos \theta$	$\sin \theta$	$\cot \theta$	$\tan \theta$	$\csc \theta$	$\sec \theta$	θ (radians)	θ (degrees)

TABLE 2 COMMON LOGARITHMS ∎

359

TABLE 2 Common Logarithms

x	0	1	2	3	4	5	6	7	8	9
1.0	.0000	.0043	.0086	.0128	.0170	.0212	.0253	.0294	.0334	.0374
1.1	.0414	.0453	.0492	.0531	.0569	.0607	.0645	.0682	.0719	.0755
1.2	.0792	.0828	.0864	.0899	.0934	.0969	.1004	.1038	.1072	.1106
1.3	.1139	.1173	.1206	.1239	.1271	.1303	.1335	.1367	.1399	.1430
1.4	.1461	.1492	.1523	.1553	.1584	.1614	.1644	.1673	.1703	.1732
1.5	.1761	.1790	.1818	.1847	.1875	.1903	.1931	.1959	.1987	.2014
1.6	.2041	.2068	.2095	.2122	.2148	.2175	.2201	.2227	.2253	.2279
1.7	.2304	.2330	.2355	.2380	.2405	.2430	.2455	.2480	.2504	.2529
1.8	.2553	.2577	.2601	.2625	.2648	.2672	.2695	.2718	.2742	.2765
1.9	.2788	.2810	.2833	.2856	.2878	.2900	.2923	.2945	.2967	.2989
2.0	.3010	.3032	.3054	.3075	.3096	.3118	.3139	.3160	.3181	.3201
2.1	.3222	.3243	.3263	.3284	.3304	.3324	.3345	.3365	.3385	.3404
2.2	.3424	.3444	.3464	.3483	.3502	.3522	.3541	.3560	.3579	.3598
2.3	.3617	.3636	.3655	.3674	.3692	.3711	.3729	.3747	.3766	.3784
2.4	.3802	.3820	.3838	.3856	.3874	.3892	.3909	.3927	.3945	.3962
2.5	.3979	.3997	.4014	.4031	.4048	.4065	.4082	.4099	.4116	.4133
2.6	.4150	.4166	.4183	.4200	.4216	.4232	.4249	.4265	.4281	.4298
2.7	.4314	.4330	.4346	.4362	.4378	.4393	.4409	.4425	.4440	.4456
2.8	.4472	.4487	.4502	.4518	.4533	.4548	.4564	.4579	.4594	.4609
2.9	.4624	.4639	.4654	.4669	.4683	.4698	.4713	.4728	.4742	.4757
3.0	.4771	.4786	.4800	.4814	.4829	.4843	.4857	.4871	.4886	.4900
3.1	.4914	.4928	.4942	.4955	.4969	.4983	.4997	.5011	.5024	.5038
3.2	.5051	.5065	.5079	.5092	.5105	.5119	.5132	.5145	.5159	.5172
3.3	.5185	.5198	.5211	.5224	.5237	.5250	.5263	.5276	.5289	.5302
3.4	.5315	.5328	.5340	.5353	.5366	.5378	.5391	.5403	.5416	.5428
3.5	.5441	.5453	.5465	.5478	.5490	.5502	.5514	.5527	.5539	.5551
3.6	.5563	.5575	.5587	.5599	.5611	.5623	.5635	.5647	.5658	.5670
3.7	.5682	.5694	.5705	.5717	.5729	.5740	.5752	.5763	.5775	.5786
3.8	.5798	.5809	.5821	.5832	.5843	.5855	.5866	.5877	.5888	.5899
3.9	.5911	.5922	.5933	.5944	.5955	.5966	.5977	.5988	.5999	.6010
4.0	.6021	.6031	.6042	.6053	.6064	.6075	.6085	.6096	.6107	.6117
4.1	.6128	.6138	.6149	.6160	.6170	.6180	.6191	.6201	.6212	.6222
4.2	.6232	.6243	.6253	.6263	.6274	.6284	.6294	.6304	.6314	.6325
4.3	.6335	.6345	.6355	.6365	.6375	.6385	.6395	.6405	.6415	.6425
4.4	.6435	.6444	.6454	.6464	.6474	.6484	.6493	.6503	.6513	.6522
x	0	1	2	3	4	5	6	7	8	9

TABLE 2   Common Logarithms (continued)

x	0	1	2	3	4	5	6	7	8	9
4.5	.6532	.6542	.6551	.6561	.6571	.6580	.6590	.6599	.6609	.6618
4.6	.6628	.6637	.6646	.6656	.6665	.6675	.6684	.6693	.6702	.6712
4.7	.6721	.6730	.6739	.6749	.6758	.6767	.6776	.6785	.6794	.6803
4.8	.6812	.6821	.6830	.6839	.6848	.6857	.6866	.6875	.6884	.6893
4.9	.6902	.6911	.6920	.6928	.6937	.6946	.6955	.6964	.6972	.6981
5.0	.6990	.6998	.7007	.7016	.7024	.7033	.7042	.7050	.7059	.7067
5.1	.7076	.7084	.7093	.7101	.7110	.7118	.7126	.7135	.7143	.7152
5.2	.7160	.7168	.7177	.7185	.7193	.7202	.7210	.7218	.7226	.7235
5.3	.7243	.7251	.7259	.7267	.7275	.7284	.7292	.7300	.7308	.7316
5.4	.7324	.7332	.7340	.7348	.7356	.7364	.7372	.7380	.7388	.7396
5.5	.7404	.7412	.7419	.7427	.7435	.7443	.7451	.7459	.7466	.7474
5.6	.7482	.7490	.7497	.7505	.7513	.7520	.7528	.7536	.7543	.7551
5.7	.7559	.7566	.7574	.7582	.7589	.7597	.7604	.7612	.7619	.7627
5.8	.7634	.7642	.7649	.7657	.7664	.7672	.7679	.7686	.7694	.7701
5.9	.7709	.7716	.7723	.7731	.7738	.7745	.7752	.7760	.7767	.7774
6.0	.7782	.7789	.7796	.7803	.7810	.7818	.7825	.7832	.7839	.7846
6.1	.7853	.7860	.7868	.7875	.7882	.7889	.7896	.7903	.7910	.7917
6.2	.7924	.7931	.7938	.7945	.7952	.7959	.7966	.7973	.7980	.7987
6.3	.7993	.8000	.8007	.8014	.8021	.8028	.8035	.8041	.8048	.8055
6.4	.8062	.8069	.8075	.8082	.8089	.8096	.8102	.8109	.8116	.8122
6.5	.8129	.8136	.8142	.8149	.8156	.8162	.8169	.8176	.8182	.8189
6.6	.8195	.8202	.8209	.8215	.8222	.8228	.8235	.8241	.8248	.8254
6.7	.8261	.8267	.8274	.8280	.8287	.8293	.8299	.8306	.8312	.8319
6.8	.8325	.8331	.8338	.8344	.8351	.8357	.8363	.8370	.8376	.8382
6.9	.8388	.8395	.8401	.8407	.8414	.8420	.8426	.8432	.8439	.8445
7.0	.8451	.8457	.8463	.8470	.8476	.8482	.8488	.8494	.8500	.8506
7.1	.8513	.8519	.8525	.8531	.8537	.8543	.8549	.8555	.8561	.8567
7.2	.8573	.8579	.8585	.8591	.8597	.8603	.8609	.8615	.8621	.8627
7.3	.8633	.8639	.8645	.8651	.8657	.8663	.8669	.8675	.8681	.8686
7.4	.8692	.8698	.8704	.8710	.8716	.8722	.8727	.8733	.8739	.8745
7.5	.8751	.8756	.8762	.8768	.8774	.8779	.8785	.8791	.8797	.8802
7.6	.8808	.8814	.8820	.8825	.8831	.8837	.8842	.8848	.8854	.8859
7.7	.8865	.8871	.8876	.8882	.8887	.8893	.8899	.8904	.8910	.8915
7.8	.8921	.8927	.8932	.8938	.8943	.8949	.8954	.8960	.8965	.8971
7.9	.8976	.8982	.8987	.8993	.8998	.9004	.9009	.9015	.9020	.9025
x	0	1	2	3	4	5	6	7	8	9

TABLE 3 NATURAL LOGARITHMS AND POWERS OF e ■

361

TABLE 2   Common Logarithms (continued)

x	0	1	2	3	4	5	6	7	8	9
8.0	.9031	.9036	.9042	.9047	.9053	.9058	.9063	.9069	.9074	.9079
8.1	.9085	.9090	.9096	.9101	.9106	.9112	.9117	.9122	.9128	.9133
8.2	.9138	.9143	.9149	.9154	.9159	.9165	.9170	.9175	.9180	.9186
8.3	.9191	.9196	.9201	.9206	.9212	.9217	.9222	.9227	.9232	.9238
8.4	.9243	.9248	.9253	.9258	.9263	.9269	.9274	.9279	.9284	.9289
8.5	.9294	.9299	.9304	.9309	.9315	.9320	.9325	.9330	.9335	.9340
8.6	.9345	.9350	.9355	.9360	.9365	.9370	.9375	.9380	.9385	.9390
8.7	.9395	.9400	.9405	.9410	.9415	.9420	.9425	.9430	.9435	.9440
8.8	.9445	.9450	.9455	.9460	.9465	.9469	.9474	.9479	.9484	.9489
8.9	.9494	.9499	.9504	.9509	.9513	.9518	.9523	.9528	.9533	.9538
9.0	.9542	.9547	.9552	.9557	.9562	.9566	.9571	.9576	.9581	.9586
9.1	.9590	.9595	.9600	.9605	.9609	.9614	.9619	.9624	.9628	.9633
9.2	.9638	.9643	.9647	.9652	.9657	.9661	.9666	.9671	.9675	.9680
9.3	.9685	.9689	.9694	.9699	.9703	.9708	.9713	.9717	.9722	.9727
9.4	.9731	.9736	.9741	.9745	.9750	.9754	.9759	.9763	.9768	.9773
9.5	.9777	.9782	.9786	.9791	.9795	.9800	.9805	.9809	.9814	.9818
9.6	.9823	.9827	.9832	.9836	.9841	.9845	.9850	.9854	.9859	.9863
9.7	.9868	.9872	.9877	.9881	.9886	.9890	.9894	.9899	.9903	.9908
9.8	.9912	.9917	.9921	.9926	.9930	.9934	.9939	.9943	.9948	.9952
9.9	.9956	.9961	.9965	.9969	.9974	.9978	.9983	.9987	.9991	.9996
x	0	1	2	3	4	5	6	7	8	9

TABLE 3   Natural Logarithms and Powers of e

x	e^x	e^{-x}	$\ln x$	x	e^x	e^{-x}	$\ln x$
0.00	1.00000	1.00000		1.60	4.95302	0.20189	0.4700
0.01	1.01005	0.99004	−4.6052	1.70	5.47394	0.18268	0.5306
0.02	1.02020	0.98019	−3.9120	1.80	6.04964	0.16529	0.5878
0.03	1.03045	0.97044	−3.5066	1.90	6.68589	0.14956	0.6419
0.04	1.04081	0.96078	−3.2189	2.00	7.38905	0.13533	0.6931
0.05	1.05127	0.95122	−2.9957				
0.06	1.06183	0.94176	−2.8134	2.10	8.16616	0.12245	0.7419
0.07	1.07250	0.93239	−2.6593	2.20	9.02500	0.11080	0.7885
0.08	1.08328	0.92311	−2.5257	2.30	9.97417	0.10025	0.8329
0.09	1.09417	0.91393	−2.4079	2.40	11.02316	0.09071	0.8755
1.10	1.10517	0.90483	−2.3026	2.50	12.18248	0.08208	0.9163
				2.60	13.46372	0.07427	0.9555

TABLE 3   Natural Logarithms and Powers of e (continued)

x	e^x	e^{-x}	$\ln x$	x	e^x	e^{-x}	$\ln x$
0.11	1.11628	0.89583	−2.2073	2.70	14.87971	0.06720	0.9933
0.12	1.12750	0.88692	−2.1203	2.80	16.44463	0.06081	1.0296
0.13	1.13883	0.87810	−2.0402	2.90	18.17412	0.05502	1.0647
0.14	1.15027	0.86936	−1.9661	3.00	20.08551	0.04978	1.0986
0.15	1.16183	0.86071	−1.8971				
0.16	1.17351	0.85214	−1.8326	3.50	33.11545	0.03020	1.2528
0.17	1.18530	0.84366	−1.7720	4.00	54.59815	0.01832	1.3863
0.18	1.19722	0.83527	−1.7148	4.50	90.01713	0.01111	1.5041
0.19	1.20925	0.82696	−1.6607				
				5.00	148.41316	0.00674	1.6094
0.20	1.22140	0.81873	−1.6094	5.50	224.69193	0.00409	1.7047
0.30	1.34985	0.74081	−1.2040				
0.40	1.49182	0.67032	−0.9163	6.00	403.42879	0.00248	1.7918
0.50	1.64872	0.60653	−0.6931	6.50	665.14163	0.00150	1.8718
0.60	1.82211	0.54881	−0.5108				
0.70	2.01375	0.49658	−0.3567	7.00	1096.63316	0.00091	1.9459
0.80	2.22554	0.44932	−0.2231	7.50	1808.04241	0.00055	2.0149
0.90	2.45960	0.40656	−0.1054				
				8.00	2980.95799	0.00034	2.0794
				8.50	4914.76884	0.00020	2.1401
1.00	2.71828	0.36787	0.0000				
1.10	3.00416	0.33287	0.0953	9.00	8103.08392	0.00012	2.1972
1.20	3.32011	0.30119	0.1823	9.50	13359.72683	0.00007	2.2513
1.30	3.66929	0.27253	0.2624				
1.40	4.05519	0.24659	0.3365	10.00	22026.46579	0.00005	2.3026
1.50	4.48168	0.22313	0.4055				

Additional Natural Logarithms

x	$\ln x$	x	$\ln x$	x	$\ln x$
11	2.3979	20	2.9957	70	4.2485
12	2.4849	25	3.2189	75	4.3175
13	2.5649	30	3.4012	80	4.3820
14	2.6391	35	3.5553	85	4.4427
		40	3.6889	90	4.4998
15	2.7081				
16	2.7726	45	3.8067	95	4.5539
17	2.8332	50	3.9120	100	4.6052
18	2.8904	55	4.0073		
19	2.9444	60	4.0943	1000	6.9078
		65	4.1744		

ANSWERS TO SELECTED EXERCISES

TO THE STUDENT

If you need further help with trigonometry, you may want to obtain a copy of the *Student's Solution Manual* that goes with this book. It contains solutions to all the odd-numbered exercises and all the chapter test exercises. Your college bookstore either has the *Manual* or can order it for you.

In this section we provide the answers that we think most students will obtain when they work the exercises using the methods explained in the text. If your answer does not look exactly like the one given here, it is not necessarily wrong. In many cases there are equivalent forms of the answer. For example, if the answer section shows 3/4 and your answer is .75, you have obtained the correct answer but written it in a different (yet equivalent) form. Unless the directions specify otherwise, .75 is just as valid an answer as 3/4. In general, if your answer does not agree with the one given in the text, see whether it can be transformed into the other form. If it can, then it is the correct answer. If you still have doubts, talk with your instructor.

■ CHAPTER 1   THE TRIGONOMETRIC FUNCTIONS

1.1 EXERCISES   (PAGE 8)

1. II   **3.** III   **5.** none   **7.** none   **9.** II   **11.** IV   **13.** I and III   **15.** $3\sqrt{2}$ **17.** $5\sqrt{2}$ **19.** $\sqrt{34}$
21. $\sqrt{29}$ **23.** 4   **25.** $\sqrt{133}$ **27.** yes   **29.** yes   **31.** no   **33.** yes   **35.** yes   **39.** 5, 12, 13;
$5^2 + 12^2 = 169$ and $13^2 = 169$ **41.** 8, 6, 10; $8^2 + 6^2 = 100$ and $10^2 = 100$ **43.** 15, 8, 17; $15^2 + 8^2 = 289$
and $17^2 = 289$ **45.** yes   **47.** yes   **49.** no   **51.** 3, -5 **53.** 13, -3

55. $(x + 5)^2 + (y - 6)^2 = 9$

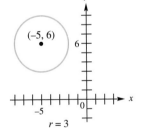

57. \$74.52   **59.** 217.9 m   **61.** $(9, \infty)$ **63.** $(-\infty, -8)$ **65.** $[10, \infty)$ **67.** $[8, 13]$ **69.** $(2, \infty)$
71. $(1, 4]$ **75.** -10 **77.** 0   **79.** -64 **81.** $-2m^2 - 4m + 6$ **83.** $-2p^2 + 4p + 6$
In Exercises 85–101, we give the domain, then the range, and then whether it is a function.   **85.** $(-\infty, \infty)$; $(-\infty, \infty)$;
function   **87.** $(-\infty, \infty)$; $[-5, \infty)$;   function   **89.** $(-\infty, \infty)$; $[-5, \infty)$;   function   **91.** $(-\infty, 0]$; $(-\infty, \infty)$;
not a function   **93.** $[2, \infty)$; $[0, \infty)$;   function   **95.** $[-1, 1]$; $[0, 1]$;   function   **97.** $[-5, \infty)$; $[0, \infty)$;   function
99. $(-\infty, -3] \cup [3, \infty)$; $(-\infty, \infty)$;   not a function   **101.** $(-\infty, \infty)$; $(-\infty, \infty)$;   function   **103.** $(-\infty, -1) \cup (-1, \infty)$
105. $(-\infty, -8/3) \cup (-8/3, -1/5) \cup (-1/5, \infty)$ **107.** $(-\infty, \infty)$ **109.** $(5/2, \infty)$

1.2 EXERCISES (PAGE 18)

1. 70°; 110° **3.** 55°; 35° **5.** 100°; 80° **7.** 90 − x degrees **9.** 45° **11.** 158° 47′ **13.** 112° 42′
15. 38° 32′ **17.** 27° 17′ **19.** 53° 41′ 13″ **21.** 59° 17′ 23″ **23.** 20.9° **25.** 91.598° **27.** 274.316°
29. 31° 25′ 47″ **31.** 89° 54′ 01″ **33.** 178° 35′ 58″ **35.** 320° **37.** 235° **39.** 90° **41.** 179°
43. 130° **45.** 94.5937° **47.** 30° + $n \cdot 360°$ **49.** 60° + $n \cdot 360°$ **51.** 135° + $n \cdot 360°$
53. −90° + $n \cdot 360°$ Angles other than the ones given are possible in Exercises 57–71.

57. 435°; −285° quadrant I   **59.** 482°; −238°; quadrant II   **61.** 594°; −126°; quadrant III

63. 660° −60°; quadrant IV   **65.** 78°; −282°; quadrant I   **67.** 152°; −208°; quadrant II

69. 308°; −412°; quadrant IV   **71.** 201°; −519°; quadrant III   **73.** $r = 3\sqrt{2}$

75. $r = \sqrt{34}$ **77.** $r = 4$ **79.** 1800°
81. 12.5 rotations per hr

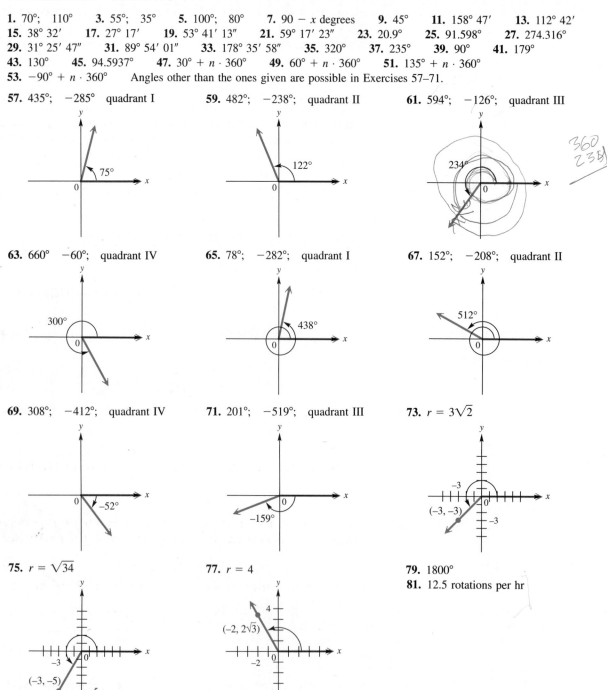

1.3 EXERCISES (PAGE 25)

1. 51°; 51° **3.** 50°; 60°; 70° **5.** 60°; 60°; 60° **7.** 65°; 115° **9.** 49°; 49° **11.** 48°; 132°
13. 91° **15.** 2° 29′ **17.** 24° 48′ 56″ **21.** Answers are given in numerical order: 55°, 65°, 60°, 65°, 60°, 120°,
60°, 60°, 55°, 55°. **23.** right; scalene **25.** acute; equilateral **27.** right; scalene **29.** right; isosceles
31. obtuse; scalene **33.** acute; isosceles **39.** A and P; C and R; B and Q; AC and PR;
CB and RQ; AB and PQ **41.** H and F; K and E; HGK and FGE; HK and FE; GK and GE; HG and FG
43. $P = 78°$; $M = 46°$; $A = N = 56°$ **45.** $T = 74°$; $Y = 28°$; $Z = W = 78°$ **47.** $T = 20°$; $V = 64°$;
$R = U = 96°$ **49.** $a = 20$; $b = 15$ **51.** $a = 6$; $b = 7\ 1/2$ **53.** $x = 6$ **55.** 30 m **57.** 500 m, 700 m
59. 112.5 ft **61.** $x = 110$ **63.** $c = 111.1$ **65.** unknown side in first quadrilateral is 40 cm; unknown sides
in second quadrilateral are 27 cm and 36 cm. **67.** $x = 10$; $y = 5$

1.4 EXERCISES (PAGE 34)

1.

3.

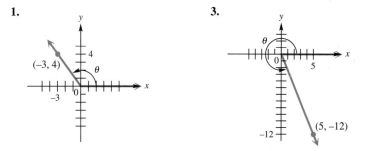

In Exercises 5–29, we give, in order, sine, cosine, tangent, cotangent, secant, and cosecant.
5. 4/5; $-3/5$; $-4/3$; $-3/4$; $-5/3$; 5/4 **7.** $-12/13$; 5/13; $-12/5$; $-5/12$; 13/5; $-13/12$ **9.** 4/5;
3/5; 4/3; 3/4; 5/3; 5/4 **11.** 24/25; $-7/25$; $-24/7$; $-7/24$; $-25/7$; 25/24 **13.** 1; 0; undefined; 0;
undefined; 1 **15.** 0; 1; 0; undefined; 1; undefined **17.** $\sqrt{3}/2$; 1/2; $\sqrt{3}$; $\sqrt{3}/3$; 2; $2\sqrt{3}/3$
19. $-1/2$; $\sqrt{3}/2$; $-\sqrt{3}/3$; $-\sqrt{3}$; $2\sqrt{3}/3$; -2 **21.** $-\sqrt{2}/2$; $\sqrt{2}/2$; -1; -1; $\sqrt{2}$; $-\sqrt{2}$
23. $-2/3$; $\sqrt{5}/3$; $-2\sqrt{5}/5$; $-\sqrt{5}/2$; $3\sqrt{5}/5$; $-3/2$ **25.** $\sqrt{3}/4$; $-\sqrt{13}/4$; $-\sqrt{39}/13$; $-\sqrt{39}/3$;
$-4\sqrt{13}/13$; $4\sqrt{3}/3$ **27.** $-\sqrt{10}/5$; $\sqrt{15}/5$; $-\sqrt{6}/3$; $-\sqrt{6}/2$; $\sqrt{15}/3$; $-\sqrt{10}/2$
29. $-.34727$; .93777; $-.37031$; -2.7004; 1.0664; -2.8796 **35.** positive **37.** negative **39.** positive
41. positive **43.** negative **45.** negative
In Exercises 47–51, we give, in order, sine, cosine, tangent, cotangent, secant, cosecant.
47. $-2\sqrt{5}/5$; $\sqrt{5}/5$; -2;   **49.** $-4\sqrt{65}/65$; $-7\sqrt{65}/65$; 4/7;   **51.** $5\sqrt{34}/34$; $-3\sqrt{34}/34$; $-5/3$;
$-1/2$; 5; $-5/2$ 7/4; $-\sqrt{65}/7$; $-\sqrt{65}/4$ $-3/5$; $-\sqrt{34}/3$; $\sqrt{34}/5$

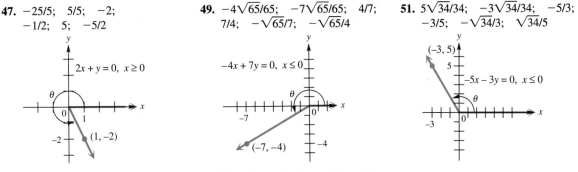

53. -3 **55.** -3 **57.** 5 **59.** 1 **61.** -1 **63.** 0 **65.** undefined

1.5 EXERCISES (PAGE 44)

1. 1/3   **3.** −5   **5.** $2\sqrt{2}$ **7.** $-3\sqrt{5}/5$ **9.** .70069071   **11.** 2.2778902   **15.** $\cot\theta = 1/\tan\theta$; $\cot\theta\tan\theta = 1$ **17.** 1/2   **19.** $\sqrt{3}$ **21.** −100   **23.** 5°   **25.** 3°   **27.** 4°   **29.** II   **31.** III   **33.** IV   **35.** II or IV   **37.** I or II   **39.** I or III   In Exercises 41–51, we give, in order, sine and cosecant, cosine and secant, and tangent and cotangent.   **41.** +; +; +   **43.** −; −; +   **45.** −; +; −   **47.** +; +; +   **49.** −; +; −   **51.** −; −; +   **53.** impossible   **55.** possible   **57.** impossible   **59.** possible   **61.** possible   **63.** impossible   **65.** possible   **67.** impossible   **69.** $-\sqrt{5}/3$ **71.** $-\sqrt{5}/2$ **73.** −4/3   **75.** $-\sqrt{3}/2$ **77.** $2\sqrt{11}/7$ **79.** 3.44701905   **81.** −.56616682   In Exercises 83–95, we give, in order, sine, cosine, tangent, cotangent, secant, and cosecant.   **83.** −4/5; −3/5; 4/3; 3/4; −5/3; −5/4   **85.** 7/25; −24/25; −7/24; −24/7; −25/24; 25/7   **87.** 1/2; $-\sqrt{3}/2$; $-\sqrt{3}/3$; $-\sqrt{3}$; $-2\sqrt{3}/3$; 2   **89.** $8\sqrt{67}/67$; $\sqrt{201}/67$; $8\sqrt{3}/3$; $\sqrt{3}/8$; $\sqrt{201}/3$; $\sqrt{67}/8$ **91.** $3\sqrt{13}/13$; $2\sqrt{13}/13$; 3/2; 2/3; $\sqrt{13}/2$; $\sqrt{13}/3$ **93.** .164215; −.986425; −.166475; −6.00691; −1.01376; 6.08958   **95.** a; $\sqrt{1-a^2}$; $a\sqrt{1-a^2}/(1-a^2)$; $\sqrt{1-a^2}/a$; $\sqrt{1-a^2}/(1-a^2)$; $1/a$ **99.** false;   for example, if $\theta = 180°$, $\sin 180° + \cos 180° = 0 + (-1) = -1 \neq 1$

CHAPTER 1 REVIEW EXERCISES (PAGE 48)

1. $(-\infty, -4]$ **3.** 5   **5.** yes   **7.** 2   **9.** $-a^2 + 3a + 2$ In Exercises 11–17, we give the domain, then the range, and then tell whether it is a function.   **11.** $(-\infty, \infty)$; $(-\infty, \infty)$; function   **13.** $(-\infty, \infty)$; $[0, \infty)$; function   **15.** $[-1, \infty)$; $(-\infty, \infty)$; not a function   **17.** $[-3, 3]$; $[-5, 5]$; not a function   **19.** 309°   **21.** 72°   **23.** 1280°   **25.** 47.420°   **27.** 74° 17′ 54″   **29.** 183° 05′ 50″   **31.** 58°; 58°   **33.** $V = 41°$; $Z = 32°$; $Y = U = 107°$ **35.** $m = 45$; $n = 60$ **37.** $r = 108/7$ **39.** proportional; equal   In Exercises 41–51 and 59–63, we give, in order, sine, cosine, tangent, cotangent, secant, cosecant.   **41.** $-\sqrt{2}/2$; $-\sqrt{2}/2$; 1; 1; $-\sqrt{2}$; $-\sqrt{2}$ **43.** 0; −1; 0; undefined; −1; undefined   **45.** 15/17; −8/17; −15/8; −8/15; −17/8; 17/15   **47.** $-5\sqrt{26}/26$; $\sqrt{26}/26$; −5; −1/5; $\sqrt{26}$; $-\sqrt{26}/5$ **49.** −1/2; $\sqrt{3}/2$; $-\sqrt{3}/3$; $-\sqrt{3}$; $2\sqrt{3}/3$; −2   **51.** $5\sqrt{34}/34$; $3\sqrt{34}/34$; 5/3; 3/5; $\sqrt{34}/3$; $\sqrt{34}/5$ **53.** −6   **55.** possible   **57.** possible   **59.** $\sqrt{3}/5$; $-\sqrt{22}/5$; $-\sqrt{66}/22$; $-\sqrt{66}/3$; $-5\sqrt{22}/22$; $5\sqrt{3}/3$ **61.** $-2\sqrt{5}/5$; $-\sqrt{5}/5$; 2; 1/2; $-\sqrt{5}$; $-\sqrt{5}/2$ **63.** −2/5; $-\sqrt{21}/5$; $2\sqrt{21}/21$; $\sqrt{21}/2$; $-5\sqrt{21}/21$; −5/2   **65.** IV; negative

■ CHAPTER 2 ACUTE ANGLES AND RIGHT TRIANGLES

2.1 EXERCISES (PAGE 58)

In Exercises 1–9, we give, in order, sine, cosine, tangent, cotangent, secant, and cosecant.   **1.** 3/5; 4/5; 3/4; 4/3; 5/4; 5/3   **3.** 21/29; 20/29; 21/20; 20/21; 29/20; 29/21   **5.** n/p; m/p; n/m; m/n; p/m; p/n **7.** .7593; .6508; 1.1667; .8571; 1.5365; 1.3170   **9.** .8717; .4901; 1.7785; .5623; 2.0403; 1.1472   In Exercise 11, we give, in order, the unknown side, sine, cosine, tangent, cotangent, secant, and cosecant.   **11.** $a = \sqrt{95}$; 7/12; $\sqrt{95}/12$; $7\sqrt{95}/95$; $\sqrt{95}/7$; $12\sqrt{95}/95$; 12/7   **15.** cot 40°   **17.** sec 43°   **19.** cos 51° 31′   **21.** $\cos(90° - \gamma)$ **23.** $\sin(70° - \alpha)$ **25.** 30°   **27.** 20°   **29.** 12°   **31.** 70°   **33.** true   **35.** false   **37.** true   **39.** true   **41.** $\sqrt{3}/3$ **43.** 1/2   **45.** $2\sqrt{3}/3$ **47.** $\sqrt{2}$ **49.** $\sqrt{2}/2$ **51.** 1   **53.** $\sqrt{3}/2$ **55.** $\sqrt{3}$ **57.** 2   **59.** tangent and cotangent   **61. (a)** 60° **(b)** k **(c)** $k\sqrt{3}$ **(d)** 2; $\sqrt{3}$; 30°; 60°   **63.** $a = 12$; $b = 12\sqrt{3}$; $d = 12\sqrt{3}$; $c = 12\sqrt{6}$ **65.** $m = 7\sqrt{3}/3$; $a = 14\sqrt{3}/3$; $n = 14\sqrt{3}/3$; $q = 14\sqrt{6}/3$ **67.** $s^2\sqrt{3}/4$

2.2 Exercises (Page 67)

1. 60° **3.** 45° **5.** 60° **7.** 45° **9.** 82° **11.** 74° 30′ **13.** 60° **15.** 60° **19.** 255°
21. 98° 20′ **23.** 38° 10′ In Exercises 25–43, we give, in order, sine, cosine, tangent, cotangent, secant, and cosecant. **25.** $\sqrt{3}/2$; $-1/2$; $-\sqrt{3}$; $-\sqrt{3}/3$; -2; $2\sqrt{3}/3$ **27.** $1/2$; $-\sqrt{3}/2$; $-\sqrt{3}/3$; $-\sqrt{3}$; $-2\sqrt{3}/3$; 2 **29.** $-\sqrt{3}/2$; $-1/2$; $\sqrt{3}$; $\sqrt{3}/3$; -2; $-2\sqrt{3}/3$ **31.** $-1/2$; $\sqrt{3}/2$; $-\sqrt{3}/3$; $-\sqrt{3}$; $2\sqrt{3}/3$; -2 **33.** $\sqrt{3}/2$; $1/2$; $\sqrt{3}$; $\sqrt{3}/3$; 2; $2\sqrt{3}/3$ **35.** $1/2$; $-\sqrt{3}/2$; $-\sqrt{3}/3$; $-\sqrt{3}$; $-2\sqrt{3}/3$; 2 **37.** $1/2$; $\sqrt{3}/2$; $\sqrt{3}/3$; $\sqrt{3}$; $2\sqrt{3}/3$; 2 **39.** $\sqrt{3}/2$; $1/2$; $\sqrt{3}$; $\sqrt{3}/3$; 2; $2\sqrt{3}/3$
41. $-1/2$; $\sqrt{3}/2$; $-\sqrt{3}/3$; $-\sqrt{3}$; $2\sqrt{3}/3$; -2 **43.** $\sqrt{3}/2$; $1/2$; $\sqrt{3}$; $\sqrt{3}/3$; 2; $2\sqrt{3}/3$
45. $\sqrt{3}/3$; $\sqrt{3}$ **47.** $\sqrt{3}/2$; $\sqrt{3}/3$; $2\sqrt{3}/3$ **49.** -1; -1 **51.** $-\sqrt{3}/2$; $-2\sqrt{3}/3$ **53.** 30°; 150°
55. 60°; 240° **57.** 120°; 240° **59.** 240°; 300° **61.** 135°; 315° **63.** 90°; 270° **65.** 0°; 180°
67. yes **69.** positive **71.** negative **73.** negative **75.** 1 **77.** 23/4 **79.** $1/2 + \sqrt{3}$ **81.** $-29/12$
83. false; $\dfrac{1 + \sqrt{3}}{2} \neq 1$ **85.** true **87.** false; $\sqrt{3}/2 \neq 0$ **89.** true

2.3 Exercises (Page 72)

1. .62478851 **3.** 1.1369414 **5.** .85264016 **7.** .31210365 **9.** −.79543592 **11.** .01163527
13. −.41204452 **15.** −.32281638 **17.** −.99691733 **19.** 1.1415183 **21.** −3.1791978
23. −.32170674 **25.** .86288438 **27.** −6.0483717 **29.** −.20092405 **31.** −.35830784 **33.** .27256013
35. −.72306685 **37.** −.26675715 **39.** −1.8926337 **43.** 57.997172° **45.** 30.502748° **47.** 46.173581°
49. 81.168073° **51.** 46.593881°; 313.40612° **53.** 24.392576°; 155.60742° **55.** 41.248183°; 221.24818°
57. −1 **59.** 1 **61.** 2×10^{8} m per sec **63.** 19° **65.** 48.7° **67.** false **69.** true **71.** false
73. false

Trigonometric Function Values: Exact and Approximate (Page 73)

1. $-\sqrt{3}/2$ **3.** −.64278761 **5.** $-\sqrt{3}$ **7.** $-\sqrt{3}/2$ **9.** −.17364818 **11.** $\sqrt{3}/3$ **13.** $-2\sqrt{3}/3$
15. 1.5557238 **17.** −1 **19.** −.17632698

2.4 Exercises (Page 78)

1. 28,999.5 to 29,000.5 **3.** 1649.5 to 1650.5; 159.5 to 160.5 **5.** Both 2 and 65 are exact measurements.
7. $B = 53°\ 40'$; $a = 571$ m; $b = 777$ m **9.** $M = 38.8°$; $n = 154$ m; $p = 198$ m **11.** $A = 47.9108°$;
$c = 84.816$ cm; $a = 62.942$ cm **15.** $B = 62°\ 00'$; $a = 8.17$ ft; $b = 15.4$ ft **17.** $A = 17°\ 00'$;
$a = 39.1$ in; $c = 134$ in **19.** $c = 85.9$ yd; $A = 62°\ 50'$; $B = 27°\ 10'$ **21.** $b = 42.3$ cm; $A = 24°\ 10'$;
$B = 65°\ 50'$ **23.** $B = 36°\ 36'$; $a = 310.8$ ft; $b = 230.8$ ft **25.** $A = 50°\ 51'$; $a = .4832$ m; $b = .3934$ m
27. $A = 71°\ 36'$; $B = 18°\ 24'$; $a = 7.413$ m **29.** $A = 47.568°$; $b = 143.97$ m; $c = 213.38$ m
31. $B = 32.791°$; $a = 156.77$ cm; $b = 101.00$ cm **33.** $a = 115.072$ m; $A = 33°\ 29.40'$; $B = 56°\ 30.60'$
37. 9.3 m **39.** 67° 10′ **41.** 19.46 ft **43.** 33.4 m **45.** 26.92 in **47.** 13.3 ft **49.** 37° 40′
51. 42,600 ft **53.** 26° 20′ **55.** 54° 40′ **57.** 6.993792×10^{9} mi **59.** 79°

2.5 EXERCISES (PAGE 86)

1. 31° 20′   **3.** 3.3 ft   **5.** 10.8 ft   **7.** 84.7 m   **9.** 5.29° or 5° 20′   **11.** 446

13. 114 ft   **15.** 5.18 m

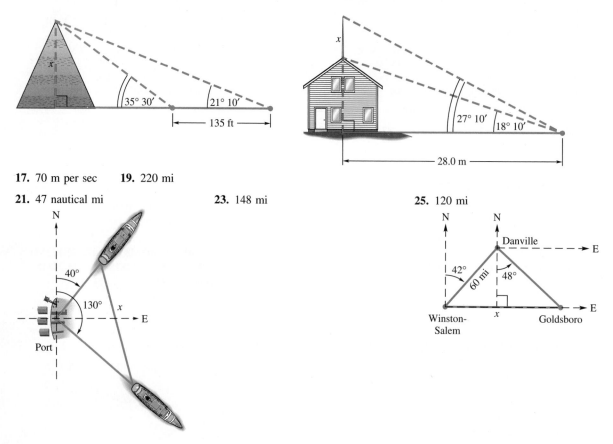

17. 70 m per sec   **19.** 220 mi

21. 47 nautical mi   **23.** 148 mi   **25.** 120 mi

CHAPTER 2 REVIEW EXERCISES (PAGE 91)

In Exercises 1 and 17–21, we give, in order, the sine, cosine, tangent, cotangent, secant, and cosecant.   **1.** 60/61; 11/61; 60/11; 11/60; 61/11; 61/60   **3.** 10°   **5.** 7°   **7.** true   **9.** true   **11.** 39°   **13.** 38.94°
17. $\sqrt{3}/2$; $-1/2$; $-\sqrt{3}$; $-\sqrt{3}/3$; -2; $2\sqrt{3}/3$ **19.** $-\sqrt{3}/2$; $1/2$; $-\sqrt{3}$; $-\sqrt{3}/3$; 2; $-2\sqrt{3}/3$
21. $\sqrt{2}/2$; $-\sqrt{2}/2$; -1; -1; $-\sqrt{2}$; $\sqrt{2}$ **23.** 210°; 330°   **25.** 135°; 315°   **27.** 1   **29.** $-5/3$
31. .95371695   **33.** $-.71592968$ **35.** 1.5576601   **37.** 1.9362132   **39.** .99984900   **43.** 55.673870°
45. 12.733938°   **47.** 63.008286°   **49.** 47.1°; 132.9°   **51.** false; 1.4088321 ≠ 1   **53.** true
55. $B = 31° 30′$; $a = 638$; $b = 391$ **57.** $B = 50.28°$; $a = 32.38$ m; $c = 50.66$ m; $C = 90°$ **59.** 73.7 ft
61. 18.75 cm   **63.** 1200 m   **65.** 140 mi

∎ CHAPTER 3   RADIAN MEASURE AND THE CIRCULAR FUNCTIONS

3.1 EXERCISES   (PAGE 100)

1. $\pi/3$ **3.** $\pi/2$ **5.** $5\pi/6$ **7.** $7\pi/6$ **9.** $5\pi/3$ **11.** $5\pi/2$ **13.** $\pi/9$ **17.** 60°   **19.** 315°
21. 330°   **23.** −30°   **25.** 288°   **27.** 132°   **29.** 63°   **31.** 66°   **33.** .68   **35.** .742   **37.** 2.43
39. 1.122   **41.** .9847   **43.** .832391   **45.** −.518537   **47.** −3.66119   **49.** 114° 35′   **51.** 99° 42′
53. 5° 14′   **55.** 564° 14′   **57.** −198° 55′   **61.** $\sqrt{3}/2$ **63.** 1   **65.** $2\sqrt{3}/3$ **67.** 1   **69.** $-\sqrt{3}$
71. 1/2   **73.** −1   **75.** $-\sqrt{3}/2$ **77.** 1   **79.** 0   **81.** $-\sqrt{3}$ **83.** 1/2   **85.** We begin the answers with
the blank next to 30°, and then proceed counterclockwise from there: $\pi/6$;   45°; $\pi/3$;   120°;   135°; $5\pi/6$; π;
$7\pi/6$; $5\pi/4$;   240°;   300°; $7\pi/4$; $11\pi/6$.   **87. (a)** 4π **(b)** $2\pi/3$

3.2 EXERCISES   (PAGE 105)

1. 25.1 in   **3.** 25.8 cm   **5.** 318 m   **7.** 5.05 m   **9.** 53.429 m   **11.** $s = r\gamma\pi/180$ **13.** 1200 km
15. 3500 km   **17.** 5900 km   **19.** 7300 km   **21. (a)** 11.6 in **(b)** 37° 05′   **23.** 38.5°   **25.** 146 in
27. 21 m   **29.** .20 km   **31.** 2100 mi   **33.** 850 ft   **35.** 1120 m^2 **37.** 1300 cm^2 **39.** 114 cm^2
41. 359 m^2 **43.** 760.34 m^2 **45.** $A = r^2\,\gamma\,\pi/360$ **47. (a)** 11.25°; $\pi/16$ **(b)** 480 ft **(c)** 15 ft **(d)** 570 ft^2
49. 7800 mi   **51.** $V = \dfrac{1}{2}\theta(r_1{}^2 - r_2{}^2)h$ (θ in radians)

3.3 EXERCISES   (PAGE 115)

1. .48775041   **3.** .53171474   **5.** .73135046   **7.** .80036052   **9.** .99813420   **11.** 1.0170372
13. .96364232   **15.** .42442278   **17.** 1.2131367   **19.** −.44357977   **21.** −.75469733   **23.** −.99668945
25. −3.8665127   **27.** −.30904176   **29.** .31728667   **31.** 2.2812466   **33.** 14.333769   **35.** 4.6358445
37. $\pi/3$ **39.** $\pi/6$ **41.** $\pi/3$ **43.** $\pi/4$ **45.** $\pi/6$ **47.** $\pi/12$ **49.** .28376614   **51.** .41352850
53. −1/2   **55.** −1   **57.** $-\sqrt{3}/2$ **59.** −2   **61.** $-\sqrt{3}$ **63.** −1/2   **65.** $2\sqrt{3}/3$ **67.** $-\sqrt{2}/2$
69. .20952066   **71.** 1.4429646   **73.** 1.0151896   **75.** .95409991   **77.** 1.4747226   **79.** 1.0181269
81. $\pi/3$ **83.** $5\pi/6$ **85.** $4\pi/3$ **87.** $7\pi/4$ **89.** $3\pi/4$ **91.** (−.80114362, .59847214)
93. (.43854733, −.89870810)   **95.** I   **97.** IV   **99. (a)** 30° **(b)** 60° **(c)** 75° **(d)** 86° **(e)** 86° **(f)** 60°

3.4 EXERCISES   (PAGE 119)

1. $5\pi/4$ radians   **3.** $\pi/25$ radians per sec   **5.** 9 min   **7.** 10.768 radians   **9.** $72\pi/5$ cm per sec
11. 6 radians per sec   **13.** 9.29755 cm per sec   **15.** 18π cm   **17.** 12 sec   **19.** $3\pi/32$ radians per sec
23. $\pi/6$ radians per hr   **25.** $\pi/30$ radians per sec   **27.** $7\pi/30$ cm per min   **29.** 168π m per min
31. 1500π m per min   **33.** about 29 sec   **35. (a)** 2π radians per day; $\pi/12$ radians per hr   **(b)** 0
(c) $12{,}800\pi$ km per day   or   about 533π km per hr   **(d)** 9050π km per day   or   about 377π km per hr
37. .24 radian per sec   **39.** .303 m

CHAPTER 3 REVIEW EXERCISES   (PAGE 123)

1. $\pi/4$ **3.** $4\pi/9$ **5.** $11\pi/6$ **7.** $17\pi/3$ **9.** 225°   **11.** 480°   **13.** −110°   **15.** 168°   **17.** $\sqrt{3}$
19. −1/2   **21.** $-\sqrt{3}$ **23.** 2   **25.** 35.8 cm   **27.** 7.683 cm   **29.** 273 m^2 **31.** 4500 km
33. .86602663   **35.** .97030688   **37.** 1.6755332   **39.** 1.1311944   **41.** $\pi/6$ **43.** $\pi/3$ **45.** 1.2246633
47. $\sqrt{3}$ **49.** −1/2   **51.** $-\sqrt{3}$ **53.** 2   **55.** .38974894   **57.** .51489440   **59.** 1.1053762   **61.** $\pi/4$
63. $7\pi/6$ **65.** 15/32 sec   **67.** $\pi/20$ radians per sec   **69.** 285.3 cm

4.1 EXERCISES   (PAGE 135)

1. 2

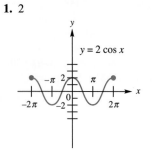

3. 2/3

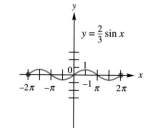

5. 1

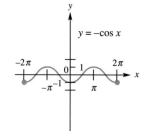

7. 2

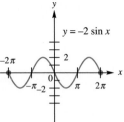

9. 4π;   1

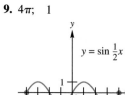

11. 6π;   1

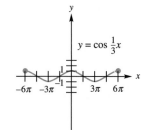

13. $2\pi/3$;   1

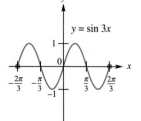

15. π;   1

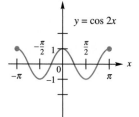

17. $\pi/2$;   1

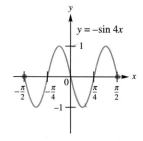

19. 8π;   2

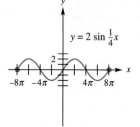

21. $2\pi/3$;   2

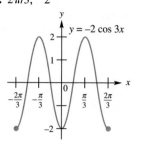

23. 2;   1

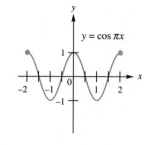

25. G   **27.** E   **29.** B   **31.** F

35.   **37.**

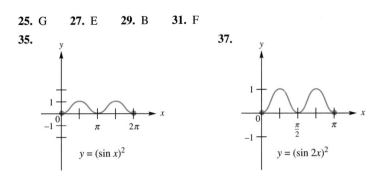

$y = (\sin x)^2$ $y = (\sin 2x)^2$

39. (a) 20   **(b)** 75   **41. (a)** about 2 hr   **(b)** 1 year   **43.** 1;   240°   or $4\pi/3$

45. (a) 5;   1/60   **(b)** 60   **(c)** 5;   1.545;   −4.045;   −4.045;   1.545   **(d)**

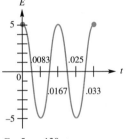

$E = 5 \cos 120\pi t$

4.2 EXERCISES (PAGE 144)

1. D   **3.** H   **5.** B   **7.** F   **9.** 1; $\pi/2$; $\pi/2$;   left;   4;   down   **11.** 2; 2π;   none; π to the right

13. 4; 4π;   none; π to the left   **15.** 3; π;   none; $\pi/4$ to the right   **17.** 1; $2\pi/3$;   up 2; $\pi/15$ to the right

19.   **21.**   **23.**

$y = \cos\left(x - \frac{\pi}{2}\right)$ $y = \sin\left(x + \frac{\pi}{4}\right)$ $y = 2\cos\left(x - \frac{\pi}{3}\right)$

25.   **27.**   **29.**

$y = \frac{3}{2}\sin 2\left(x + \frac{\pi}{4}\right)$ $y = -4\sin(2x - \pi)$ $y = \frac{1}{2}\cos\left(\frac{1}{2}x - \frac{\pi}{4}\right)$

31.

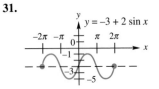
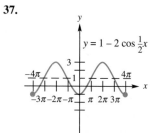

$y = -3 + 2 \sin x$

33.

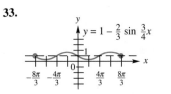
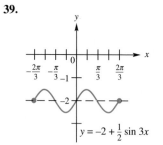

$y = 1 - \frac{2}{3} \sin \frac{3}{4}x$

35.

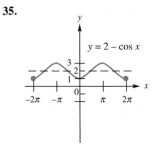
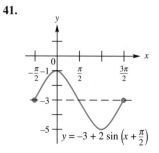

$y = 2 - \cos x$

37.

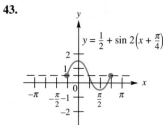

$y = 1 - 2 \cos \frac{1}{2}x$

39.

$y = -2 + \frac{1}{2} \sin 3x$

41.

$y = -3 + 2 \sin \left(x + \frac{\pi}{2}\right)$

43.

$y = \frac{1}{2} + \sin 2\left(x + \frac{\pi}{4}\right)$

4.3 EXERCISES (PAGE 155)

1. B   **3.** E   **5.** D

9.

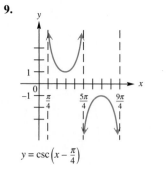

$y = \csc \left(x - \frac{\pi}{4}\right)$

11.

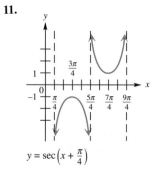

$y = \sec \left(x + \frac{\pi}{4}\right)$

13.

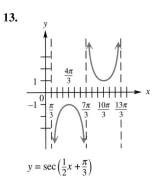

$$y = \sec\left(\tfrac{1}{2}x + \tfrac{\pi}{3}\right)$$

15.

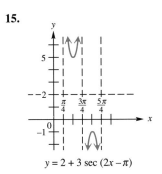

$$y = 2 + 3\sec(2x - \pi)$$

17.

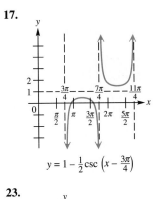

$$y = 1 - \tfrac{1}{2}\csc\left(x - \tfrac{3\pi}{4}\right)$$

19.

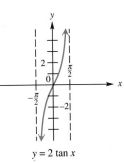

$$y = 2\tan x$$

21.

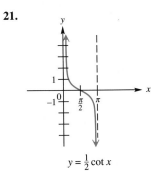

$$y = \tfrac{1}{2}\cot x$$

23.

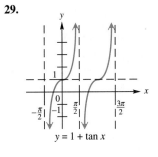

$$y = \cot 3x$$

25.

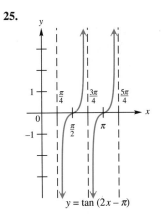

$$y = \tan(2x - \pi)$$

27.

$$y = \cot\left(3x + \tfrac{\pi}{4}\right)$$

29.

$$y = 1 + \tan x$$

31.

$$y = 1 - \cot x$$

33.

$$y = -1 + 2\tan x$$

35.

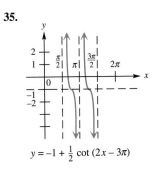

$$y = -1 + \tfrac{1}{2}\cot(2x - 3\pi)$$

37.

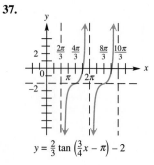

$$y = \frac{2}{3} \tan \left(\frac{3}{4} x - \pi \right) - 2$$

39. (a) 0 m   **(b)** −2.9 m   **(c)** −12.3 m   **(d)** 12.3 m   **(e)** It leads to tan $\pi/2$, which is undefined.
(f) $\{t | t \neq .25n \text{ where } n \text{ is an integer}\}$

CHAPTER 4 REVIEW EXERCISES   (PAGE 158)

1. 2; 2π; none; none   **3.** 1/2; $2\pi/3$; none; none   **5.** 2; 8π; 1; none   **7.** 3; 2π; none;
$\pi/2$ units to the left   **9.** not applicable; π; none; $\pi/8$ units to the right   **11.** not applicable; $\pi/3$; none;
$\pi/9$ units to the right   **13.** tangent   **15.** cosine   **17.** cotangent

19.

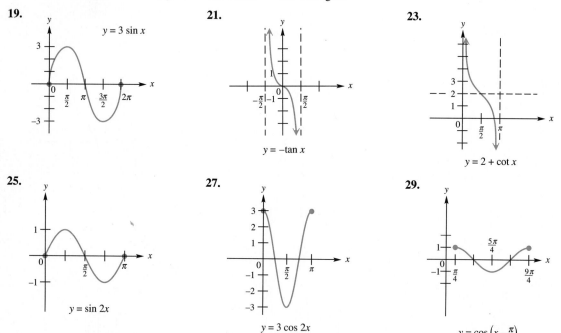

$y = 3 \sin x$

21.

$y = -\tan x$

23.

$y = 2 + \cot x$

25.

$y = \sin 2x$

27.

$y = 3 \cos 2x$

29.

$y = \cos \left(x - \frac{\pi}{4} \right)$

31.

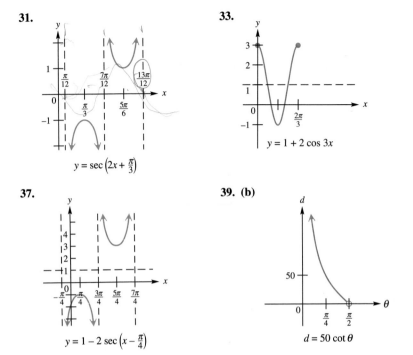

$y = \sec\left(2x + \frac{\pi}{3}\right)$

33.

$y = 1 + 2\cos 3x$

35.

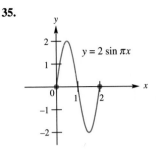

$y = 2\sin \pi x$

37.

$y = 1 - 2\sec\left(x - \frac{\pi}{4}\right)$

39. (b)

$d = 50\cot\theta$

41. (a) about 20 years
(b) from about 5000 to about 150,000

▪ CHAPTER 5   TRIGONOMETRIC IDENTITIES

5.1 EXERCISES   (PAGE 170)

1. $\sqrt{7}/4$ **3.** $-2\sqrt{5}/5$ **5.** $-\sqrt{105}/11$ **9.** $\sqrt{21}/2$ **11.** $\cos\theta = -\sqrt{5}/3$; $\tan\theta = -2\sqrt{5}/5$;
$\cot\theta = -\sqrt{5}/2$; $\sec\theta = -3\sqrt{5}/5$; $\csc\theta = 3/2$ **13.** $\sin\theta = -\sqrt{17}/17$; $\cos\theta = 4\sqrt{17}/17$; $\cot\theta = -4$;
$\sec\theta = \sqrt{17}/4$; $\csc\theta = -\sqrt{17}$ **15.** $\sin\theta = 2\sqrt{2}/3$; $\cos\theta = -1/3$; $\tan\theta = -2\sqrt{2}$; $\cot\theta = -\sqrt{2}/4$;
$\csc\theta = 3\sqrt{2}/4$ **17.** $\sin\theta = 3/5$; $\cos\theta = 4/5$; $\tan\theta = 3/4$; $\sec\theta = 5/4$; $\csc\theta = 5/3$
19. $\sin\theta = -\sqrt{7}/4$; $\cos\theta = 3/4$; $\tan\theta = -\sqrt{7}/3$; $\cot\theta = -3\sqrt{7}/7$; $\csc\theta = -4\sqrt{7}/7$ **21.** (b)   **23.** (e)

25. (a)   **27.** (a)   **29.** (d)   **33.** $\dfrac{\pm\sqrt{1 + \cot^2\theta}}{1 + \cot^2\theta}$; $\dfrac{\pm\sqrt{\sec^2\theta - 1}}{\sec\theta}$ **35.** $\dfrac{\pm\sin\theta\sqrt{1 - \sin^2\theta}}{1 - \sin^2\theta}$;

$\dfrac{\pm\sqrt{1 - \cos^2\theta}}{\cos\theta}$; $\pm\sqrt{\sec^2\theta - 1}$; $\dfrac{\pm\sqrt{\csc^2\theta - 1}}{\csc^2\theta - 1}$ **37.** $\dfrac{\pm\sqrt{1 - \sin^2\theta}}{1 - \sin^2\theta}$; $\pm\sqrt{\tan^2\theta + 1}$; $\dfrac{\pm\sqrt{1 + \cot^2\theta}}{\cot\theta}$;

$\dfrac{\pm\csc\theta\sqrt{\csc^2\theta - 1}}{\csc^2\theta - 1}$ **39.** $\sin\theta = \dfrac{\pm\sqrt{2x + 1}}{x + 1}$ **41.** 1   **43.** $-\sin\alpha$ **45.** 0   **47.** $\dfrac{1 + \sin\theta}{\cos\theta}$ **49.** 1

51. -1 **53.** $\dfrac{\sin^2\theta}{\cos^4\theta}$ **55.** 1   **57.** $\dfrac{\cos^2\alpha + 1}{\sin^2\alpha\cos^2\alpha}$ **59.** $\dfrac{\sin^2 s - \cos^2 s}{\sin^4 s}$ **61.** $4\sec\theta$; $\dfrac{3x}{\sqrt{16 + 9x^2}}$;

$\dfrac{4\sqrt{16 + 9x^2}}{16 + 9x^2}$ **63.** $\sin^3\theta$; $\sqrt{1 - x^2}$; $\dfrac{\sqrt{1 - x^2}}{x}$ **65.** $\dfrac{\tan^2\theta\sec\theta}{16}$; $\dfrac{4x\sqrt{1 + 16x^2}}{1 + 16x^2}$; $\dfrac{\sqrt{1 + 16x^2}}{1 + 16x^2}$

67. $\dfrac{25\sqrt{6} - 60}{12}$; $\dfrac{-25\sqrt{6} - 60}{12}$ **69. (a)** odd   **(b)** even   **(c)** odd   **(d)** odd   **(e)** odd   **(f)** even

71. $y = \cos 4x$

73. $y = -3 \sin 4x$

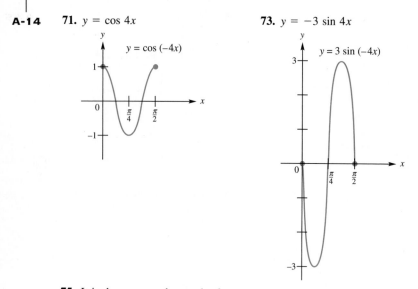

75. It is the same as the graph of $y = \sec x$.

77. The graph of $y = \tan (-x)$ is a reflection across the x-axis as compared to the graph of $y = \tan x$.

5.2 EXERCISES (PAGE 177)

1. $\dfrac{1}{\sin \theta \cos \theta}$ or $\csc \theta \sec \theta$ **3.** $1 + \cos s$ **5.** 1 **7.** 1 **9.** $2 + 2 \sin t$ **11.** $\dfrac{-2 \cos x}{(1 + \cos x)(1 - \cos x)}$ or

$\dfrac{-2 \cos x}{\sin^2 x}$ **13.** $(\sin \gamma + 1)(\sin \gamma - 1)$ **15.** $4 \sin x$ **17.** $(2 \sin x + 1)(\sin x + 1)$ **19.** $(4 \sec x - 1)(\sec x + 1)$

21. $(\cos^2 x + 1)^2$ **23.** $(\sin x - \cos x)(1 + \sin x \cos x)$ **25.** $\sin \theta$ **27.** 1 **29.** $\tan^2 \beta$ **31.** $\tan^2 x$

33. $\sec^2 x$ **83.** identity **85.** not an identity **87.** not an identity **89.** not an identity

5.3 EXERCISES (PAGE 185)

1. $\dfrac{\sqrt{6} - \sqrt{2}}{4}$ **3.** $\dfrac{\sqrt{6} - \sqrt{2}}{4}$ **5.** $\dfrac{\sqrt{2} - \sqrt{6}}{4}$ **7.** $\dfrac{\sqrt{6} + \sqrt{2}}{4}$ **9.** 0 **11.** 0 **13.** $\cot 3°$

15. $\sin 5\pi/12$ **17.** $\sec 104° 24'$ **19.** $\cos (-\pi/8)$ **21.** $\csc (-56° 42')$ **23.** $\tan (-86.9814°)$ **25.** $\tan$

27. $\cos$ **29.** $\csc$ **31.** $15°$ **33.** $\dfrac{140°}{3}$ **35.** $20°$ **37.** $\cos \theta$ **39.** $-\cos \theta$ **41.** $\cos \theta$ **43.** $-\cos \theta$

45. $\dfrac{4 - 6\sqrt{6}}{25}$; $\dfrac{4 + 6\sqrt{6}}{25}$ **47.** $16/65$; $-56/65$ **49.** $-77/85$; $-13/85$ **51.** $\dfrac{2\sqrt{638} - \sqrt{30}}{56}$;

$\dfrac{2\sqrt{638} + \sqrt{30}}{56}$ **53.** true **55.** false **57.** true **59.** true **73.** $\dfrac{-\sqrt{6} + \sqrt{2}}{4}$ **75.** $\dfrac{\sqrt{6} - \sqrt{2}}{4}$

77. $\dfrac{-\sqrt{6} - \sqrt{2}}{4}$

5.4 EXERCISES (PAGE 191)

1. $\dfrac{\sqrt{6} - \sqrt{2}}{4}$ **3.** $2 - \sqrt{3}$ **5.** $\dfrac{-\sqrt{6} - \sqrt{2}}{4}$ **7.** $\dfrac{\sqrt{6} + \sqrt{2}}{4}$ **9.** $2 - \sqrt{3}$ **11.** $\dfrac{-\sqrt{6} - \sqrt{2}}{4}$

13. $\sqrt{2}/2$ **15.** -1 **17.** 0 **19.** 1 **21.** $\dfrac{\sqrt{3} \cos \theta - \sin \theta}{2}$ **23.** $\dfrac{\cos \theta - \sqrt{3} \sin \theta}{2}$

25. $\dfrac{\sqrt{2}(\sin x - \cos x)}{2}$ **27.** $\dfrac{\sqrt{2}(\sin \theta + \cos \theta)}{2}$ **29.** $\dfrac{\sqrt{3}\tan \theta + 1}{\sqrt{3} - \tan \theta}$ **31.** $\dfrac{1 + \tan x}{1 - \tan x}$ **33.** $\sin \theta$ **35.** $\tan \theta$

37. $-\sin \theta$ **41.** 63/65; 33/65; 63/16; 33/56; I; I   **43.** $\dfrac{4\sqrt{2} + \sqrt{5}}{9}$; $\dfrac{4\sqrt{2} - \sqrt{5}}{9}$; $\dfrac{-8\sqrt{5} - 5\sqrt{2}}{20 - 2\sqrt{10}}$

or $\dfrac{4\sqrt{2} + \sqrt{5}}{2 - 2\sqrt{10}}$; $\dfrac{-8\sqrt{5} + 5\sqrt{2}}{20 + 2\sqrt{10}}$ or $\dfrac{-4\sqrt{2} + \sqrt{5}}{2\sqrt{10} + 2}$; II; II   **45.** 77/85; 13/85; −77/36; 13/84; II; I

47. −33/65; −63/65; 33/56; 63/16; III; III   **49.** 1; −161/289; undefined; −161/240; quadrantal; IV

51. $\dfrac{-(3\sqrt{22} + \sqrt{21})}{20}$; $\dfrac{-3\sqrt{22} + \sqrt{21}}{20}$; $\dfrac{-(66\sqrt{7} + 7\sqrt{66})}{154 - 3\sqrt{462}}$; $\dfrac{-66\sqrt{7} + 7\sqrt{66}}{154 + 3\sqrt{462}}$; IV; IV   **67.** $\dfrac{\sqrt{6} - \sqrt{2}}{4}$

69. $2 + \sqrt{3}$ **71.** $\dfrac{-\sqrt{6} - \sqrt{2}}{4}$ **73.** $\dfrac{\sqrt{6} - \sqrt{2}}{4}$ **75.** $-2 + \sqrt{3}$

79. $\sin (A + B + C) = \sin A \cos B \cos C + \cos A \sin B \cos C + \cos A \cos B \sin C - \sin A \sin B \sin C$ **83.** 18.4°

5.5 EXERCISES (PAGE 197)

1. $\cos \theta = 2\sqrt{5}/5$; $\sin \theta = \sqrt{5}/5$; $\tan \theta = 1/2$; $\sec \theta = \sqrt{5}/2$; $\csc \theta = \sqrt{5}$; $\cot \theta = 2$
3. $\cos x = -\sqrt{42}/12$; $\sin x = \sqrt{102}/12$; $\tan x = -\sqrt{119}/7$; $\sec x = -2\sqrt{42}/7$; $\csc x = 2\sqrt{102}/17$;
$\cot x = -\sqrt{119}/17$ **5.** $\cos 2\theta = 17/25$; $\sin 2\theta = -4\sqrt{21}/25$; $\tan 2\theta = -4\sqrt{21}/17$; $\sec 2\theta = 25/17$;
$\csc 2\theta = -25\sqrt{21}/84$; $\cot 2\theta = -17\sqrt{21}/84$ **7.** $\tan 2x = -4/3$; $\sec 2x = -5/3$; $\cos 2x = -3/5$;
$\cot 2x = -3/4$; $\sin 2x = 4/5$; $\csc 2x = 5/4$ **9.** $\sin 2\alpha = -4\sqrt{55}/49$; $\cos 2\alpha = 39/49$; $\tan 2\alpha = -4\sqrt{55}/39$;
$\cot 2\alpha = -39\sqrt{55}/220$; $\sec 2\alpha = 49/39$; $\csc 2\alpha = -49\sqrt{55}/220$ **11.** $\sqrt{3}/2$ **13.** $\sqrt{3}/3$ **15.** $\sqrt{3}/2$

17. $\sqrt{2}/2$ **19.** $-\sqrt{2}/2$ **21.** $\sqrt{2}/4$ **23.** $\dfrac{1}{2}\tan 102°$ **25.** $\dfrac{1}{4}\cos 94.2°$ **27.** $-\cos 4\pi/5$ **29.** $\sin 10x$

31. $\cos 4\alpha$ **33.** $\tan 2x/9$ **37.** 1   **39.** −1/2   **41.** −1/2   **43.** $\sqrt{3}$ **45.** $\sqrt{3}$ **47.** 0

In Exercises 73–79, other forms may be possible.   **73.** $\tan^2 2x = \dfrac{4\tan^2 x}{1 - 2\tan^2 x + \tan^4 x}$

75. $\cos 3x = 4\cos^3 x - 3\cos x$ **77.** $\tan 3x = \dfrac{3\tan x - \tan^3 x}{1 - 3\tan^2 x}$ **79.** $\tan 4x = \dfrac{4(\tan x - \tan^3 x)}{1 - 6\tan^2 x + \tan^4 x}$ **81.** 1

83. $\csc^2 3r$ **85.** 980.799 cm per sec²

5.6 EXERCISES (PAGE 203)

1. $\dfrac{\sqrt{2 - \sqrt{3}}}{2}$ **3.** $\dfrac{\sqrt{2 + \sqrt{2}}}{2}$ **5.** $1 + \sqrt{2}$ **7.** $\dfrac{\sqrt{2 + \sqrt{2}}}{2}$ **9.** $\dfrac{-\sqrt{2 + \sqrt{3}}}{2}$ **11.** $\dfrac{-\sqrt{2 + \sqrt{3}}}{2}$

15. $\sqrt{10}/4$ **17.** 3   **19.** $\sqrt{50 - 10\sqrt{5}}/10$ **21.** $-\sqrt{7}$ **23.** $\sqrt{5}/5$ **25.** $-\sqrt{42}/12$ **27.** $\sin 20°$
29. $\tan 73.5°$ **31.** $\tan 29.87°$ **33.** $\cos 9x$ **35.** $\tan 4\theta$ **37.** $\cos x/8$ **55.** 84°   **57.** 60°   **59.** 3.9

5.7 EXERCISES (PAGE 208)

1. $\dfrac{1}{2}(\sin 70° - \sin 20°)$ **3.** $\dfrac{3}{2}(\cos 8x + \cos 2x)$ **5.** $\dfrac{1}{2}[\cos 2\theta - \cos (-4\theta)] = \dfrac{1}{2}(\cos 2\theta - \cos 4\theta)$
7. $-4[\cos 9y + \cos (-y)] = -4(\cos 9y + \cos y)$ **9.** $2\cos 45° \sin 15°$ **11.** $2\cos 95° \cos (-53°) =$
$2\cos 95° \cos 53°$ **13.** $2\cos \dfrac{15\beta}{2}\sin \dfrac{9\beta}{2}$ **15.** $-6\cos \dfrac{7x}{2}\sin \dfrac{-3x}{2} = 6\cos \dfrac{7x}{2}\sin \dfrac{3x}{2}$

CHAPTER 5 REVIEW EXERCISES (PAGE 211)

1. $\sin x = -4/5$; $\tan x = -4/3$; $\sec x = 5/3$; $\csc x = -5/4$; $\cot x = -3/4$ **3. (a)** $\sin \pi/12 = \dfrac{\sqrt{6} - \sqrt{2}}{4}$;

$\cos \pi/12 = \dfrac{\sqrt{6} + \sqrt{2}}{4}$; $\tan \pi/12 = 2 - \sqrt{3}$ **(b)** $\sin \pi/12 = \dfrac{\sqrt{2 - \sqrt{3}}}{2}$; $\cos \pi/12 = \dfrac{\sqrt{2 + \sqrt{3}}}{2}$;

$\tan \pi/12 = 2 - \sqrt{3}$ **5.** (e)   **7.** (j)   **9.** (i)   **11.** (h)   **13.** (g)   **15.** (a)   **17.** (f)   **19.** (e)   **21.** 1

23. $\dfrac{1}{\cos^2 \theta}$ or $\sec^2 \theta$ **25.** $\dfrac{1}{\sin^2 \theta \cos^2 \theta}$ **27.** $\dfrac{4 + 3\sqrt{15}}{20}$; $\dfrac{4\sqrt{15} + 3}{20}$; $\dfrac{192 + 25\sqrt{15}}{231}$; I

29. $\dfrac{4 - 9\sqrt{11}}{50}$; $\dfrac{12\sqrt{11} - 3}{50}$; $\dfrac{\sqrt{11} - 16}{21}$; IV   **31.** $\sin \theta = \sqrt{14}/4$; $\cos \theta = \sqrt{2}/4$ **33.** $\sin 2x = 3/5$;

$\cos 2x = -4/5$ **35.** 1/2   **37.** $\dfrac{\sqrt{5} - 1}{2}$ **73.** $\dfrac{\sqrt{6} - \sqrt{2}}{4}$

■ CHAPTER 6 INVERSE TRIGONOMETRIC FUNCTIONS AND TRIGONOMETRIC EQUATIONS

6.1 EXERCISES (PAGE 225)

1. $-\pi/6$ **3.** $\pi/4$ **5.** 0   **7.** $\pi/3$ **9.** $3\pi/4$ **11.** $5\pi/6$ **13.** $-45°$ **15.** $-60°$ **17.** $120°$ **19.** $-30°$
21. .83798122   **23.** 2.3154725   **25.** 1.1900238   **27.** $-7.6713835°$ **29.** $113.50097°$ **31.** $30.987961°$
33. $(-\infty, \infty)$; $(0, \pi)$

35. $(-\infty, -1] \cup [1, \infty)$; $[0, \pi/2) \cup (\pi/2, \pi]$

41. $\sqrt{7}/3$ **43.** $\sqrt{5}/5$ **45.** $-\sqrt{5}/2$ **47.** $\sqrt{34}/3$ **49.** 1/2   **51.** -1 **53.** 2   **55.** $\pi/4$ **57.** $\pi/3$

59. 120/169   **61.** $-7/25$ **63.** $4\sqrt{6}/25$ **65.** $-24/7$ **67.** $\dfrac{\sqrt{10} - 3\sqrt{30}}{20}$ **69.** $-16/65$ **71.** .89442719

73. .12343998   **75.** $\sqrt{1 - u^2}$ **77.** $\dfrac{\sqrt{u^2 + 1}}{u}$ **79.** $\dfrac{\sqrt{1 - u^2}}{u}$ **81.** $\dfrac{\sqrt{u^2 - 4}}{u}$ **83.** $\dfrac{u\sqrt{2}}{2}$

85. $\dfrac{\pm 2\sqrt{4 - u^2}}{4 - u^2}$ **87. (a)** $113°$ **(b)** $84°$ **(c)** $60°$ **(d)** $47°$

6.2 EXERCISES (PAGE 232)

1. $3\pi/4, 7\pi/4$ **3.** $\pi/6, 5\pi/6$ **5.** $\pi/3, 2\pi/3, 4\pi/3, 5\pi/3$ **7.** $\pi/6, 7\pi/6, 4\pi/3, 5\pi/3$ **9.** $\pi/4, 2\pi/3, 5\pi/4, 5\pi/3$
11. π **13.** $7\pi/6, 3\pi/2, 11\pi/6$ **15.** $90°, 210°, 330°$ **17.** $45°, 135°, 225°, 315°$ **19.** $45°, 225°$
21. $45°, 90°, 135°, 225°, 270°, 315°$ **23.** $0°, 30°, 150°, 180°$ **25.** $0°, 45°, 135°, 180°, 225°, 315°$ **27.** $0°, 90°$
29. $90°, 221.8°, 318.2°$ **31.** $135°, 315°, 71.6°, 251.6°$ **33.** $71.6°, 90°, 251.6°, 270°$ **35.** $53.6°, 126.4°, 187.9°,$
$352.1°$ **37.** $149.6°, 329.6°, 106.3°, 286.3°$ **39.** no solution   **41.** $57.7°, 159.2°$ **43.** $360° \cdot n$,
$120° + 360° \cdot n, 240° + 360° \cdot n$, where n is an integer   **47. (a)** 1/4 sec **(b)** 1/6 sec **(c)** .21 sec   **49.** $14°$

6.3 EXERCISES (PAGE 236)

1. $\pi/12, 11\pi/12, 13\pi/12, 23\pi/12$ **3.** $\pi/2, 7\pi/6, 11\pi/6$ **5.** $\pi/18, 7\pi/18, 13\pi/18, 19\pi/18, 25\pi/18, 31\pi/18$
7. $3\pi/8, 5\pi/8, 11\pi/8, 13\pi/8$ **9.** $\pi/2, 3\pi/2$ **11.** $0, \pi/4, \pi/2, 3\pi/4, \pi, 5\pi/4, 3\pi/2, 7\pi/4$ **13.** $\pi/8, 3\pi/8, 5\pi/8,$
$7\pi/8, 9\pi/8, 11\pi/8, 13\pi/8, 15\pi/8$ **15.** $\pi/2$ **17.** $\pi/3, \pi, 5\pi/3$ **19.** $15°, 45°, 135°, 165°, 255°, 285°$ **21.** $0°$
23. $120°, 240°$ **25.** $30°, 150°, 270°$ **27.** $0°, 30°, 150°, 180°$ **29.** $60°, 300°$ **31.** $11.8°, 78.2°, 191.8°, 258.2°$
33. $30°, 90°, 150°, 210°, 270°, 330°$ **35.** $0°, 120°, 240°$ **37.** $\pi/12, 5\pi/12, \pi/2, 13\pi/12, 17\pi/12, 3\pi/2$
39. $0, \pi/4, 3\pi/4, \pi, 5\pi/4, 7\pi/4$ **41.** $\pi/6, \pi/2, 5\pi/6, 3\pi/2$ **45. (a)** 91.3 days after March 21, on June 20
(b) 273.8 days after March 21, on December 19 **(c)** 228.7 days after March 21, on November 4, and again after
318.8 days, on February 2 **47.** .001 sec **49.** .004 sec

6.4 EXERCISES (PAGE 240)

1. $x = \operatorname{Arccos} \dfrac{y}{5}$ **3.** $x = \dfrac{1}{3} \operatorname{Arccot} 2y$ **5.** $x = \dfrac{1}{2} \operatorname{Arctan} \dfrac{y}{3}$ **7.** $x = 4 \operatorname{Arccos} \dfrac{y}{6}$ **9.** $x = \dfrac{1}{5} \operatorname{Arccos}\left(-\dfrac{y}{2}\right)$

11. $x = -3 + \operatorname{Arccos} y$ **13.** $x = \operatorname{Arcsin}(y + 2)$ **15.** $x = \operatorname{Arcsin}\left(\dfrac{y+4}{2}\right)$ **19.** $2\sqrt{2}$ **21.** $\pi - 3$

23. $3/5$ **25.** $4/5$ **27.** 0 **29.** $1/2$ **31.** $-1/2$ **33.** 0 **35.** $t = \dfrac{50}{\pi} \operatorname{Arccos}\left(\dfrac{d-550}{450}\right)$

37. (a) $t = \dfrac{1}{2\pi f} \operatorname{Arcsin} \dfrac{e}{E_{\max}}$ **(b)** .00068 sec

39. (a) $x = \operatorname{Sin} u, -\pi/2 \le u \le \pi/2$ **(b)**

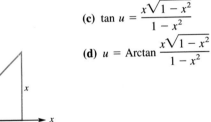

(c) $\tan u = \dfrac{x\sqrt{1 - x^2}}{1 - x^2}$

(d) $u = \operatorname{Arctan} \dfrac{x\sqrt{1 - x^2}}{1 - x^2}$

CHAPTER 6 REVIEW EXERCISES (PAGE 243)

1. $\pi/4$ **3.** $-\pi/3$ **5.** $3\pi/4$ **7.** $2\pi/3$ **9.** $3\pi/4$ **11.** $-60°$ **13.** $60.679245°$ **15.** $36.489508°$
17. $73.262206°$ **21.** $(-\infty, \infty)$ **23.** $1/2$ **25.** -1 **27.** $3\pi/4$ **29.** $\pi/4$ **31.** $\sqrt{7}/4$ **33.** $\sqrt{3}/2$
35. $\dfrac{294 + 125\sqrt{6}}{92}$ **37.** $\sqrt{1 - u^2}$ **39.** $[-1, 1]; \quad [-\pi/2, \pi/2]$ **41.** $(-\infty, \infty); \quad (0, \pi)$

43. .46364761, 3.6052403 **45.** $\pi/4$, $3\pi/4$, $5\pi/4$, $7\pi/4$ **47.** $\pi/8$, $3\pi/8$, $5\pi/8$, $7\pi/8$, $9\pi/8$, $11\pi/8$, $13\pi/8$, $15\pi/8$ **49.** $\pi/3$, π, $5\pi/3$ **51.** 270° **53.** 45°, 90°, 225°, 270° **55.** 70.5°, 180°, 289.5° **57.** 0°, 60°, 90°, 120°, 180°, 240°, 270°, 300° **59.** $x = \text{Arcsin } 2y$ **61.** $x = \left(\dfrac{1}{3}\text{Arctan } 2y\right) - \dfrac{2}{3}$ **63.** no solution **65.** $x = -1/2$ **67.** 48.8°

■ CHAPTER 7 APPLICATIONS OF TRIGONOMETRY AND VECTORS

7.1 EXERCISES (PAGE 253)

1. $C = 95°$, $b = 13$ m, $a = 11$ m **3.** $C = 80°\ 40'$, $a = 79.5$ mm, $c = 108$ mm **5.** $B = 37.3°$, $a = 38.5$ ft, $b = 51.0$ ft **7.** $C = 57.36°$, $b = 11.13$ ft, $c = 11.55$ ft **9.** $B = 18.5°$, $a = 239$ yd, $c = 230$ yd **11.** $A = 56°\ 00'$, $AB = 361$ ft, $BC = 308$ ft **13.** $B = 110.0°$, $a = 27.01$ m, $c = 21.36$ m **15.** $A = 34.72°$, $a = 3326$ ft, $c = 5704$ ft **17.** $C = 97°\ 34'$, $b = 283.2$ m, $c = 415.2$ m **23.** 118 m **25.** 1.93 mi **27.** 10.4 in **29.** 111° **31.** first location: 5.1 mi; second location: 7.2 mi **33.** 46.4 m^2 **35.** 356 cm^2 **37.** 722.9 in^2 **39.** 1071 cm^2 **41.** 100 m^2

7.2 EXERCISES (PAGE 259)

1. $B_1 = 49.1°$, $C_1 = 101.2°$, $B_2 = 130.9°$, $C_2 = 19.4°$ **3.** $B = 26°\ 30'$, $A = 112°\ 10'$ **5.** no such triangle **7.** $B = 27.19°$, $C = 10.68°$ **9.** $A = 43°\ 50'$, $B = 6°\ 52'$ **11.** $B = 20.6°$, $C = 116.9°$, $c = 20.6$ ft **13.** no such triangle **15.** $B_1 = 49°\ 20'$, $C_1 = 92°\ 00'$, $c_1 = 15.5$ km, $B_2 = 130°\ 40'$, $C_2 = 10°\ 40'$, $c_2 = 2.88$ km **17.** $A_1 = 52°\ 10'$, $C_1 = 95°\ 00'$, $c_1 = 9520$ cm, $A_2 = 127°\ 50'$, $C_2 = 19°\ 20'$, $c_2 = 3160$ cm **19.** $B = 37.77°$, $C = 45.43°$, $c = 4.174$ ft **21.** $A_1 = 53.23°$, $C_1 = 87.09°$, $c_1 = 37.16$ m, $A_2 = 126.77°$, $C_2 = 13.55°$, $c_2 = 8.719$ m **23.** 1; 90°; a right triangle **27.** does not exist

7.3 EXERCISES (PAGE 265)

1. 257 m **3.** 281 km **5.** 22 ft **7.** 18 ft **9.** 2000 km **11.** 163.5° **13.** 25.24983 mi **17.** $c = 2.83$ in, $A = 44.9°$, $B = 106.8°$ **19.** $c = 6.46$ m, $A = 53.1°$, $B = 81.3°$ **21.** $a = 156$ cm, $B = 64°\ 50'$, $C = 34°\ 30'$ **23.** $b = 9.529$ in, $A = 64.59°$, $C = 40.61°$ **25.** $a = 15.7$ m, $B = 21.6°$, $C = 45.6°$ **27.** $c = 139.0$ m, $A = 49°\ 20'$, $B = 105°\ 51'$ **29.** $A = 30°$, $B = 56°$, $C = 94°$ **31.** $A = 82°$, $B = 37°$, $C = 61°$ **33.** $A = 42°\ 00'$, $B = 35°\ 50'$, $C = 102°\ 10'$ **35.** $A = 47°\ 50'$, $B = 44°\ 50'$, $C = 87°\ 20'$ **37.** 78 m^2 **39.** 12,600 cm^2 **41.** 3650 ft^2 **43.** 1921 ft^2 **45.** 33 cans **47.** 392,000 mi^2

7.4 EXERCISES (PAGE 272)

Sketches of vectors are in proportion but not necessarily the exact length as shown in the exercises. **1.** **m** and **p**; **n** and **r** **3.** **m** and **p** equal 2**t**, or **t** is one half **m** or **p**; also, **m** = 1**p** and **n** = 1**r** **5.**

7.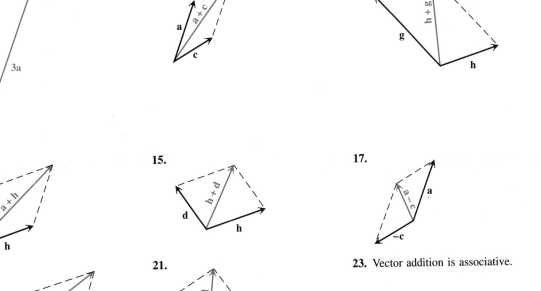

3a

9.

a + c

a

c

11.

h + g

g

h

13.

a + h

a

h

15.

h + d

d

h

17.

a − c

a

−c

19.

a + (b + c)

a

c

b + c

b

21.

c + d

d

c

23. Vector addition is associative.

25.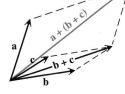

12

27°

20

27.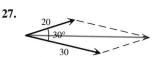

20

30°

30

29.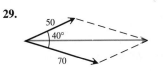

50

40°

70

31. 9.5, 7.4   **33.** 17, 20   **35.** 13.7, 7.11   **37.** 123, 155   **39.** 530 newtons   **41.** 27.2 lb   **43.** 88.2 lb
51. −22   **53.** −50   **57.** 151°

7.5 EXERCISES (PAGE 279)

1. 93.9°   **3.** 18°   **5.** 2.4 tons   **7.** 2640 lb at an angle of 167.2° with the 1480-lb force   **9.** weight 64.8 lb;
tension 61.9 lb   **11.** 190, 283 lb, respectively   **13.** 173.1°   **15.** 39.2 km   **17.** 237°; 470 mph   **19.** 358°;
170 mph   **21.** 230 km per hr; 167°   **23.** Turn at 3:21 P.M. on a bearing of 152°.

CHAPTER 7 REVIEW EXERCISES (PAGE 283)

1. 63.7 m   **3.** 41° 40′   **5.** 54° 20′ or 125° 40′   **9.** 19.87° or 19° 52′   **11.** 55.5 m   **13.** 148 cm
15. $B = 17° 10′$, $C = 137° 40′$, $c = 11.0$ yd   **17.** $c = 18.7$ cm, $A = 88° 20′$, $B = 49° 10′$ **19.** 153,600 m²

21. .234 km²   **23.** Each expression is equal to $\dfrac{1 + \sqrt{3}}{2}$.   **25.** about 2.5 cans   **27.** 13 m   **29.** 10.8 mi

31. 115 km   **33.** 5500 m   **35.** 438.14 ft

37.   **39.**

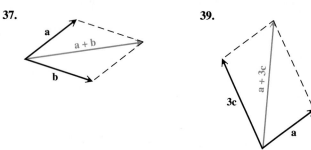

41. false   **43.** 17.9, 66.8   **45.** 28 lb   **47.** 135 newtons   **49.** 280 newtons, 30.4°   **51.** 3° 50′
53. speed 21 km per hr, bearing 118°

■ CHAPTER 8   COMPLEX NUMBERS AND POLAR EQUATIONS

8.1 EXERCISES   (PAGE 292)

1. $2i$ **3.** $\dfrac{5}{3}i$ **5.** $5i\sqrt{6}$ **7.** $4i\sqrt{5}$ **9.** -3 **11.** $-\sqrt{30}$ **13.** $\sqrt{6}/2$ **15.** $\dfrac{\sqrt{3}}{3}i$ **17.** $4i, -4i$

19. $2i\sqrt{3}, -2i\sqrt{3}$ **21.** $-\dfrac{2}{3}+\dfrac{\sqrt{2}}{3}i, -\dfrac{2}{3}-\dfrac{\sqrt{2}}{3}i$ **23.** $3+i\sqrt{5}, 3-i\sqrt{5}$ **25.** $\dfrac{1}{2}+\dfrac{\sqrt{6}}{2}i, \dfrac{1}{2}-\dfrac{\sqrt{6}}{2}i$

27. $-\dfrac{1}{2}+\dfrac{\sqrt{3}}{2}i, -\dfrac{1}{2}-\dfrac{\sqrt{3}}{2}i$ **33.** $5-3i$ **35.** $-5+2i$ **37.** $-4+i$ **39.** $8-i$ **41.** $-14+2i$

43. 5   **45.** $-5i$ **47.** $\dfrac{7}{25}-\dfrac{24}{25}i$ **49.** $\dfrac{13}{20}-\dfrac{1}{20}i$ **51.** 1   **53.** -1 **55.** i **57.** -1

61. $x=2, y=-3$ **63.** $x=\dfrac{1}{2}, y=15$ **65.** $x=14, y=8$ **67.** $E=30+60i$ **69.** $Z=\dfrac{233}{37}+\dfrac{119}{37}i$

8.2 EXERCISES   (PAGE 298)

1.   **3.**   **5.**

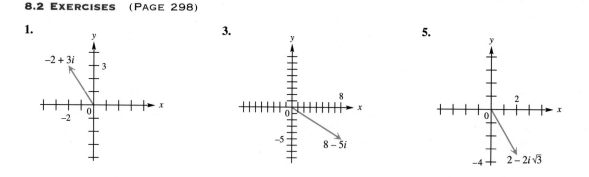

7.

9.

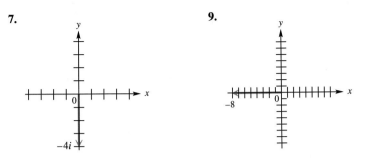

13. 0   **15.** $3 - i$ **17.** $3 - 3i$ **19.** $-3 + 3i$ **21.** $2 + 4i$ **23.** $7 + 9i$ **25.** $\sqrt{2} + i\sqrt{2}$ **27.** $10i$

29. $-2 - 2i\sqrt{3}$ **31.** $\dfrac{\sqrt{3}}{2} + \dfrac{1}{2}i$ **33.** $\dfrac{5}{2} - \dfrac{5\sqrt{3}}{2}i$ **35.** $-\sqrt{2}$ **37.** $3\sqrt{2}(\cos 315° + i \sin 315°)$

39. $6(\cos 240° + i \sin 240°)$ **41.** $2(\cos 330° + i \sin 330°)$ **43.** $5\sqrt{2}(\cos 225° + i \sin 225°)$

45. $2\sqrt{2}(\cos 45° + i \sin 45°)$ **47.** $4(\cos 180° + i \sin 180°)$ **49.** $2(\cos 270° + i \sin 270°)$

51. $\sqrt{13}(\cos 56.31° + i \sin 56.31°)$ **53.** $-1.0260604 - 2.8190779i$ **55.** $12(\cos 90° + i \sin 90°)$

57. $\sqrt{34}(\cos 59.04° + i \sin 59.04°)$

8.3 EXERCISES (PAGE 303)

1. $-3\sqrt{3} + 3i$ **3.** $-4i$ **5.** $12\sqrt{3} + 12i$ **7.** $-\dfrac{15\sqrt{2}}{2} + \dfrac{15\sqrt{2}}{2}i$ **9.** $-3i$ **11.** $\sqrt{3} - i$

13. $-1 - i\sqrt{3}$ **15.** $-\dfrac{1}{6} - \dfrac{\sqrt{3}}{6}i$ **17.** $2\sqrt{3} - 2i$ **19.** $-\dfrac{1}{2} - \dfrac{1}{2}i$ **21.** $\sqrt{3} + i$

23. $.65366807 + 7.4714602i$ **25.** $30.858023 + 18.541371i$ **27.** $.20905693 + 1.9890438i$

29. $-3.7587705 - 1.3680806i$ **33.** $1.2 - .14i$

8.4 EXERCISES (PAGE 308)

1. $27i$ **3.** 1 **5.** $\dfrac{27}{2} - \dfrac{27\sqrt{3}}{2}i$ **7.** $-16\sqrt{3} + 16i$ **9.** $-128 + 128i\sqrt{3}$ **11.** $128 + 128i$

13. $(\cos 0° + i \sin 0°)$,   **15.** $2 \text{ cis } 20°$,   **17.** $2(\cos 90° + i \sin 90°)$,
$(\cos 120° + i \sin 120°)$, $2 \text{ cis } 140°$, $2(\cos 210° + i \sin 210°)$,
$(\cos 240° + i \sin 240°)$ $2 \text{ cis } 260°$ $2(\cos 330° + i \sin 330°)$

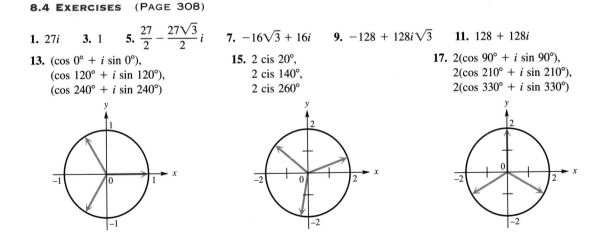

19. 4(cos 60° + *i* sin 60°),
4(cos 180° + *i* sin 180°),
4(cos 300° + *i* sin 300°)

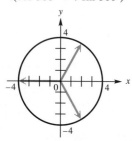

21. $\sqrt[3]{2}$(cos 20° + *i* sin 20°),
$\sqrt[3]{2}$(cos 140° + *i* sin 140°),
$\sqrt[3]{2}$(cos 260° + *i* sin 260°)

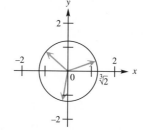

23. $\sqrt[3]{4}$(cos 50° + *i* sin 50°),
$\sqrt[3]{4}$(cos 170° + *i* sin 170°),
$\sqrt[3]{4}$(cos 290° + *i* sin 290°)

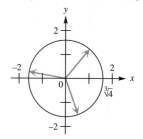

25. (cos 0° + *i* sin 0°),
(cos 180° + *i* sin 180°)

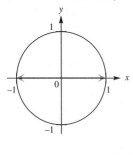

27. (cos 0° + *i* sin 0°),
(cos 60° + *i* sin 60°),
(cos 120° + *i* sin 120°),
(cos 180° + *i* sin 180°),
(cos 240° + *i* sin 240°),
(cos 300° + *i* sin 300°)

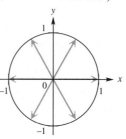

29. (cos 45° + *i* sin 45°),
(cos 225° + *i* sin 225°)

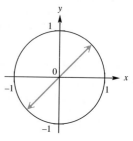

33. false   **35.** (cos 0° + *i* sin 0°), (cos 120° + *i* sin 120°), (cos 240° + *i* sin 240°)
37. (cos 90° + *i* sin 90°), (cos 210° + *i* sin 210°), (cos 330° + *i* sin 330°)   **39.** 2(cos 0° + *i* sin 0°),
2(cos 120° + *i* sin 120°), 2(cos 240° + *i* sin 240°)   **41.** (cos 45° + *i* sin 45°), (cos 135° + *i* sin 135°),
(cos 225° + *i* sin 225°), (cos 315° + *i* sin 315°)   **43.** (cos 22.5° + *i* sin 22.5°), (cos 112.5° + *i* sin 112.5°),
(cos 202.5° + *i* sin 202.5°), (cos 292.5° + *i* sin 292.5°)   **45.** 2(cos 20° + *i* sin 20°), 2(cos 140° + *i* sin 140°),
2(cos 260° + *i* sin 260°)   **47.** 1.3606 + 1.2637*i*, −1.7747 + .5464*i*, .4141 − 1.8102*i*
49. $1, -\frac{1}{2} + \frac{\sqrt{3}}{2}i, -\frac{1}{2} - \frac{\sqrt{3}}{2}i$

8.5 Exercises (Page 315)

Answers may vary in Exercises 1–9.
1. (1, 405°), (−1, 225°)   **3.** (−2, 495°), (2, 315°)
5. (5, 300°), (−5, 120°)   **7.** (−3, 150°), (3, −30°)
9. (3, 660°), (−3, 120°)   **11.** quadrantal

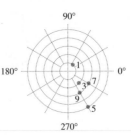

13.

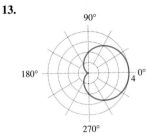

$r = 2 + 2 \cos \theta$

15.

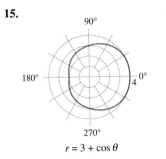

$r = 3 + \cos \theta$

17.

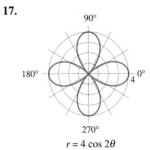

$r = 4 \cos 2\theta$

19.

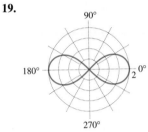

$r^2 = 4 \cos 2\theta$

21.

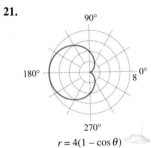

$r = 4(1 - \cos \theta)$

23.

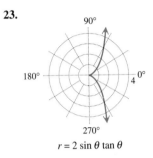

$r = 2 \sin \theta \tan \theta$

27. $x^2 + (y - 1)^2 = 1$

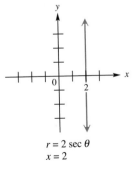

$r = 2 \sin \theta$
$x^2 + (y - 1)^2 = 1$

29. $y^2 = 4(x + 1)$

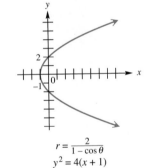

$r = \dfrac{2}{1 - \cos \theta}$
$y^2 = 4(x + 1)$

31. $(x + 1)^2 + (y + 1)^2 = 2$

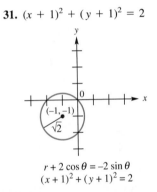

$r + 2 \cos \theta = -2 \sin \theta$
$(x + 1)^2 + (y + 1)^2 = 2$

33. $x = 2$

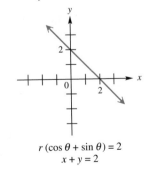

$r = 2 \sec \theta$
$x = 2$

35. $x + y = 2$

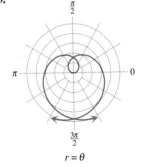

$r (\cos \theta + \sin \theta) = 2$
$x + y = 2$

37. $r(\cos \theta + \sin \theta) = 4$
39. $r = 4$
41. $r = 2 \csc \theta$ or $r \sin \theta = 2$
43.

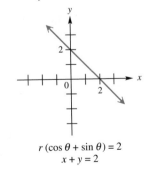

$r = \theta$

A-24   **CHAPTER 8 REVIEW EXERCISES**   (PAGE 318)

1. $3i$ **3.** $-9i, 9i$ **5.** $-2 - 3i$ **7.** $5 + 4i$ **9.** $29 + 37i$ **11.** $-32 + 24i$ **13.** $-2 - 2i$

15. $\dfrac{8}{5} + \dfrac{6}{5}i$ **17.** $-\dfrac{3}{26} + \dfrac{11}{26}i$ **19.** i **21.** $-30i$ **23.** $-\dfrac{1}{8} + \dfrac{\sqrt{3}}{8}i$ **25.** $8i$ **27.** $-\dfrac{1}{2} - \dfrac{\sqrt{3}}{2}i$ **29.** x

31.   **33.**   **35.** $5 + 4i$

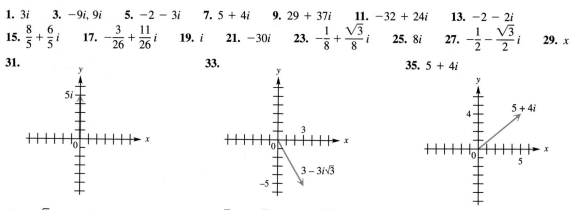

37. $2\sqrt{2}(\cos 135° + i \sin 135°)$ **39.** $-\sqrt{2} - i\sqrt{2}$ **41.** $\sqrt{2}(\cos 315° + i \sin 315°)$

43. $4(\cos 270° + i \sin 270°)$ **45.** $\sqrt[10]{8}(\cos 27° + i \sin 27°)$, $\sqrt[10]{8}(\cos 99° + i \sin 99°)$, $\sqrt[10]{8}(\cos 171° + i \sin 171°)$,

$\sqrt[10]{8}(\cos 243° + i \sin 243°)$, $\sqrt[10]{8}(\cos 315° + i \sin 315°)$ **47.** one   **49.** $5(\cos 60° + i \sin 60°)$,

$5(\cos 180° + i \sin 180°)$, $5(\cos 300° + i \sin 300°)$ **51.** $(\cos 135° + i \sin 135°)$, $(\cos 315° + i \sin 315°)$

53.   **55.**

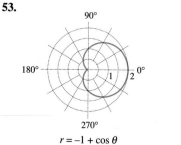
$r = -1 + \cos \theta$

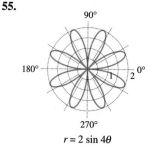
$r = 2 \sin 4\theta$

57. $y^2 + 6x - 9 = 0$ **59.** $x^2 + y^2 = x + y$ **61.** $r \cos \theta = -3$ **63.** $r = \tan \theta \sec \theta$ or $r = \dfrac{\tan \theta}{\cos \theta}$

■ CHAPTER 9   LOGARITHMS

9.1 EXERCISES   (PAGE 327)

1. 3^{11} **3.** 7 **5.** 6^3 **7.** 8^{18} **9.** 2^8 **11.** $1/2^3$ **13.** 3 **15.** $1/9^2$ **17.** 5 **19.** 8

21.   **23.**   **25.**   **27.**

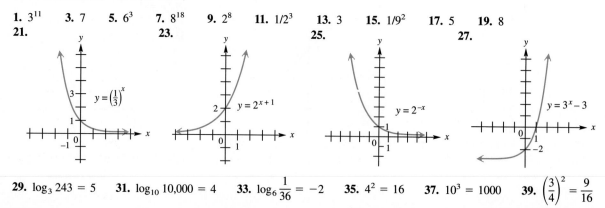

29. $\log_3 243 = 5$ **31.** $\log_{10} 10{,}000 = 4$ **33.** $\log_6 \dfrac{1}{36} = -2$ **35.** $4^2 = 16$ **37.** $10^3 = 1000$ **39.** $\left(\dfrac{3}{4}\right)^2 = \dfrac{9}{16}$

45. **47.** **49.** **51.**

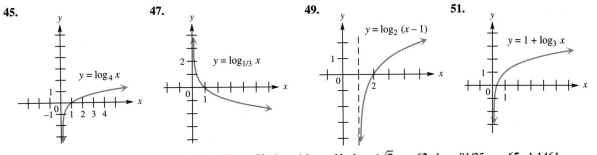

53. $\log_6 45$ **55.** $\log_5 12/7$ **57.** $\log_2 28/5$ **59.** $\log_3 16$ **61.** $\log_8 \sqrt{7}$ **63.** $\log_2 81/25$ **65.** 1.1461
67. 1.4471 **69.** 2.4080 **71.** 3.1242 **73.** .3471

9.2 EXERCISES (PAGE 335)

1. 2.4440 **3.** .5159 **5.** .9926 **7.** −.4855 **9.** −3.1726 **11.** 5.8293 **13.** 1.6094 **15.** 6.5468
17. .5423 **19.** 4.5809 **21.** 3.1 **23.** 3.3863 **27.** (a) **29.** 3.4198 **31.** 454.0462 **33.** .0762
35. 46.9649 **37.** 4.0398 **39.** .6592 **41.** 1.4899 **43.** .5794 **45.** 1.0892 **47.** 1.6083 **49.** .8803
51. 6.4 **53.** 1.8 **55.** 3.2×10^{-6} **57.** 2.0×10^{-3} **59.** 527 years **61.** 82% **63.** 1 **65.** (a) 2
(b) 3 **(c)** 3 **(d)** 1 **67.** (a) 11% **(b)** 36% **(c)** 84%

9.3 EXERCISES (PAGE 342)

1. 3 **3.** 1/5 **5.** 3/2 **7.** 4/3 **9.** 8 **11.** 3 **13.** .477 **15.** −.824 **17.** 1.285 **19.** 1.263
21. 1.781 **23.** −1.433 **25.** 2.993 **27.** 16 **29.** 5 **31.** 2.154 **33.** .01
35. Let $x = 4$ and $y = 5$. Then $1^4 = 1^5$, but $4 \neq 5$. **39.** x must be positive. **41.** (a) about 961,000
(b) about 7 years **(c)** about 17.3 years **43.** (a) $11,260.96 **(b)** $11,416.64 **(c)** $11,497.99 **(d)** $11,580.90
(e) $11,581.83 **45.** $606.53

CHAPTER 9 REVIEW EXERCISES (PAGE 345)

1. 27 **3.** 2^{11} **5.** 8

7. **9.** **11.** **13.**

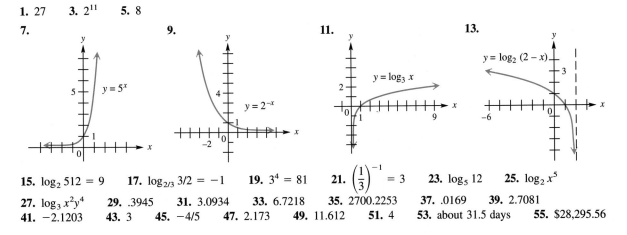

15. $\log_2 512 = 9$ **17.** $\log_{2/3} 3/2 = -1$ **19.** $3^4 = 81$ **21.** $\left(\dfrac{1}{3}\right)^{-1} = 3$ **23.** $\log_5 12$ **25.** $\log_2 x^5$
27. $\log_3 x^2 y^4$ **29.** .3945 **31.** 3.0934 **33.** 6.7218 **35.** 2700.2253 **37.** .0169 **39.** 2.7081
41. −2.1203 **43.** 3 **45.** −4/5 **47.** 2.173 **49.** 11.612 **51.** 4 **53.** about 31.5 days **55.** $28,295.56

INDEX